2D Materials-Based Sensors

2D Materials-Based Sensors: Technology and Applications is a comprehensive book covering the latest innovations on sensors using the unique properties of 2D materials. The primary aim of the book is to present up-to-date information on these sensors in a concise form for students, scientists, and technologists engaged in sensor research and development. Commencing from an introductory survey of the 2D materials family, including graphene, graphitic carbon nitride, hexagonal boron nitride, layered metal chalcogenides, 2D-Xenes and MXenes, layered double hydroxides, metal-organic frameworks, and metal oxides, the book describes the impressive properties of these materials, and their top-down and bottom-up synthesis techniques. The succeeding chapters present detailed descriptions of 2D materials-based mechanical sensors, photodetectors, magnetic sensors together with quantum sensors, heavy-metal ion sensors, gas sensors, and biological sensors. The book concludes with the research challenges, emerging trends, and future outlook providing a balanced perspective of the advances in 2D material sensors, including their limitations.

Key Features:
- Describes 2D materials process technology and integration of these materials into sensor fabrication.
- Provides an in-depth coverage of the underlying physics, operation, and performance metrics of 2D materials-based sensors.
- Highlights state-of-the-art developments in 2D material sensors vis-à-vis challenges faced in their fabrication.
- Presents glimpses of emerging 2D materials-based sensor technologies that will shape the future.
- Expedites understanding and learning through plain, lucid explanations, and clear, insightful diagrams.

The author, **Vinod Kumar Khanna, Ph.D. (Physics)**, is an independent researcher at Chandigarh, India; a former emeritus scientist, Council of Scientific and Industrial Research (CSIR), India, and Emeritus Professor, Academy of Scientific and Innovative Research (AcSIR), Ghaziabad, India; a retired chief scientist, CSIR-Central Electronics Engineering Research Institute, Pilani, India, and professor, AcSIR, India. He has worked for more than 37 years on the design, fabrication, and characterization of power semiconductor devices, MEMS, and nanotechnology-based sensors. He has published 194 research papers in refereed journals and conference proceedings, 20 books, and six chapters in his credit, two US and three Indian.

Emerging Materials and Technologies
Series Editor: Boris I. Kharissov

The *Emerging Materials and Technologies* series is devoted to highlighting publications centered on emerging advanced materials and novel technologies. Attention is paid to those newly discovered or applied materials with potential to solve pressing societal problems and improve quality of life, corresponding to environmental protection, medicine, communications, energy, transportation, advanced manufacturing, and related areas.

The series takes into account that, under present strong demands for energy, material, and cost savings, as well as heavy contamination problems and worldwide pandemic conditions, the area of emerging materials and related scalable technologies is a highly interdisciplinary field, with the need for researchers, professionals, and academics across the spectrum of engineering and technological disciplines. The main objective of this book series is to attract more attention to these materials and technologies and invite conversation among the international R&D community.

Computational Studies
From Molecules to Materials
Edited by Ambrish Kumar Srivastava

2D Semiconductors for Environmental Remediation
Edited by Honey John, Nisha T Padmanabhan, Sona Stanly and Jith C Janardhanan

Materials from Natural Sources
Structure, Properties, and Applications
Edited by Ramesh Gardas, Neha Patni, and Amita Chaudhary

Dielectric Materials for Capacitive Energy Storage
Edited By Haibo Zhang and Hua Tan

Multifunctional Coordination Materials for Green Energy Technologies
Edited by Ghulam Yasin, Anuj Kumar, Sajjad Ali, Tuan Anh Nguyen, and Saira Ajmal

Advancements in Nanomaterials for Energy Conversion and Storage
Edited by Piyush Kumar Sonkar and Vellaichamy Ganesan

2D Materials-Based Sensors
Technology and Applications
Vinod Kumar Khanna

For more information about this series, please visit:
www.routledge.com/Emerging-Materials-and-Technologies/book-series/CRCEMT

2D Materials-Based Sensors
Technology and Applications

Vinod Kumar Khanna

CRC Press
Taylor & Francis Group
Boca Raton New York London

CRC Press is an imprint of the
Taylor & Francis Group, an **informa** business

ISBN: 978-1-032-36173-4 (hbk)
ISBN: 978-1-032-36174-1 (pbk)
ISBN: 978-1-003-33058-5 (ebk)

DOI: 10.1201/9781003330585

Typeset in Times
by SPi Technologies India Pvt Ltd (Straive)

Dedication

To my parents Late Shri Amarnath Khanna and Shrimati Pushpa Khanna in respectful remembrance of their selfless sacrifices for my career and for showing me the path of life.

And to my grandson Hansh, daughter Aloka, and wife Amita for filling my life with love, joy, and happiness.

Contents

Preface

As silicon MEMS and NEMS-based sensor technologies are fast attaining maturity, there is a surging demand for new materials. The 2D materials including graphene, transition metal dichalcogenides (TMDCs), Xenes, metal oxides, and others have opened a new frontier in material science bringing forth with them a plethora of opportunities for fabrication of novel, enhanced-functionality sensors. This family of nanomaterials exhibits a wide range of electronic properties bridging from metallic or semi-metallic behavior (e.g., graphene) through semiconducting (e.g., MoS_2, WS_2) to insulating nature (e.g., hexagonal boron nitride). Thanks to their enormous surface area-to-volume ratio facilitating surface reactions, unique physisorption/chemisorption properties, and the freedom to modify the surface chemistry to target specific analyte species (functionalization) together with good optical transparency and high mechanical strength, they have become the favorite materials for designing small-size sensors with controlled properties. Additional impetus to the pace of sensor development has come from the recent advances in engineering of 2D materials through stacking, doping, and alloying.

Consequent upon the proliferation of research activity on 2D materials, the last decade has witnessed an exponential deluge of research papers on sensors made from these materials for mechanical, optical, chemical and biochemical stimuli. The aim of this book is to bring together and systematically organize the far-flung and scattered knowledge created from the rapid strides in 2D materials-based sensors and to make up-to-date information available in an easily accessible, concise, and cohesive format.

The book is organized into nine chapters. Beginning with an introduction to the key members of 2D materials family and description of their exceptional properties in Chapter 1, techniques of synthesis/deposition of these materials and their incorporation into the process flow of sensor devices are elaborated in Chapter 2. In the physical sensor category, Chapter 3 deals with devices for pressure and strain sensing, acceleration measurement, and other mechanical parameters. Among optical devices, the recent advances in photodetectors are reviewed in Chapter 4. Magnetic sensors are treated in Chapter 5. In the chemical sensor class, devices for detection of heavy metal ions are discussed in Chapter 6 and those for gases in Chapter 7. Biological sensors for health and wellness monitoring, and disease detection are covered in detail in Chapter 8. Finally, the perspectives and challenges of research on 2D material sensors are summarized in Chapter 9.

Presenting an all-inclusive overview of the latest research and developments in the area of sensors fabricated from 2D nanomaterials, this reference text will be a valuable resource for industrial and academic researchers and engineers, and advanced students in the field of applied physics, materials science, nanotechnology, biotechnology, chemical, and electronics engineering.

Welcome to the World of 2D Material Sensors!
Advancing from 3D solids to 2D planar sheets has deep implications
Because charge carriers have to follow new rules and regulations

Changing over from Newtonian classical mechanical motion
To quantum mechanical philosophies and notions
From free carrier movement
To quantum confinement
From fixed material conductivities and bandgaps
To tunable conductivities and bandgaps
Allowing tailoring of properties through defect centers and charge traps
From a small specific surface area
To an extraordinary large specific surface area
Arousing amazement and euphoria
The 2D materials portfolio opens a slew of exotic properties for low-cost sensor
benefits
For supersensitive device development with low detection limits
So together, let us explore the 2D material sensor opportunities
For designing sensors with imagination and ingenuity
Let's unveil the vast treasures of 2D material sensors
Demystify and admire their impressive properties and splendor
It is a subject offering ample scope for innovations.
Full of sensations, beyond imagination!

Vinod Kumar Khanna
Chandigarh, India

Acknowledgments

I am thankful to Almighty God for giving me the wisdom and strength to complete this work.

I deeply appreciate the great scientists and engineers whose pioneering work on 2D nanomaterials paved the way towards a global revolution in material science. I salute the diligence, dedication, and passion of all workers whose contributions have placed 2D materials and sensors at the present advanced state of knowledge. Their ground-breaking research papers, articles, webpages, and books are cited in the reference lists appended at the end of each chapter.

I acknowledge the kind cooperation of the editor and staff at CRC press all through this project.

Last but not the least, I owe a debt of gratitude to my family for their affectionate and unwavering support and continuous encouragement in my endeavors.

Vinod Kumar Khanna
Chandigarh, India

About the Author

INTRODUCTION

Vinod Kumar Khanna is an independent researcher at Chandigarh, India. He is a retired chief scientist from the Council of Scientific and Industrial Research (CSIR)-Central Electronics Engineering Research Institute (CEERI), Pilani, India, and a retired professor from the Academy of Scientific and Innovative Research (AcSIR), Ghaziabad, India. He is a former emeritus scientist, CSIR, and a professor emeritus, AcSIR, India. His broad areas of research were the design, fabrication, and characterization of power semiconductor devices and micro- and nanosensors.

ACADEMIC QUALIFICATIONS

He received the M.Sc. degree in Physics with specialization in electronics from the University of Lucknow in 1975, and Ph.D. degree in Physics from Kurukshetra University in 1988 for the thesis entitled 'Development, characterization and modeling of the porous alumina humidity sensor'.

RESEARCH/TEACHING EXPERIENCE AND ACCOMPLISHMENTS

His research experience spans over a period of 40 years from 1977 to 2017. Starting his career as a research assistant in the Department of Physics, University of Lucknow, from 1977 to1980, he joined CSIR-Central Electronics Engineering Research Institute, Pilani (Rajasthan), in April 1980. At CSIR-CEERI, he worked on several CSIR-funded as well as sponsored research and development projects. His major fields of research included power semiconductor devices and microelectronics/MEMS and nanotechnology-based sensors and dosimeters.

In the power semiconductor devices area, he worked on the high-voltage and high-current rectifier (600A, 4300V) for railway traction, high-voltage TV deflection transistor (5A, 1600V), power Darlington transistor for AC motor drives (100A, 500V), fast-switching thyristor (1300A, 1700V), power DMOSFET, and IGBT. He contributed toward the development of sealed-tube Ga/Al diffusion for deep junctions, surface electric field control techniques using edge beveling and contouring of large-area devices, and floating field limiting ring design. He carried out an extensive characterization of minority-carrier lifetime in power semiconductor devices as a function of process steps. He also contributed toward the development of P–I–N diode neutron dosimeter and PMOSFET-based gamma ray dosimeter.

In the area of sensor technology, he worked on the nanoporous aluminum oxide humidity sensor; ion-sensitive field-effect transistor-based microsensors for biomedical, food, and environmental applications; microheater-embedded gas sensor for automotive electronics; MEMS acoustic sensor for satellite launch vehicles; and capacitive MEMS ultrasonic transducer for medical applications.

As an AcSIR faculty member, he was the course coordinator of MEMS/IC Technology for advanced semiconductor electronics program (2011–2013) and taught 'MEMS Technology' to students pursuing M.Tech degree. As an adjunct faculty, BESU, Kolkata, he taught 'MEMS Technology & Design' to M.Tech (Mechatronics) students. He was invited by IIT, Jodhpur, for delivering lectures on 'Semiconductor Fundamentals and Technology' to B.Tech. students during February 2011. He guided B.Tech./M.Tech. thesis of students from BITS, Pilani; VIT, Vellore; and Kurukshetra University. He also guided a Ph.D. thesis on 'MEMS acoustic sensor', MNIT, Jaipur.

SEMICONDUCTOR FACILITY CREATION AND MAINTENANCE

He was responsible for setting up and looking after diffusion/oxidation facilities, edge beveling and contouring, reactive sputtering, and carrier lifetime measurement facilities. As the Head of MEMS and microsensors group, he looked after the maintenance of six-inch MEMS fabrication facility for R&D projects as well as augmentation of processing equipment under this facility at CSIR-CEERI.

SCIENTIFIC POSITIONS HELD

During his tenure of service at CSIR-CEERI from April 1980 till superannuation in November 2014, he was promoted to various positions including one merit promotion. He retired as a chief scientist and Professor (AcSIR: Academy of Scientific and Innovative Research), and Head of MEMS and Microsensors Group. Subsequently, he worked for three years as Emeritus Scientist, CSIR and Emeritus Professor, AcSIR, from November 2014 to November 2017. After completion of the emeritus scientist scheme, he now lives at Chandigarh. He is a passionate author and enjoys reading and writing.

MEMBERSHIP OF PROFESSIONAL SOCIETIES

He is a fellow and life member of the Institution of Electronics and Telecommunication Engineers (IETE), India. He is a life member of Indian Physics Association (IPA), Semiconductor Society, India (SSI), and Indo-French Technical Association (IFTA).

FOREIGN TRAVEL

He is widely travelled. He participated in and presented research papers at the IEEE Industry Application Society (IEEE-IAS) Annual Meeting at Denver, Colorado, USA, in September–October, 1986. His short-term research assignments include deputations to Technische Universität Darmstadt, Germany, in 1999; at Kurt-Schwabe-Institut fur Mess-und Sensortechnike e.V., Meinsberg, Germany, in 2008; and at Fondazione Bruno Kessler, Trento, Italy, in 2011 under collaborative programs. He was a member of the Indian Delegation to Institute of Chemical Physics, Novosibirsk, Russia, in 2009.

SCHOLARSHIPS AND AWARDS

He was awarded the National Scholarship by the Ministry of Education and Social Welfare, Government of India, on the basis of Higher Secondary result, 1970; CEERI Foundation Day Merit Team Award for projects on fast-switching thyristor (1986); power Darlington transistor for transportation (1988), P–I–N diode neutron dosimeter (1992); and high-voltage TV deflection transistor (1994); Dr N. G. Patel Prize for best poster presentation in 12th National Seminar on Physics and Technology of Sensors, 2007, BARC, Mumbai; CSIR-DAAD Fellowship in 2008 under Indo-German Bilateral Exchange Programme of Senior scientists, 2008. He is featured in the Stanford–Elsevier prestigious list of world top 2% scientists (2022, Elsevier Data Repository, V4, doi:10.17632/btchxktzyw.4). He is named as a Highly Ranked Scholar-Lifetime # 4 Nanoelectronics by ScholarGPS.

RESEARCH PUBLICATIONS AND BOOKS

He has published 194 research papers in leading peer-reviewed national/international journals and conference proceedings. He has authored 20 books, and has also contributed six chapters in edited books. He has five granted patents to his credit, including two US patents.

About the Book

2D materials are a class of nanomaterials made of a single or few layers of atoms with thickness of a few nanometers or less. This book is about 2D materials, their interesting properties, and the sensors made utilizing these special properties. It contains nine chapters. Following is a summary of the contents of the book highlighting the 2D materials used and their roles in sensor technology in each chapter.

Chapter 1 introduces the members of 2D materials family, namely, graphene, graphene oxide, reduced graphene oxide (rGO), graphitic carbon nitride (g-C_3N_4), hexagonal boron nitride (hBN), layered metal chalcogenides (LMDCs, e.g., MoS_2, $MoSe_2$, WS_2, WSe_2); 2D Xenes, 2D MXenes, layered double hydroxides, metal-organic frameworks, and metal oxides. Their mechanical, electronic, and optical properties used in sensors are discussed.

Chapter 2 deals with the synthesis and processing of 2D materials for sensors under two heads: top-down methods (exfoliation, ion intercalation, and aqueous acid etching) and bottom-up methods (wet-chemical synthesis, magnetron sputtering, pulsed laser deposition, and molecular beam epitaxy). 2D materials considered are graphene, graphene oxide, reduced graphene oxide, g-C_3N_4, hexagonal boron nitride, MoS_2, $MoSe_2$, WSe_2, black phosphorous, and MXenes.

Chapter 3 describes 2D material mechanical sensors including pressure sensors (piezoresistive, capacitive, squeezed film, and iontronic), accelerometers, air-flow and liquid-flow sensors, and ultrasound sensor. 2D materials used are graphene, suspended graphene, and $PtSe_2$.

Chapter 4 looks into 2D material photodetectors fabricated in photoconductor, heterostructure, Schottky junction, P–N junction, in-plane PN homojunction, and phototransistor structures, namely, FET, ferroelectric-driven FET, encapsulated FET, and dual-gate FET. 2D materials used are graphene, MoS_2, WSe_2, and black phosphorous.

Chapter 5 surveys magnetic sensors of induction coil, Hall-effect, and extraordinary magnetoresistance configurations; and quantum sensors using negatively charged boron vacancies. 2D materials used are VSe_2, graphene, hBN-encapsulated ultra-clean graphene, boron vacancy defects in hexagonal boron nitride, and $CrTe_2$-^{10}hBN.

Chapter 6 presents heavy metal ion sensors, mainly for As (III), Cd (II), Pb (II), Cu (II), and Hg (II) ions working on chemically modified electrodes (voltammetry), FET, fluorescence, and chemiluminescence principles. 2D materials used are reduced graphene oxide, graphene, MoS_2, g-C_3N_4, and Ti_3C_2 MXene.

Chapter 7 treats gas sensors for SO_2, H_2S, NO_2, CO, H_2, THF, VOCs, and moisture, operating on chemiresistive, chemically modified electrodes (voltammetry), FET, quartz crystal microbalance, surface acoustic wave, and photoelectrochemical principles. 2D materials used are graphene, graphene oxide, $MoSe_2$, MoS_2, WS_2, WSe_2, $g\text{-}C_3N_4$, $SnSe_2$, black phosphorous, LDH, and 2D $MoO_{3\text{-}x}$.

Chapter 8 outlines biological sensors for glucose, ascorbic acid, dopamine, uric acid, paracetamol, aspirin, caffeine, pesticides, lung cancer biomarker CYFRA21-1, neuron-specific enolase (NSE) cancer biomarker, PSA, COVID-19 virus, and DNA using modified electrodes (voltammetry), FET, fluorescence, and surface plasmon resonance effects. 2D materials used are MXene, MoS_2, WSe_2, conductive MOF, graphene, carboxylic graphene, graphene oxide, $g\text{-}C_3N_4$, and black phosphorous.

Chapter 9 explores the research challenges facing 2D materials and suggests viable solutions, namely, large-area synthesis of 2D materials, their doping, ohmic contacts formation, insulators for 2D nanoelectronics, and defects engineering in these materials. 2D materials discussed are graphene, hBN, MoS_2, $MoSe_2$, WS_2, $TaSe_2$, $MoTe_2$, and ReS_2.

The book will be of immense value for post-graduate and Ph.D. students, and scientists engaged in research on 2D materials as well as process engineers working on 2D materials technology development.

Abbreviations, Acronyms, and Chemical Symbols

0D, 1D, 2D 3D	Zero, one, two-, and three-dimensional
1L MoS$_2$	Monolayer molybdenum disulfide
1-T phase	Metallic octahedral crystal phase
1X PBS	A phosphate-buffered saline solution with composition: 0.27M KCl, 0.137M NaCl, 0.18M KH$_2$PO4 and 0.01M Na$_2$HPO4, adjusted to pH=7.4
1,3-BUT	1,3-butanediol
2-H phase	Semiconducting trigonal prismatic crystal phase
3L WSe$_2$	Three-layer tungsten diselenide
A	Ampere
AA	Ascorbic acid
AAE	Aqueous acid etching
AAS	Atomic absorption spectroscopy
AC	Alternating current
AChE	Acetylcholinesterase
aF	Attofarad (1.0×10^{-18} Farad)
AFM	Atomic force microscopy
Ag	Silver (Argentum)
AgCl	Silver chloride
Al	Aluminum
AlCl$_3$·xH$_2$O	Aluminum chloride hydrate
Alk-Ti$_3$C$_2$	Alkaline intercalated Ti$_3$C$_2$
Al (NO$_3$)$_3$·9H$_2$O	Aluminum nitrate nonahydrate
Al$_2$O$_3$	Aluminum oxide
AMR	Anisotropic magnetoresistance
A^{n-}	Anion
An-Br	9,10-Dibromo-anthracene
An-CH$_3$	9,10-Dimethylanthracene
AP-CVD	Ambient-pressure chemical vapor deposition
APTES	(3-Aminopropyl) triethoxysilane
Ar	Argon
As	Arsenic
ASA	Aspirin
ASLSV	Anodic stripping linear sweep voltammetry
ASV	Anodic stripping voltammetry
ATCl	Acetylthiocholine chloride
AT-cut	A cut at 35°angle from the Z-axis of a quartz crystal
atm	Atmosphere
ATR	Attenuated reflectance (method)

ATR-FTIR	Attenuated reflectance-Fourier transform infrared (spectroscopy)
Au	Aurum (Gold)
AuNCs	Gold nanoclusters
AuNPs	Gold nanoparticles
AuPd	An alloy of gold and palladium
$Au_{0.6}Pd_{0.4}$	An alloy of gold (0.6) and palladium (0.4)
B	Boron
BCN	Boron carbon nitride
B_2H_6	Diborane
BK7 glass	High optical quality borosilicate-crown glass made by SCHOTT
[Bmim] [Tf2N]	1-butyl-3-methyl-imidazolium bis(trifluoromethanesulfonyl) imide
BN	Boron nitride
BNGr	B- and N-codoped graphene
$B_3N_3H_6$	Borazine
BOX	Buried oxide
BP	Black phosphorous
BPNS	Black phosphorous nanosheet
BP-TFG	Black phosphorous-tilted fiber grating
Br^-	Bromide ion
BSA	Bovine serum albumin
C	Carbon
°C	Degree Centigrade
Ca	Calcium
CAF	Caffeine
CaF_2	Calcium fluoride
C-band	4–8 GHz
Cd	Cadmium
cDNA	Complementary deoxyribonucleic acid
CeO_2	Cerium (IV) oxide or ceric oxide
CGR	Carboxylic graphene
CGR–NF	Carboxylic graphene-Nafion
CGT	$Cr_2Ge_2Te_6$
CH_4	Methane
CHF_3	Trifluoromethane or fluoroform
C_4H_9Li	n-Butyllithium
$C_4H_{13}N_3$	Diethylenetriamine, 1,2-Ethanediamine
$C_4H_{10}Se$	Diethyl selenide
$C_{12}H_{25}SH$	1-Dodecanethiol
CL	Chemiluminescence
Cl^-	Chloride ion
ClO_2	Chlorine dioxide
ClO_4^-	Perchlorate ion
cm	Centimeter
C-MOF	Conductive metal–organic framework
CMOS	Complementary metal–oxide–semiconductor

C/N ratio	Carbon/nitrogen ratio
CO	Carbon monoxide
Co	Cobalt
C=O	Carbonyl group
CO$_2$	Carbon dioxide
CO$_3{}^{2-}$	Carbonate anion
C-O-C	Epoxide or alkoxy group
Co$_{69.25}$Fe$_{4.25}$Si$_{13}$B$_{12.5}$Nb$_1$	Melt-extracted amorphous microwire made of cobalt, iron, silicon, boron and niobium
-COOH	Carboxyl group
COVID-19	Coronavirus disease-2019
Cr	Chromium
CRG	Chemically reduced graphene
Cr$_2$Ge$_2$Te$_6$	Chromium germanium telluride
CrI$_3$	Chromium (III) iodide
CrSiTe$_3$	Chromium tellurosilicate
CrTe$_2$	Chromium ditelluride
CS	Chitosan
Cu	Copper
CuInSe$_2$	Copper indium selenide
Cu(OAc)$_2$·H$_2$O	Copper(II) acetate monohydrate
CV	Cyclic voltammetry
CVD	Chemical vapor deposition
CYFRA21-1	Cytokeratin 19 fragment
Cys	Cysteine
DA	Dopamine
DBA(Mes)$_2$	9,10-dimesityl-9,10-dihydro-9,10-diboraanthracene
DC	Direct current
DCC	N, N'-Dicyclohexylcarbodimide
DI	Deionized (water)
DLG	Double layer graphene
DMF	N, N-dimethylformamide
DNA	Deoxyribonucleic acid
DNAzyme	Deoxyribozyme or DNA enzyme or catalytic DNA
DPASV	Differential pulse anodic stripping voltammetry
DPV	Differential pulse voltammetry
DRIE	Deep reactive ion etching
ds-DNA	Double stranded deoxyribonucleic acid
DSNSA	Duplex-specific nuclease signal amplification
ds-Ti$_2$CCl$_2$	Directly synthesized-Ti$_2$CCl$_2$
EA	Ethanolamine
e-beam	Electron-beam
EDC	*N*-Ethyl-*N'*-(3-dimethylaminopropyl) carbodiimide hydrochloride
EIS	Electrochemical impedance spectroscopy
ELISA	Enzyme-linked immunosorbent assay

EMF	Electromotive force
EMR	Extraordinary magnetoresistance
EPR	Electron paramagnetic resonance (spectroscopy)
EQE	External quantum efficiency
EQE (λ)	EQE at a wavelength λ
ES	Excited state
ESFDA	Electrostatic-force-directed assembly
ESR	Electron spin resonance, same as electron paramagnetic resonance (EPR) spectroscopy
EtOH	Ethanol
eV	Electron volt
F	Farad
Fe	Iron (Ferrum)
$FeCl_3$	Iron (III) chloride
$FeCl_3$-FLG	$FeCl_3$-intercalated few-layered graphene
$FeCl_3 \cdot 6H_2O$	Ferric chloride hexahydrate or Iron (III) chloride hexahydrate
$[Fe(CN)_6]^{3-}$	Ferricyanide ion or hexacyanidoferrate (III) ion or hexacyanoferrate(III) ion
$[Fe(CN)_6]^{4-}$	Ferrocyanide ion or hexacyanidoferrate(II) ion or hexacyanoferrate(II) ion
Fe_3GeTe_2	Iron germanium telluride
$Fe(NO_3)_3 \cdot 9H_2O$	Ferric nitrate nonahydrate or iron (III) nitrate nonahydrate
Fe_2O_3	Ferric oxide or iron (III) oxide
FET	Field-effect transistor
fg	Femtogram (10^{-15} gram)
FGT	Fe_3GeTe_2
FIB	Focused ion beam
FLG	Few-layered graphene, few-layered graphite
fMLP	N-formyl-methionyl-leucyl-phenylalanine
FR	Flame retardant
FRET	Fluorescence resonance energy transfer or Förster resonance energy transfer
F4-TCNQ	2,3,5,6-Tetrafluoro-7,7,8,8-tetracyano-quinodimethane
FTIR	Fourier transform infrared (spectroscopy)
FTO	Fluorine-doped tin oxide
g	Gram
Ga	Gallium
GaAs	Gallium arsenide
GaSe	Gallium (II) selenide
Gbit	Gigabit
GCE	Glassy carbon electrode
$g\text{-}C_3N_4$	Graphitic carbon nitride
GeSe	Germanium monoselenide or germanium selenide
GHz	Gigahertz
GMR	Giant magnetoresistance
GO	Graphene oxide

GO-DCC	Graphene oxide-N, N'-dicyclohexylcarbodimide
GO_x	Glucose oxidase
GPa	GigaPascal
GR	Graphene
GR–NF	Graphene–Nafion
GS	Ground state
h	Hour
H^+	Hydrogen ion, proton
H_2	Hydrogen
HAFe/GN	Horizontally arranged Fe_2O_3/graphene nanosheets
$HAuCl_4$	Hydrogen tetrachloroaurate (III) or tetrachloroauric (III) acid or chloroauric acid
H_2BDC	Terephthalic acid or 1,4-benzenedicarboxylic acid
hBN, h-BN	Hexagonal boron nitride
HCl	Hydrogen chloride, hydrochloric acid
HCT-116	Human colorectal carcinoma cell line
He	Helium
HF	Hydrofluoric acid
HfO_2	Hafnium (IV) oxide
Hg	Mercury (Hydrargyrum)
HHTP	2,3,6,7,10,11-Hexahydroxytriphenylene
H_3NBH_3	Borazine
HNO_3	Nitric acid
H_2O	Water
H_2O_2	Hydrogen peroxide
HOPG	Highly ordered pyrolytic graphite
HPLC	High performance liquid chromatography
H_3PO_4	Phosphoric acid or orthophosphoric acid
H_2PP	Protoporphyrin IX
$H_2PtCl_6 \cdot 6H_2O$	Hydrogen hexachloroplatinate (IV) hexahydrate or hexachloro-platinic acid hexahydrate
H_2S	Hydrogen sulfide
H_2Se	Hydrogen selenide
H_2SO_4	Sulfuric acid
HSQ	Hydrogen silsesquioxane
Hz	Hertz
I^-	Iodide ion
I_+ and I_-	Current-injecting contacts
ICP-MS	Inductively coupled plasma mass spectroscopy
ICP-OES	Inductively coupled plasma optical emission spectroscopy
IDEs, IDTs	Interdigitated electrodes
In_2O_3	Indium (III) oxide
In_2Se_3	Indium selenide
IQE	Internal quantum efficiency
IQE (λ)	IQE at wavelength λ
IR	Infrared

ISC	Intersystem crossing
ITO	Indium tin oxide
J	Joule
K	Kelvin
K$^+$	Potassium ion
KCl	Potassium chloride
keV	Kilo electron volt
KMnO$_4$	Potassium permanganate
KOH	Potassium hydroxide
kPa	KiloPascal
KrF	Krypton-fluoride
L	Liter
La$_3$Ga$_5$SiO$_{14}$	Lanthanum gallium silicate (Langasite)
LC	Inductance–capacitance
LCR	Inductance (L), capacitance (C), and resistance (R)
LDH	Layered double hydroxide
LDR	Linear dynamic range
LGS	Langasite (Lanthanum gallium silicate)
Li$^+$	Lithium cation
LiOH	Lithium hydroxide
Li$_x$MoS$_2$	Lithium intercalated molybdenum disulfide
LMDC	Layered metal chalcogenide
LMH	Layered materials heterostructure
LoD or LOD	Limit of detection
LPCVD	Low-pressure chemical vapor deposition
M	Molar
m	Meter
m$^\circ$	Milli degree
M^{2+}	Divalent metal cation
M^{3+}	Trivalent metal cation
mA	Milliampere
MAX phase	Simplified form of "$M_{n+1}AX_n$" (M=an early transition metal, A=a group A element, X= C and/or N atom, n=1, 2 or 3)
Mbar	Millibar
MBE	Molecular beam epitaxy
MEMS	Microelectromechanical systems
MeV	Mega electron-volt
meV	Millielectronvolt
Mg	Magnesium
Mg$_6$Al$_2$(CO$_3$)(OH)$_{16}$•4(H$_2$O)	Magnesium-aluminum hydrotalcite or aluminum–magnesium layered double hydroxide or anionic clay
MGM	Metal-graphene-metal
Mg(OH)$_2$	Magnesium hydroxide
MG/SG	Middle gate/Side gate
MHz	Megahertz
min	Minute

MIP	Molecularly imprinted polymer
miRNA	MicroRNA or Microribonucleic acid
mJ	Millijoule
mK	Milli Kelvin
MLG	Monolayer graphene
mm	Millimeter
mM	Millimolar
MnBi$_2$Te$_4$	Manganese bismuth telluride
Mo	Molybdenum
MoCl$_5$	Molybdenum (V) chloride
MOCVD	Metal-organic chemical vapor deposition
MOF	Metal-organic framework
MoO$_3$	Molybdenum trioxide
MoS$_2$	Molybdenum disulfide
MoSe	Molybdenum monoselenide
MoSe$_2$	Molybdenum diselenide
MoTe$_2$	Molybdenum ditelluride or molybdenum telluride
MOX	Metal oxide
MPa	MegaPascal
MS	Metastable state
ms	Millisecond
MSD	Magnetron sputtering deposition
MUA	11-Mercaptoundecanoic acid
mV	Millivolt
MX$_2$	M is a transition metal, X is a chalcogen atom
N	Nitrogen
Na$^+$	Sodium ion
nA	Nanoampere (10^{-9} A)
NaBH$_4$	Sodium borohydride, or sodium tetrahydridoborate or sodium tetrahydroborate
NaAc-Hac	Sodium acetate-acetic acid buffer solution
NaCl	Sodium chloride (commonly called Table salt)
Na$_2$MoO$_4$·2H$_2$O	Sodium molybdate dihydrate
Na-NH$_2$	1,5 Naphthalenediamine
NaNO$_3$	Sodium nitrate
NaOH	Sodium hydroxide (Caustic soda)
NaYF$_4$	Sodium yttrium fluoride
NaYF$_4$: Yb, Er @SiO$_2$ UCNPs	Sodium yttrium fluoride upconversion nanoparticles doped with lanthanide metals (ytterbium, erbium) and coated with silicon dioxide
Nb	Niobium
NCM-460	A human colon epithelial cell line (A normal, non-cancerous cell line derived from normal human colon mucosa)
NCs	Nanoclusters
Nd:YAG	Neodymium-doped yttrium aluminum garnet (Nd:Y$_3$Al$_5$O$_{12}$)
NEMS	Nanoelectromechanical systems

NEP	Noise equivalent power
NF	Nafion
NFs	Nanoflowers
ng	Nanogram (10^{-9} gram)
NH_3	Ammonia
$N_2H_4.H_2O$	Hydrazine hydrate
$(NH_4)_6Mo_7O_{24}$	Ammonium heptamolybdate
NHS	N-Hydroxysuccinimide
Ni	Nickel
NIR	Near infrared
Nm	Newton-meter
nm	Nanometer
nM	Nanomolar
nmhr^{-1}	Nanometer hour^{-1}
nmol	Nanomole
NMP	N-Methyl-2-pyrrolidone
nN	Nano Newton
NO_2	Nitrogen dioxide
NO_3^-	Nitrate anion
NPA	Nanopatch antenna
NPs	Nanoparticles
NRs	Nanorods
ns	Nanosecond
NSE	Neuron-specific enolase
NSs	Nanosheets
NST	Nb_2SiTe_4
nT	Nanotesla
N-V center	Nitrogen-vacancy center
O_2	Oxygen
ODMR	Optical detection of magnetic resonance
Oe	Oersted
-OH	Hydroxyl group
OMDR	Optical-microwave double resonance
Pa	Pascal
PAR	Paracetamol
Pb	Lead (Plumbum)
PBE	Photobolometric effect
PBS	Phosphate-buffered saline
PC	Polycarbonate
PCE	Photoconductive effect
Pd	Palladium
PDMS	Polydimethylsiloxane
PEDOT: PSS	Poly (3,4-ethylenedioxythiophene): Poly(styrenesulfonate)
PEDT	Poly (3,4-ethylenedioxythiophene)
PET	Polyethylene terephthalate, Photo-induced electron transfer
PFM	Piezoresponse force microscopy

pg	Picogram (10^{-12} grams)
PGE	Photogating effect
PI	Polyimide
PL	Photoluminescence
PLD	Pulsed laser ablation
pm	Picometer
pM	Picomolar
PM2.5	Particulate matter 2.5 (concentration of particles with diameter $\leq$ 2.5 µm, expressed in micrograms per cubic meter of air ($\mu g/m^3$))
PM10	Particulate matter 10 (concentration of particles with diameter $\leq$ 10 µm, expressed in micrograms per cubic meter of air ($\mu g/m^3$))
PMMA	Polymethyl methacrylate
pmol	Picomole
ppb	Parts per billion (parts of gas per billion parts of air), billion=10 million=10^9
ppm	Parts per million (parts of gas per million parts of air), million=10^6
ps	Picosecond
PSA	Prostate-specific antigen
PSS	Poly(styrenesulfonate)
Pt	Platinum
pT	Picotesla
PTE	Photothermoelectric effect
PTFE	Polytetrafluoroethylene
PtNPs	Platinum nanoparticles
PtSe$_2$	Platinum diselenide
PVA	Polyvinyl alcohol
P(VDF-TrFE)	Poly (vinylidene fluoride-co-trifluoroethylene) copolymer
PVE	Photovoltaic effect
QCM	Quartz crystal microbalance
QR	Quenching ratio
QSHI	Quantum spin Hall insulator
QY	Quantum yield
RCOO$^-$	Carboxylate ion
ReS$_2$	Rhenium disulfide
RF	Radio frequency
rGO	Reduced graphene oxide
Rh	Rhodium
RHEED	Reflection high-energy electron diffraction
RIE	Reactive ion etching
RIU	Refractive index unit
RNA	Ribonucleic acid
RPM	Revolutions per minute
s	Second
S	Sulfur (a chemical element), Siemen (unit of electrical conductance)

SANS	Small-angle neutron scattering
SARS-CoV-2	Severe acute respiratory syndrome-coronavirus-2
SAW	Surface acoustic wave
SC	Sodium cholate
SCE	Saturated calomel electrode
SDS	Sodium dodecyl sulfate
Se	Selenium
SeO$_2$	Selenium dioxide
SGGT	Solution-gated graphene transistor
Si	Silicon
SiC	Silicon carbide
SiGe	Silicon–germanium
Si$_3$N$_4$	Silicon nitride
SiO$_2$	Silicon dioxide
SLG	Single-layer graphene
Sn	Tin
SnCl$_2$.2H$_2$O	Stannous chloride dihydrate or Tin (II) chloride dihydrate
SnCl$_4$.5H$_2$O	Tin (IV) chloride pentahydrate or Stannic chloride pentahydrate
SnI$_4$	Tin (IV) iodide
Sn-MOF	Sn-based metal organic framework
SnO$_2$	Tin (IV) oxide
S–N ratio	Signal-to-noise ratio or SNR
SnS$_2$	Tin (IV) sulfide
SnSe$_2$	Stannic selenide or Tin (IV) selenide
SnSO$_4$	Tin (II) sulfate
SO	Spin-orbit (coupling)
SO$_2$	Sulfur dioxide
SO$_3$	Sulfur trioxide
SO$_4^-$	Sulfate ion
SOG	Spin-on-glass
SOI	Silicon-on-insulator (wafer)
SPR	Surface plasmon resonance
SQUID	Superconducting quantum interference device
Sr	Strontium
ss-DNA	Single-stranded deoxyribonucleic acid
Sulfo-NHS	N-hydroxysulfosuccinimide
SW-48	Human colorectal adenocarcinoma cell line
SW-AdASV	Square wave-adsorptive anodic stripping voltammetry
SWASV	Square wave-anodic stripping voltammetry
SWCNT	Single-walled carbon nanotube
T	Tesla
Ta	Tantalum
TAC	Thermally assisted conversion
TDSO	Time differential of the signal output
Te	Tellurium
TFG	Tilted fiber grating

TFSI	Bis(trifluoromethane)sulfonimide or bis(trifluoromethylsulfonyl) amine or bis(trifluoromethanesulfonyl)amine
TGA	Thioglycolic acid
THF	Tetrahydrofuran
T-Hg (II)-T	Thymine-Hg (II)-thymine
Ti	Titanium
Ti_3AlC_2	Titanium aluminum carbide
Ti_3C_2	Titanium carbide
$TiCl_4$	Titanium tetrachloride
$Ti_3C_2T_x$	Titanium carbide MXene, T=O, -OH, -F, - Cl, etc.
TiO_2	Titanium dioxide or titanium (IV) oxide or titania
TMDCs or TMDs	Transition metal dichalcogenides
TMR	Tunnel magnetoresistance
TRG	Thermally reduced graphene
UA	Uric acid
UCNPs	Upconversion nanoparticles
ULCB	Ultralow carbon bainitic (steel)
UV	Ultraviolet
V	Volt, Vanadium (a chemical element)
V_+ and V_-	Voltage measuring contacts
VAFe/GN	Vertically arranged Fe_2O_3/graphene nanosheets
V_{B-} defects	Negatively charged boron vacancy defects
vdW	van der Waals (force)
V_2O_5	Vanadium pentoxide I
VOC	Volatile organic compound
VSe_2	Vanadium diselenide
W	Watt
WCl_6	Tungsten hexachloride
WDM	Wavelength division multiplexing
WO_3	Tungsten (VI) oxide
WS_2	Tungsten disulfide
WSe_2	Tungsten diselenide
wt%	Weight percentage
Xe	Xenon
XPS	X-ray photoelectron spectroscopy
ZFS	Zero-field splitting
Zn	Zinc
$ZnCl_2.6H_2O$	Zinc chloride hexahydrate
$Zn (NO_3)_2.6H_2O$	Zinc nitrate hexahydrate
ZnO	Zinc oxide
$ZnSO_3$	Zinc sulfite
Zr	Zirconium
ZrB_4	Zirconium (IV) bromide
Zr_2CBr_2	Zirconium-based MXene (Bromine-based)
Zr_2CCl_2	Zirconium-based MXene (Chlorine-based)
$ZrCl_4$	Zirconium (IV) chloride

Mathematical Notations

A	Area
a	Acceleration
B	Bandwidth, Magnetic field
$\mathbf{B}$	Magnetic field vector
B_Y	Applied magnetic field in the Y-direction
B_z	Projection of magnetic field along the hBN c-axis.
C	Capacitance, mass sensitivity constant
c	Concentration
$c_1, c_2, c_3,\ldots$	Concentrations
$C_{\text{Parasitic}}$	Parasitic capacitance
D	Distance, thickness, zero-field splitting, a ZFS parameter equal to half the sum of resonance frequencies in OMDR
D^*	Specific detectivity
D_{ES}	Zero-field splitting in the excited state
D_{GS}	Zero-field splitting in the ground state
E	A ZFS parameter equal to half the difference of resonance frequencies in OMDR
$\mathbf{E}$	Electric field
e	Electronic charge
E_1, E_2	Energy levels
$E_{+1/2}$	Energy of the parallel state
$E_{-1/2}$	Energy for antiparallel aligned state
$E_{\text{Deposition}}$	Electrochemical deposition potential
E_{G}	Energy bandgap of a material
F	Force, Fluorescence emission intensity
F_0	Original fluorescence emission intensity
f_0	Resonance frequency in vacuum
f_{C}	Cut-off frequency
F_{Initial}	Initial fluorescence emission intensity
F_{Final}	Final fluorescence emission intensity
$f_{\text{Resonance}}$	Resonance frequency at ambient pressure
G	Gauge factor, photoconductive gain
G_0	Original conductance
g_0	Distance between the membrane and the stationary substrate
g_{e}	Landé g-factor
H	Magnetic field
h	Planck's constant
$\hbar$	Reduced Planck's constant
I	Current
$i_1, i_2, i_3,\ldots$	Currents
I_{b}	Biasing current
I_{D}	Dark current

I_H	Hall current
I_P	Photogenerated current
$I_P{}^*$	Photocurrent at $1\,mWcm^{-2}$ optical intensity
I_X	Applied current in the X-direction
k_1, k_2	Constants
L	Length, inductance
l	Azimuthal quantum number
$L_{Effective}$	Effective inductance
m	Mass
m_e	Electron mass
m_l	Magnetic quantum number
$\mathbf{MR}$	Magnetoresistance
m_s	Spin quantum number
n	Principal quantum number, a number representing the particular harmonic of vibration
$N_{Absorbed}$	Number of photons absorbed by the photodetector
$N_{Collected}$	Number of photogenerated free charge carriers collected by the photodetector electrodes
P	Incident optical power
$P_{Ambient}$	Pressure of ambient gas
$\mathbf{PL}$	Original photoluminescence
R	Resistance, photoresponsivity, reflectance
r_a, r_b	Radii of circular discs
$R_{Baseline}$	Baseline resistance of the sensor
R_{Gas}	Resistance of device after exposure to a known concentration of analyte gas
R_I	Photoresponsivity
$R_{Non\text{-}local}$	Non-local resistance
$R_{Parasitic}$	Parasitic resistance
$R_{Two\text{-}terminal}$	Two-terminal resistance
S	Sensitivity, response
S_A	Absolute sensitivity of the Hall sensor
S_I	Current-related sensitivity of the Hall sensor
S_{Max}	Maximum sensitivity
S_{Min}	Minimum sensitivity
T	Transmittance
$U(\theta)$	Interaction energy of a magnetic dipole moment with an external magnetic field
V	Voltage
V_b	Biasing voltage
$V_{Back\ gate}$	Back gate voltage
$V_{Difference}$	Non-local voltage of a graphene magnetometer
V_{DS}	Drain–source voltage
$V_{g,\ min}$	Gate voltage of minimum conductivity
V_{GS}	Gate-source voltage
V_H	Hall voltage

V_P	Photogenerated voltage
V_{Peak}	Potential for peak current
X_C	Capacitive reactance
X_L	Inductive reactance
Z_{COIL}	Impedance of the coil

Greek Alphabets

γ_e	Electron gyromagnetic ratio
ΔE	Energy separation between spectral lines, energy difference between the parallel state and antiparallel aligned state
Δf	Change in resonant frequency of the device
ΔG	Change in conductance
Δm	Change in mass
ΔP	Change in pressure
ΔPL	Change in photoluminescence
ΔR	Change in resistance
Δt	Time delay
$\Delta V_{DA\text{-}AA}$	Peak-to-peak separation potential between dopamine and ascorbic acid
$\Delta V_{g,\,min}$	Shift in Dirac point
ΔV_{Th}	Change in threshold voltage
$\Delta V_{UA\text{-}AA}$	Peak-to-peak separation potential between uric acid and ascorbic acid
$\Delta V_{UA\text{-}DA}$	Peak-to-peak separation potential between uric acid and dopamine
Δz	Zeeman shift
ε	Mechanical strain
ε_0	Permittivity of free space
ε_r	Relative permittivity of a material
θ	Angle between the magnetic field and magnetic dipole moment
λ	Wavelength
$\lambda_{Emission}$	Emission wavelength
$\lambda_{Excitation}$	Excitation wavelength
μ	Magnetic dipole moment
μA	Microampere
μ_B	Bohr magneton
$\mu_{Effective}$	Effective permeability
μL	Microliter
μM	Micromolar
μm	Micrometer or micron
$\mu mole$	Micromole
μs	Microsecond
μV	Microvolt
ν	Frequency
ν_0	Resonance frequency
ν_- and ν_+	Resonance frequencies in OMDR spectrum
ρh	Mass per unit area of the membrane
ρ_{Piezo}	Piezoresistivity
τ_f	Fall time
τ_r	Rise time
Ω	Ohm
$K\Omega$	Kilo ohm

1 The 2D Materials Family and Properties of 2D Materials Used in Sensors

This introductory chapter provides a quick overview of the prime members of 2D materials family that have been the focus of attention in recent years. It will be a curtain raiser to their significant impact on sensor technology and their vast contribution in developing novel devices. Starting from graphene and transition metal dichalcogenides (TMDCs) through black phosphorous, the description moves to Xenes, Mxenes, layered double hydroxides (LDHs), metal-organic frameworks (MOFs), and metal oxides (MOXs). Emphasis is laid on their remarkable properties for making sensors with improved characteristics over the conventional bulk materials.

1.1 NANOMATERIALS: 0D, 1D, AND 2D

1.1.1 Nanomaterial

A material that has one or more external dimensions smaller than 100 nm but larger than 1 nm is referred to as a nanomaterial. It is a material that has at least one external dimension in the range 1–100 nm (Figure 1.1).

1.1.2 0D Nanomaterial

If all three dimensions of a material are smaller than 100 nm, it is called a zero-dimensional or 0D material, e.g., a nanoparticle or a quantum dot.

1.1.3 1D Nanomaterial

If two dimensions of a material are <100 nm in size while the third dimension is >100 nm, it is known as a one-dimensional material or 1D material, e.g., a nanowire or carbon nanotube.

DOI: 10.1201/9781003330585-1

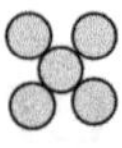

(a) 0D Nanomaterials (Noble metal nanoparticles, magnetic nanoparticles, nanoclusters, fullerenes, quantum dots) : All dimensions in nanoscale, no dimension in micro/macroscale.

(b) 1D Nanomaterials (Nanotubes, nanorods, nanowires, nanofibers): Two dimensions in nanoscale, one dimension in micro/macroscale.

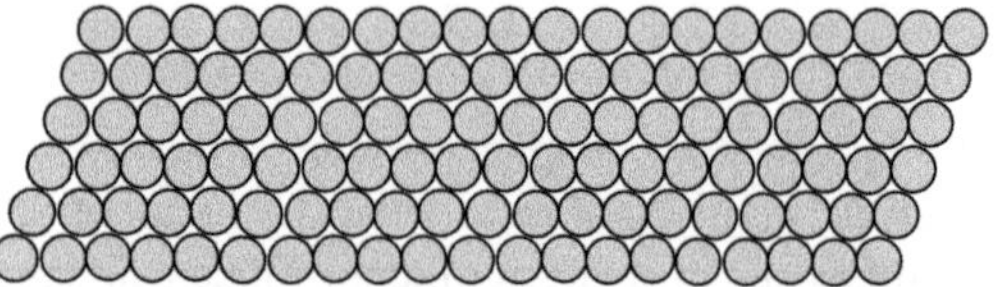

(c) 2D Nanomaterials (Nanofilms, nanolayers, nanosheets, all 2D materials, e.g., graphene, LMDCs, Xenes, MXenes, LDHs, MoFs, MOXs) : One dimension in nanoscale, two dimensions in micro/macroscale.

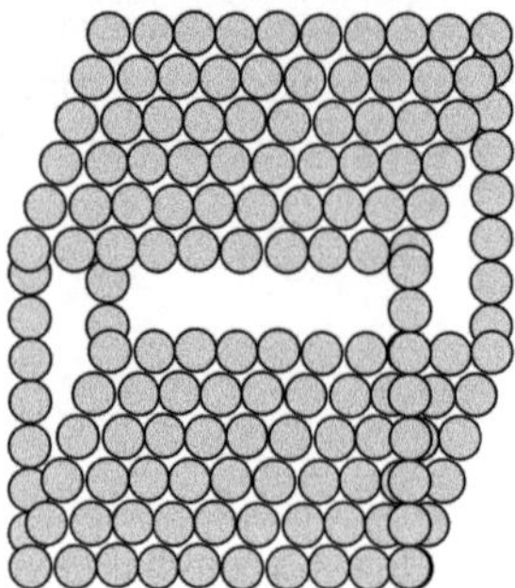
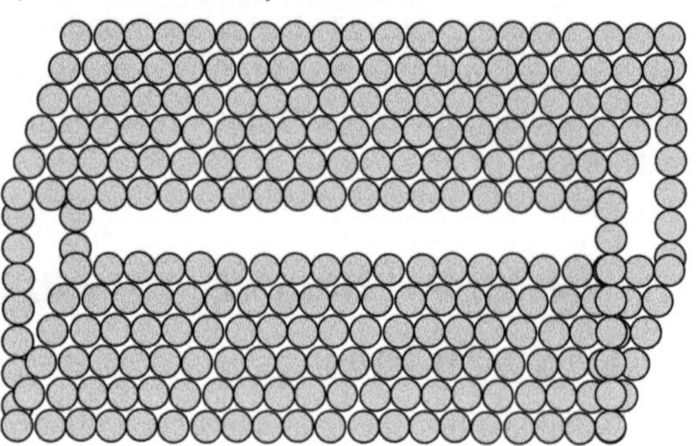

(d) 3D Materials (Graphite, 3D polycrystals, ...) : No dimension in nanoscale, all dimensions in micro/macroscale.

FIGURE 1.1   0d, 1D, 2D nanomaterials, and 3D bulk materials. Part (a) shows particles of different sizes having all dimensions in the nano regime and gives specific examples of such particles. Part (b) shows line-shaped or linear geometry materials with length dimension only exceeding the nanoscale limit. Examples of wires, rods, etc. are mentioned. Part (c) shows planar geometry materials whose thickness dimension only lies in the nanoscale. Examples of thin films, graphene, etc. are cited. Part (d) shows solid materials made of multiple layers with all dimensions larger than the nanoscale.

1.1.4   2D Nanomaterial

If one dimension of a material is <100 nm whereas the other two dimensions are larger than 100 nm, it is said to be a two-dimensional material or 2D material. The 2D materials or single-layer materials are crystalline nanomaterials comprising single or few atomic layers and having two dimensions, namely, those in X- and Y-directions, i.e., along the horizontal X–Y plane, greater than the nanometric range (100 nm) but the third dimension in the vertical or Z-direction is in the nanometric range (1–100 nm). Hence, they are thin films of atomic-scale thickness.

Because of the confinement of electron motion along the Z-direction, they act as quantum wells, which are thin layers confining the electrons in one dimension between two energy barriers.

2D materials are obtained from respective 3D materials of stratified structure, in which the in-plane atomic bonding is much stronger than along the stacking of strata. So, these 3D materials have the tendency of delamination resulting in the formation of stable layers of atomic thickness.

1.1.5 GRAPHENE: THE FIRST 2D NANOMATERIAL

The first 2D material isolated was graphene (Novoselov et al. 2004). It was described as a stable, few atoms thick, monocrystalline graphitic film with semi-metallic character in which electron and hole concentrations $\sim10^{13}$ cm^{-2} with extraordinarily high mobilities $\sim10^4$ cm^2 V^{-1} s^{-1} could be induced by the application of a gate voltage.

Ensuing the discovery of graphene in 2004, there was a boom in research on 2D materials leading to a vast library of such materials displaying a broad spectrum of properties from conducting to insulating behavior and showing mechanically softest to toughest nature.

1.2 GRAPHENE AND RELATED MATERIALS

1.2.1 GRAPHENE

It is an allotrope of carbon extracted from graphite (Figure 1.2(a)). It is a single layer of graphite. It consists of a plane of 0.14-nm thick single layer of sp^2-bonded carbon atoms (Figure 1.2(b)). These atoms are organized in a hexagonal lattice. The sp^2 hybridization occurs during covalent bonding. The bond length from a carbon atom to another carbon atom is 0.142 nm.

Pristine graphene has a breaking strength of 42 Nm^{-1} equating to an intrinsic strength of 130 GPa (Lee et al. 2008). For comparison, pure tungsten shows an ultimate tensile strength of 900 MPa (Prasad and Annamalai 2021), which is the highest for any metal. The strength of ultralow carbon bainitic (ULCB) steel ranges from 600 to 1000 MPa (Zhang et al. 2016).

The optical transparency of graphene is 97.7% (Sheehy and Schmalian 2009) meaning that a single layer of graphene absorbs 2.3% incident light so that two layers will absorb 4.6% light giving a transparency of 95.4%.

Achievement of electron mobility $>2\times10^5$ cm^2 V^{-1} s^{-1} has been reported in a single-layer suspended graphene having an electron density of $\sim2 \times 10^{11}$ cm^{-2} (Bolotin et al. 2008). The drawback of graphene is its zero bandgap, hampering its use in electronic switching. Its bandgap is tunable by the application of an electric field or strain but carrier mobility is sacrificed in this process, e.g., the mobility reduces to 200 cm^2 V^{-1} s^{-1} for 150 meV bandgap (Mir et al. 2020). Graphene is N-doped using boron and P-doped with nitrogen.

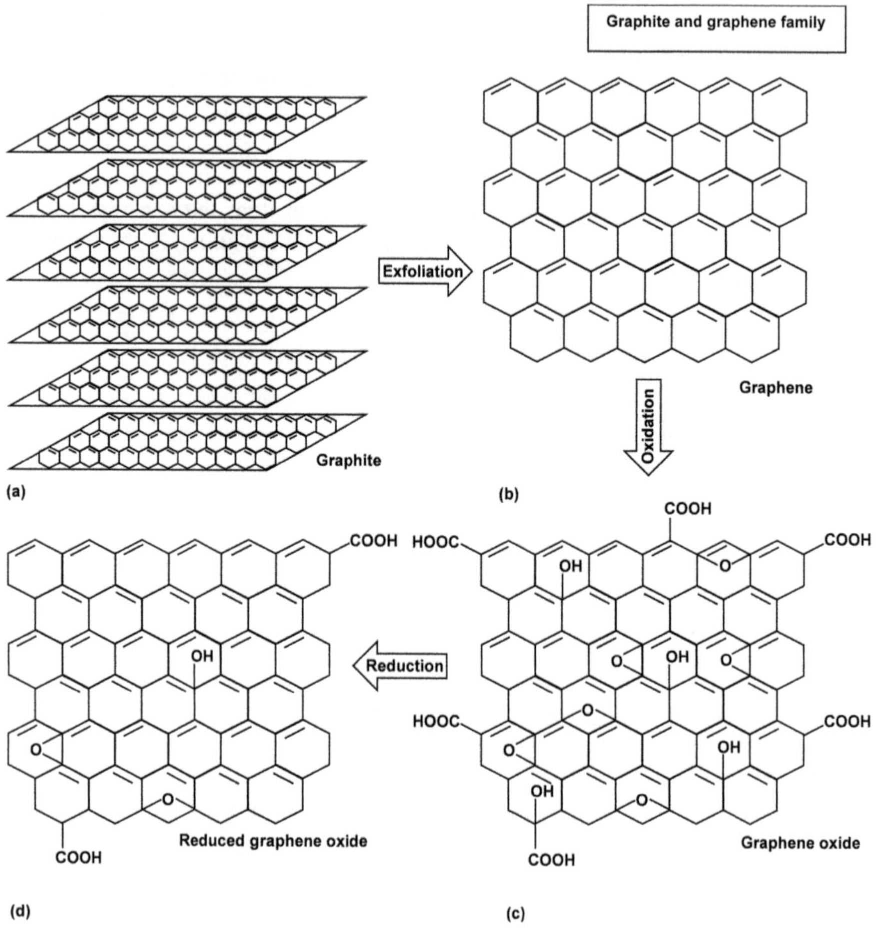

FIGURE 1.2 Graphite and the three leading members of the graphene family: (a) graphite, (b) graphene, (c) graphene oxide, and (d) reduced graphene oxide. Part (a) shows the layered structure of graphite in which different layers are attached to each other by weak van der Waals forces. Each layer comprises a hexagonal pattern of carbon atoms. Each carbon atom in a hexagon is covalently bonded to three other carbon atoms. Part (b) shows the graphene structure as a single layer of graphite. In this monolayer, the carbon atoms are bonded via sp^2-hybridized orbitals in a hexagonal honeycomb lattice. The molecular bond length is 0.142 nm. Part (c) shows graphene oxide as a layered carbon structure with a graphene-like hexagonal lattice and having oxygen functional groups (=O, –O–, –OH, –COOH) on basal planes and edges, looking like an oxygenated planar molecular material. Part (d) shows the structure of reduced graphene oxide which is similar in appearance to graphene oxide but has a smaller number of oxygen-containing functional groups than graphene oxide.

Besides its exceptional electronic properties, isotopically pure graphene shows an extremely high intrinsic thermal conductivity >4000 W $(mK)^{-1}$ as found from opto-thermal Raman measurements at 320 K (Chen et al. 2012). Pure natural diamond has a thermal conductivity of 2400–2500 $W(mK)^{-1}$ at 300 K while the same for copper is 400 $W(mK)^{-1}$ (Graebner 1995).

The specific surface area of a solid is defined as the total area of its surface per unit mass. For graphene, it is theoretically estimated to be 2630 m^2 g^{-1}.

1.2.2 GRAPHENE OXIDE

Graphene oxide (GO), Figure 1.2(c), is a monomolecular layer of graphite with attached oxygen-containing functional groups such as hydroxyl (–OH), carboxyl (–COOH), carbonyl (C=O), and epoxide or alkoxy (C–O–C, a three-membered ring structure involving two carbon atoms and one oxygen atom) groups (Jiříčková et al. 2022). The presence of these oxygenated groups imparts special properties to GO. These properties are associated with advantages over graphene such as higher solubility and enhanced dispersion in polymeric solutions and possibilities of func-tionalizing their surfaces. GO is synthesized by Hummers' method and its modified protocols, which are highly reproducible techniques for producing good-quality gra-phene oxide, frequently followed in engineering and research laboratories (Hummers and Offeman 1958, Kang et al. 2016). It will be described in Chapter 2, Sec.2.6.1.

1.2.3 REDUCED GRAPHENE OXIDE (rGO)

GO can be treated by several methods to produce rGO (Smith et al. 2019). The rGO, Figure 1.2(d), resembles graphene in its properties. However, it differs from gra-phene due to the presence of residual oxygen and other heteroatoms. It also contains structural defects. Reduction of GO to rGO provides an attractive route to obtain graphene-like films.

As graphene oxide undergoes reaction, its electrical behavior changes from insu-lating to semiconducting and then semi-metallic (Eda et al. 2009). Its apparent trans-port gap varies from 10–50 meV and, finally, falls to zero with progressive reduction.

1.3 GRAPHITIC CARBON NITRIDE (g-C$_3$N$_4$)

Graphitic carbon nitride (Figure 1.3) is a polymeric N-type visible light-active semi-conductor made of carbon and nitrogen with C/N ratio = 3/4, and a small amount of hydrogen impurity. Its unique 2D delocalized π–π-conjugated structure is formed by an infinite extension of triazine ring or tri-s-triazine ring as basic units (Dong et al.2021).

It is a layered material with the planar layers held together by van der Waals (vdW) forces. The layers are made of covalent-bonded carbon and nitrogen atoms. It is the analog of graphite with sp^2 type of hybridization but it is made of alternate carbon–nitrogen bonds. Because of its stacked construction, it is often looked upon as sp^2-hybridized nitrogen-substituted graphene.

It is extremely chemically and thermally stable. It shows high intrinsic photo-absorption and photo-responsiveness making it suitable for photophysical and photo-chemical applications, although it has drawbacks of a narrow visible light response ~450 nm, and fast recombination of optically generated charge carriers. It has a bandgap of 2.7 eV. With respect to normal hydrogen electrode (NHE), its conduction band is located at -1.4 eV and valence band at $+1.3$ eV.

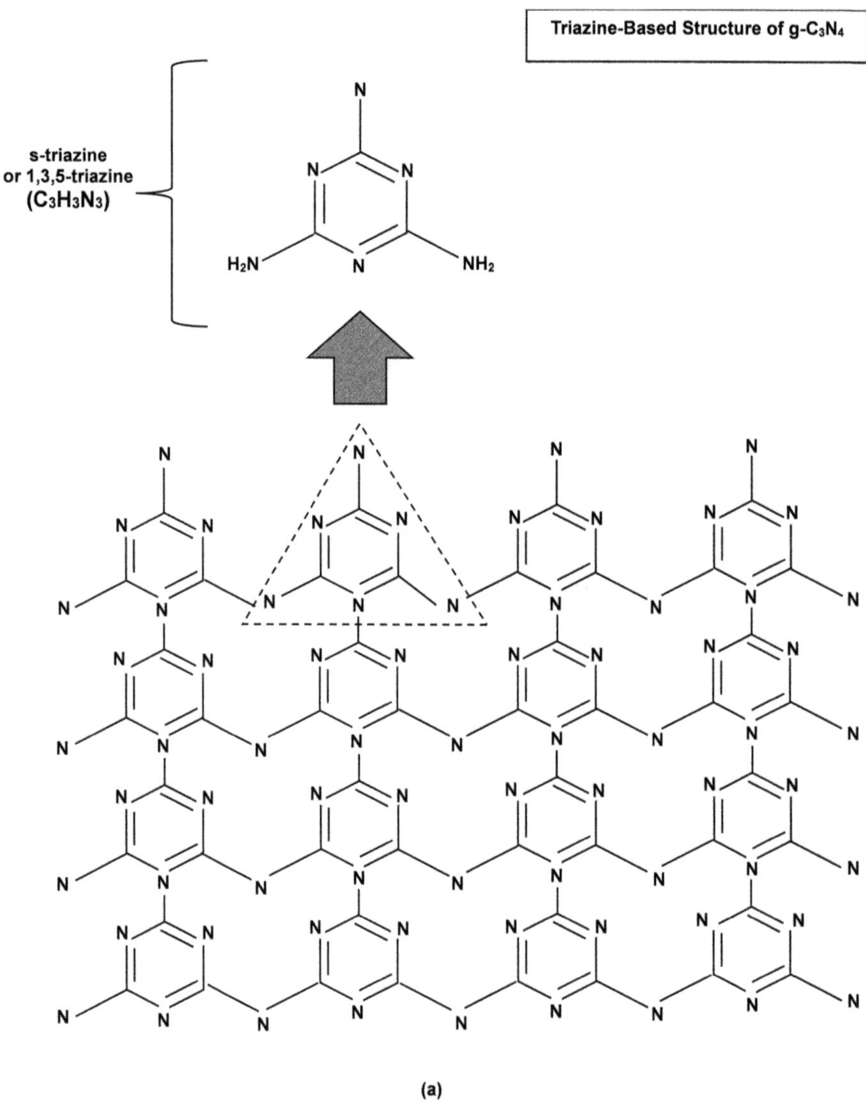

FIGURE 1.3   Two structures of g-C_3N_4: (a) triazine-based and (b) tri-s-triazine (heptazine)-based. Part (a) shows triazine-based structure of g-C_3N_4. The g-C_3N_4 has the appearance of an infinite extension of triazine ring (C_3N_3), which is shown enclosed in a dashed triangle, composed of three nitrogen and three carbon atoms in a six-membered ring, as the basic structural unit.

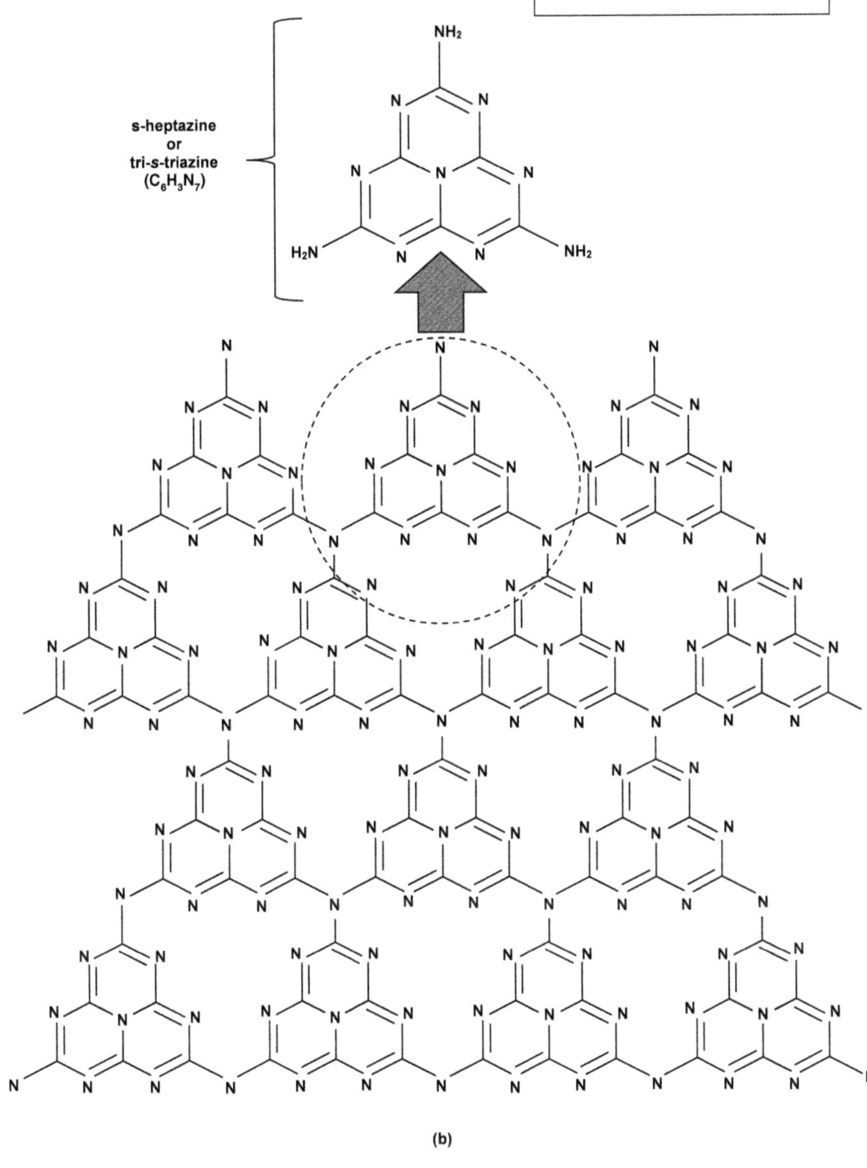

FIGURE 1.3 (Continued) Part (b) shows heptazine-based structure of g-C$_3$N$_4$ as an infinite extension of heptazine or tri-s-triazine ring (C$_6$N$_7$), which is shown enclosed in a dashed circle, made of a combination of three triazine rings, with three hydrogen atoms at the corners, as the basic structural unit.

1.4 HEXAGONAL BORON NITRIDE (hBN)

It is an isomorph of graphene having the same crystal structure as graphene (Figure 1.4). Its lattice is similar to the graphene lattice with alternating boron and nitrogen atoms in the place of carbon atoms in the graphene lattice. But its lattice constant is 1.8% longer than that of graphene (Bhimanapati et al. 2016). Further,

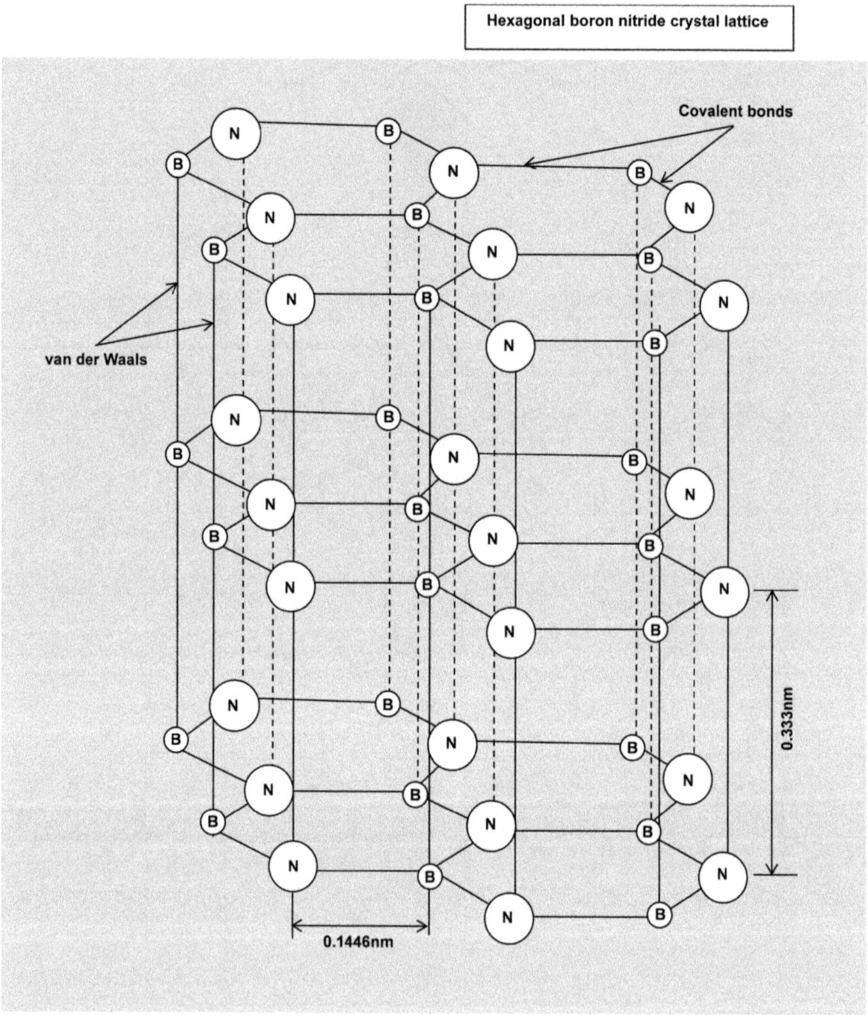

FIGURE 1.4 Crystal lattice structure of hexagonal boron nitride. The diagram shows the layered structure of hBN resembling graphite and graphene. It has a honeycomb lattice structure with similar dimensions to graphene; hence, it is often called 'white graphene'. Hexagonal rings forming parallel planes are bound to each other by van der Waals forces. These rings consist of boron and nitrogen atoms attached to each other in the plane by strong covalent bonds with an angle of 120° between the bonds. Each boron atom is attached to three nitrogen atoms. Similarly, each nitrogen atom is bound to three boron atoms. In hBN, the B–N bond length is 0.1446 nm and the interlayer spacing is 0.333 nm.

unlike graphene which is a conductor, h-BN is a wide-bandgap semiconductor with a bandgap of 5.9 eV, which makes it useful for deep UV emitters and detectors. Defects hosted by it can be engineered to achieve single-photon emission at room temperature (Caldwell et al. 2019). In comparison to using an SiO_2 substrate for graphene, the use of h-BN as a substrate greatly improves the mobility of carriers in graphene.

1.5 LAYERED METAL CHALCOGENIDES (LMDCs)

1.5.1 COMPOSITION

These materials are composed of layers of metal chalcogenides, compounds containing at least one chalcogen metal ion (Butler et al. 2013). Chalcogens are the chemical elements such as sulfur, selenium, and tellurium. The atoms in any layer of an LMDC are bound by covalent bonds while interlayer binding is through vdW forces. The weak interaction between layers allows the exfoliation of a monolayer or stack of a few monolayers from a bulk LMDC crystal.

1.5.2 CRYSTAL STRUCTURE

TMDCs, a family of nanomaterials with a three-atom thick unit cell, consist of a layer of transition metal atoms packed between layers of chalcogen atoms on both sides. Threefold symmetry is observed in the hexagonal honeycomb lattice. A crystal with an even number of layers has an inversion center. However, a crystal with an odd number of layers may or may not have an inversion center. Figure 1.5 illustrates the crystal structure of TMDCs. Within each layer of the layered 2D structure, the hexagonally packed MX_6 octahedra or trigonal prisms share edges with their six nearest MX_6 neighbors; the octahedra are for d^0, d^3 and some d^1 metals, and prisms are for d^1 and d^2 metals.

1.5.3 PROPERTIES

TMDCs display interesting properties bolstering their technological applications. They are semiconducting materials represented by the stoichiometry formula MX_2 where M is a transition metal such as Mo, W, and Ta and X =S, Se, or Te. MoS_2 is the most common and extensively studied material of TMDCs. More examples of TMDCs are $MoSe_2$, $MoTe_2$ WS_2, and WSe_2. Besides atomic scale thickness, TMDC monolayers show direct bandgaps promoting their usefulness for optoelectronic applications. They exhibit strong spin–orbit (SO) coupling (Manzeli et al. 2017). The strong SO coupling splits the spin states in the valence band by hundreds of meV. The splitting in the conduction band is up to few tens of meV (Wang et al. 2015). Hence, the electron spin can be controlled by tuning the photon energy and handedness of the excitation laser.

In-plane stiffness of monolayer MoS_2 is 180 Nm^{-1} (Bertolazzi et al. 2011). The corresponding Young's modulus = 270 GPa is at par with that of steel. A strain of 6–11% causes breaking. The breaking strength is 15 Nm^{-1} (23 GPa). Chemical vapor

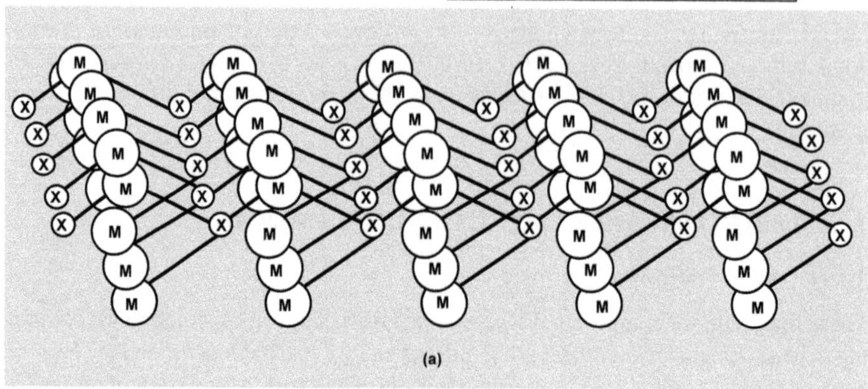

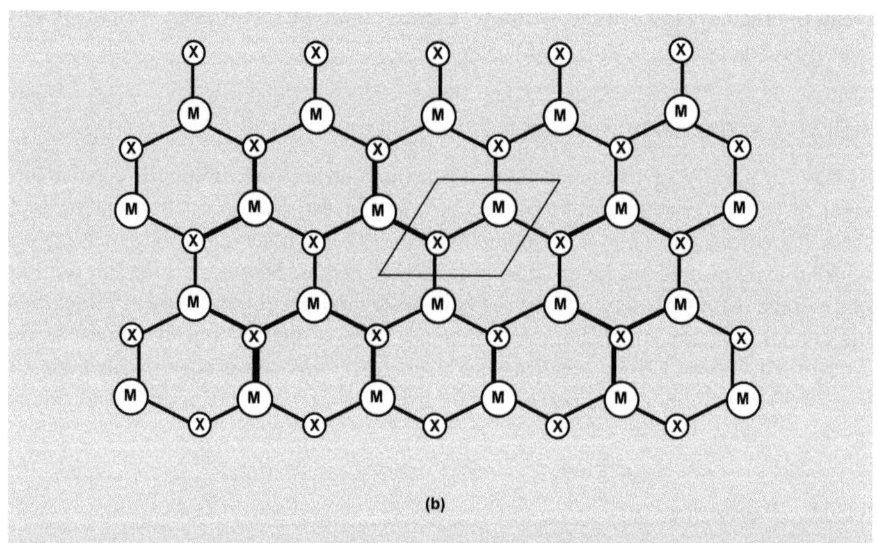

FIGURE 1.5 Schematic representation of 2D transition-metal dichalcogenides: (a) side view and (b) top view. The diagram displays the MX_2 structure (M= metal atom=Mo, W, Ta, Nb, Ti, Re; X= chalcogen atom=S, Se, Te) of transition-metal dichalcogenides. The MX_2 structure consists of an atomic layer of transition metal M crammed between two atomic layers of chalcogen atoms X. Part (a) shows the covalently bonded X–M–X layer sequence, as viewed from the side. Part (b) shows the hexagonal crystal lattice, as viewed from the top.

deposition (CVD) monolayer MoS_2 and WS_2 show high elastic moduli ~170 GPa comparable to that of exfoliated MoS_2 monolayer (Liu et al. 2014).

A major impediment to the progress of 2D TMDCs is their low electron/hole mobility (Cheng and Liu 2018), probably because they are extremely thin and, therefore, are more susceptible to damage. MoS_2 shows a low mobility~ 1 cm^2 V^{-1} s^{-1} without gate dielectric deposition. But the mobility increases to 150 cm^{2}V^{-1} s^{-1} when the HfO$_2$ gate layer is deposited over it (Mir et al. 2020). Hexagonal boron nitride coating has a similar effect on mobility.

1.6 THE 2D-XENES

1.6.1 NAMING OF XENES

These are mono-elemental materials, and are presently of 12 kinds. They include Xenes that are prepared by experimental synthesis or theoretically predicted. The 12 Xenes are listed below according to the positions of their constituent elements (enclosed within brackets for each Xene) in the periodic table (Li et al. 2022a):

Group III: Borophene (boron) and gallenene (gallium).
Group IV: Silicene (silicon), germanene (germanium), stanene (stannum meaning tin), and plumbene (plumbum meaning lead).
Group V: Phosphorene (phosphorous), arsenene (arsenic), antimonene (antimony), and bismuthene (bismuth).
Group VI: selenene (selerium) and tellurene (tellurium).

It may be noted that the naming of Xenes is done in such a manner to rhyme with graphene. Like graphene, Xenes have a hexagonal honeycomb lattice. But unlike graphene, they have a non-planar puckered structure buckled to different degrees. Figure 1.6 shows the structure of silicine and Figure 1.7 compares its structure with that of graphene.

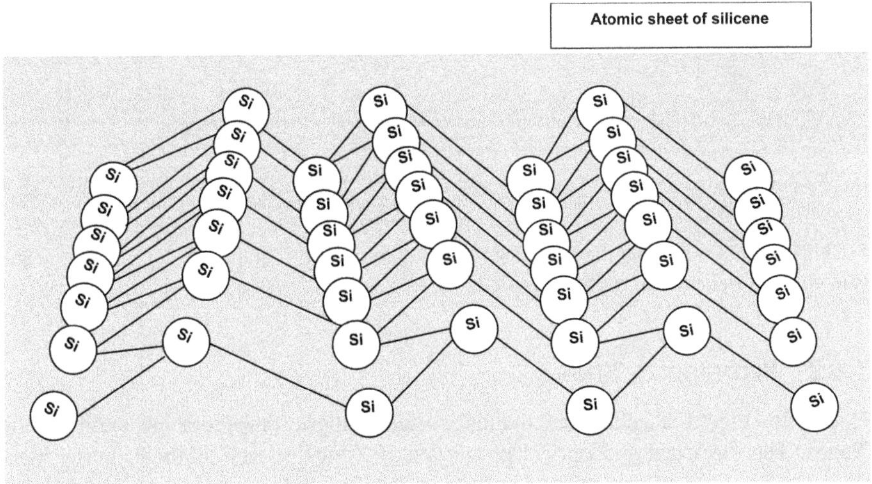

FIGURE 1.6 Structure of silicene, the 2D equivalent of graphene and 2D allotrope of silicon. The diagram shows the appearance of atomic sheet of silicine with the silicon atoms bound in a buckled hexagonal honeycomb lattice. The buckling of the lattice happens due to the large ionic radius of silicon atoms. As a result, the lower and upper sublattices sit in parallel planes separated along the vertical direction.

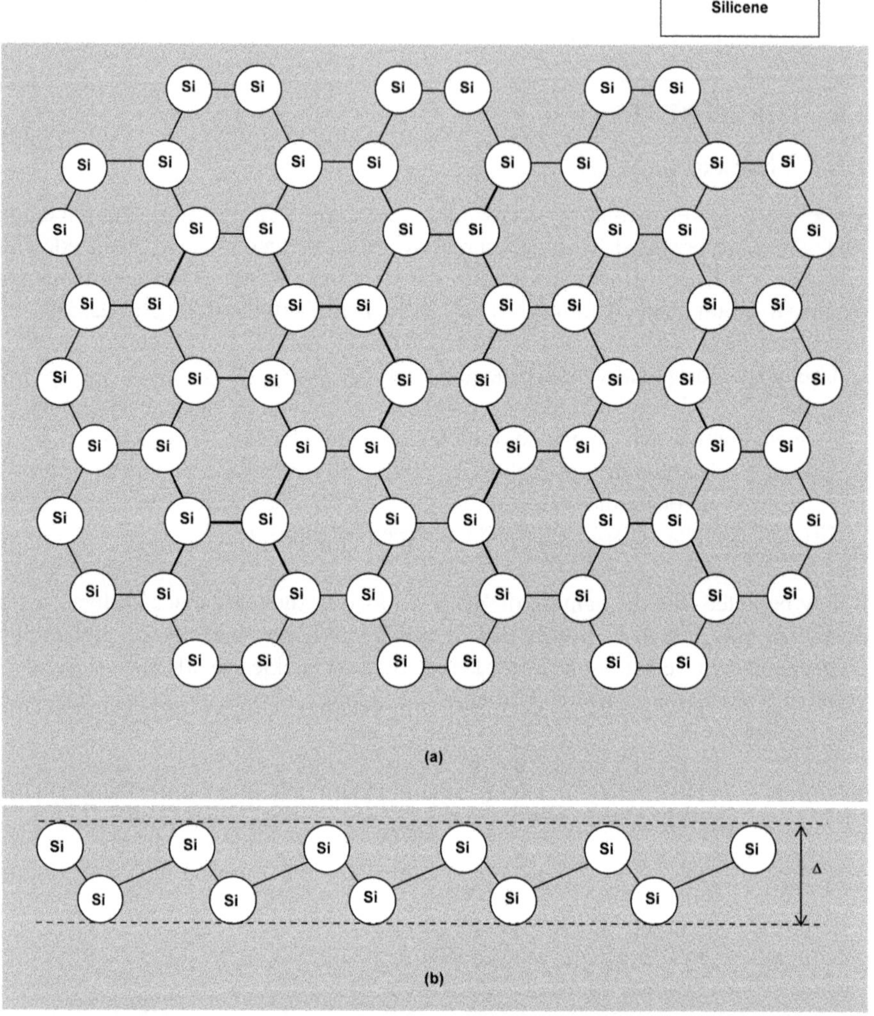

FIGURE 1.7 Comparing the crystal structure of silicine with graphene: (a) and (b) top and side views of the buckled silicine atomic layer.

1.6.2 Properties of Xenes

Distinctive electrical, photonic, magnetic, and catalytic properties are exhibited by Xenes. The electrical and optical properties of Xenes as well as their energy band structure are alterable by thickness control. Further, as a consequence of their out-of-plane thickness, Xenes display exceptional physical and chemical reactivities. Their electrical, optical, and catalytic properties are easily adjustable by chemical modification.

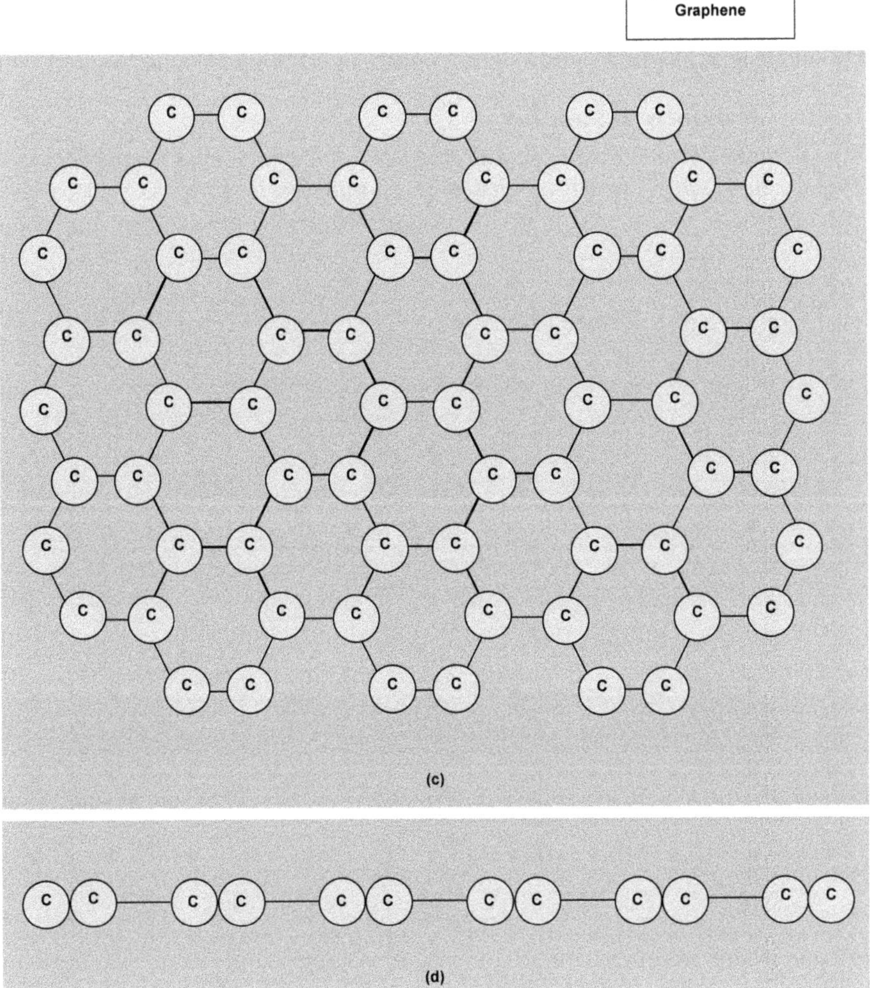

FIGURE 1.7 (Continued)   (c) and (d) top and side views of the planar graphene mono-layer. The diagram shows that in striking dissimilarity to graphene, silicene lacks flatness showing a periodically buckled topology. For silicine, the buckling distance in the out-of-plane direction is Δ.

Borophene and gallenene have metallic character. Silicene, germanene, stanene, and plumbene are quantum spin Hall insulators (QSHIs) arising from interaction of topological band inversion and intense SO coupling. QSHIs have large tunable energy gaps but gapless edge or surface states. These states are topologically protected. They are invulnerable to impurities or geometrical perturbations. Phosphorene, arsenene, antimonene, selenene and tellurene are semiconductors (Grazianetti and Martella 2021).

1.6.3 PHOSPHORENE

Phosphorene or black phosphorous is an important 2D-Xene material.

1.6.3.1 Phosphorene Structure

It is a single-elemental material with a layered structure similar to graphene and TMDCs; the single element in this material is phosphorous (Figure 1.8). So, it consists

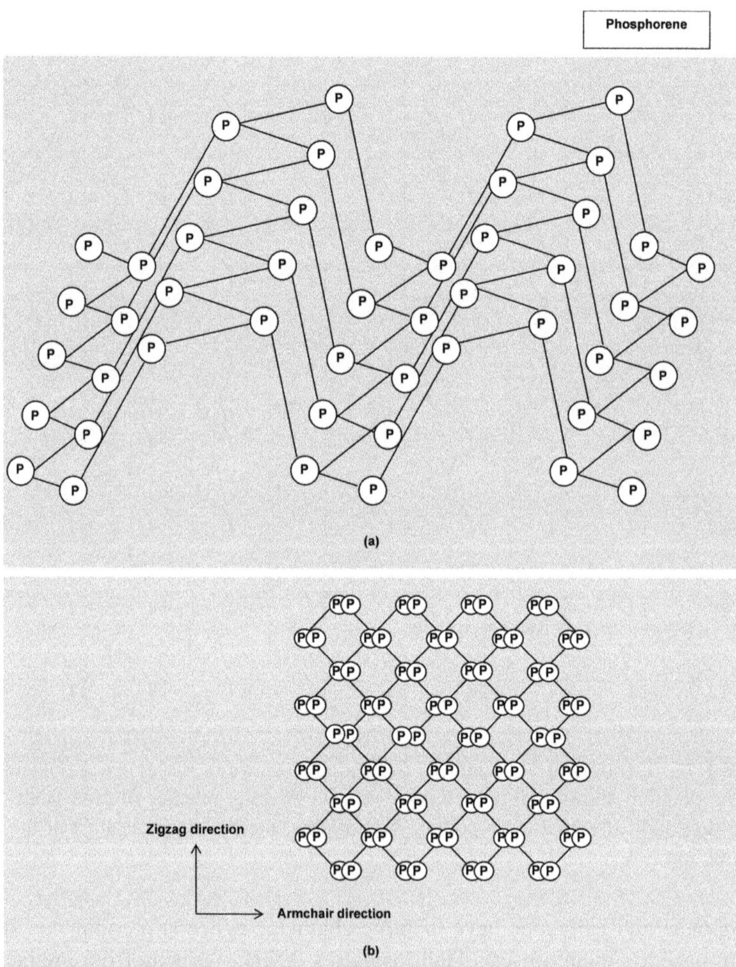

FIGURE 1.8   Phosphorene, a single layer of black phosphorous, has the structure of a quadrangular pyramid. It has a tetragonal lattice with P–P bond length = 0.222 nm in the horizontal direction and 0.226 nm in the other direction. Different views of phosphorene are given in parts (a), (b), and (c). Part (a) shows the perspective view of phosphorene consisting of phosphorous atoms forming sp^3 bonds with a lone pair of valence electrons to produce a hexagonal puckered pattern resulting in an orthorhombic pleated honeycomb structure. Part (b) shows the top view of phosphorene and part (c) its side view. Armchair and zigzag directions are indicated by arrows.

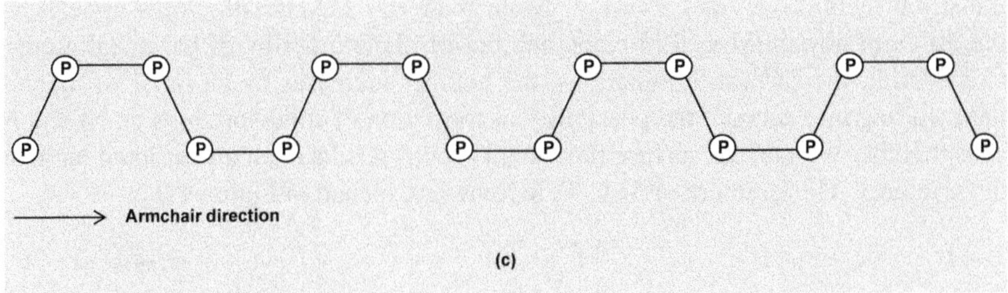

FIGURE 1.8 (Continued)

of phosphorous atoms only. It has a puckered geometry, a corrugated orthorhombic crystal organization, and shows distinctive crystalline symmetries (Ling et al. 2015).

1.6.3.2 Phosphorene Properties

The carrier mobility in black phosphorous is high, up to 1000 cm^2 V^{-1}s^{-1}, and the on/off ratio is 10^4–10^5 (Shi et al. 2021). These values fall outside the range enclosed by graphene or TMDCs. So, it is favored for this slot, especially for gigahertz electronics.

It has a direct bandgap in thin film form; its bandgap is 0.3 eV. This value is for thickness exceeding 4 nm or 8 layers. The bandgap of single-layer black phosphorous is 2.0 eV. The thickness dependence of bandgap allows easy bandgap variation by choosing the suitable thickness. The ease of bandgap tuning in the 0.3–2.0 eV range makes it very useful for different applications.

Another important property of black phosphorous is in-plane anisotropy (Xia et al. 2019). Due to this property, it shows physical effects such as linear dichroism (LD) or diattenuation, and anisotropic plasmons. LD is the difference between absorption of light that is polarized parallel to an orientation axis and the light that is polarized perpendicular to that axis, i.e., the differential absorption of light between two mutually orthogonal, linearly polarized states. A theory for the collective plasmon modes of a system of interacting electrons is formulated in the presence of explicit mass or velocity anisotropy (Ahn and Das Sarma 2021). Considerable anisotropy of plasmon dispersion itself builds up in such systems.

Furthermore, the physical properties of phosphorene can be tailored for new applications, e.g., designing polarization-sensitive photodetectors and high-gain digital inverters (Li et al. 2019).

1.7 THE 2D-MXENES

1.7.1 GENERAL FORMULA

The 2D-MXenes are 2D carbides and nitrides of transition metals, spoken as 'maxenes', and described by the general formula (Gogotsi and Huang 2021):

$$MXene = M_{n+1}X_nT_x \qquad (1.1)$$

where M is a transition metal such as Cr, V, Ti, Mo, Nb, Zr, Ta, and Sr; X = C or N; and T_x symbolizes surface terminations, e.g., =O, –OH, –F, and –Cl, on the outermost transition metal layer, and $n = 1$–4. More than 100 MXene structures arise from the different possibilities of in-plane and out-of-plane ordering of the metal atoms. Considering the surface terminations, the number increases by an order of magnitude. Taking into account the possibility of formation of solid solutions on M and X sites together with mixed surface functional groups results in an unrestricted number of structures. The structure of $Ti_3C_2T_x$ MXene is sketched in Figure 1.9.

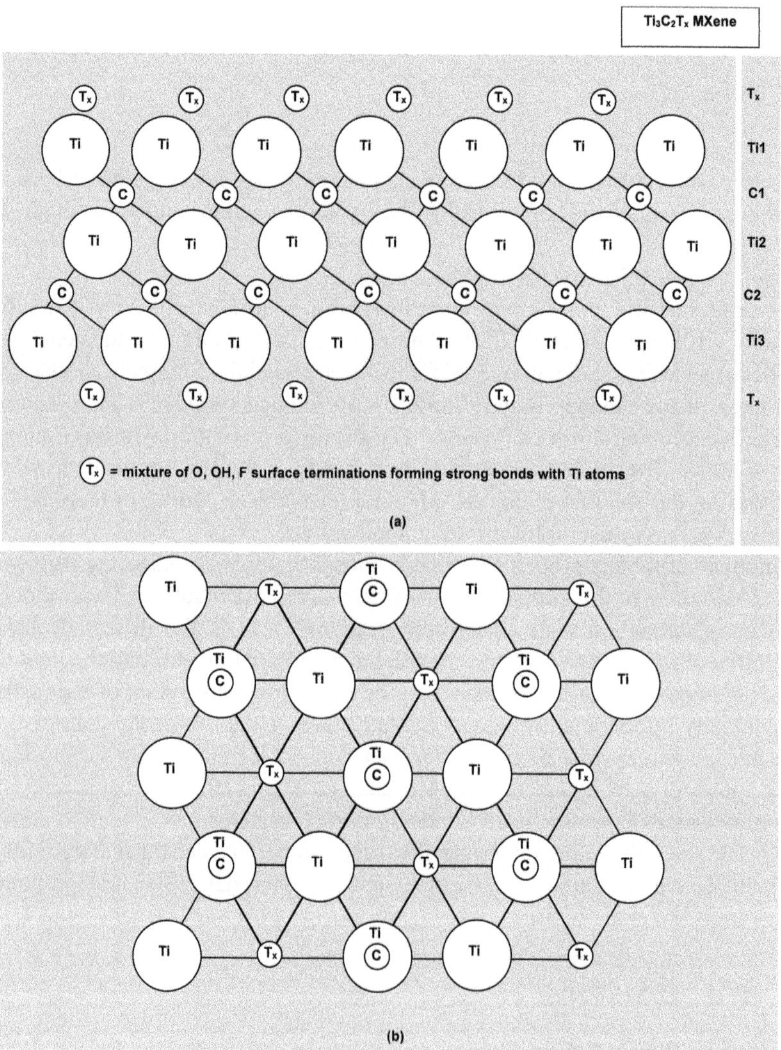

FIGURE 1.9 $Ti_3C_2T_x$ MXene: (a) side view showing the layered structure and (b) top view depicting the hexagonal lattice. Part (a) shows a central Ti_3C_2 layer constructed from three layers of titanium atoms and two layers of carbon atoms assembled together with the surface titanium atoms strongly bonded with O, OH, and F functional groups. Part (b) shows the two-dimensional structure of Ti_3C_2 consisting of two Ti_3C_2 sheets oriented in the (0, 0, 1) direction with the Ti atoms tied to O, OH, and F surface terminations through strong bonds.

1.7.2 Properties

MXenes show metallic electrical conductivity with conductivities up to 20,000 $S\text{-cm}^{-1}$ (Gogotsi and Huang 2021). This behavior distinguishes them from other 2D materials which are usually semi-metallic, semiconducting, or insulating. Their specific surface area ranges from 250 to 1000 $m^2 g^{-1}$ (Donga et al. 2020). They are hydrophilic in nature, form stable aqueous colloidal solutions (Gogotsi and Anasori 2019), show high photothermal conversion efficiencies, and absorb electromagnetic waves efficiently. Their useful properties have been exploited in catalysis, energy storage, water purification, electromagnetic interference shielding, plasmonics, etc. (Vahidmohammadi et al. 2021).

1.8 LAYERED DOUBLE HYDROXIDES

1.8.1 Structure and Formula

LDHs are a group of lamellar inorganic solids. They have a brucite $[Mg\,(OH)_2]$-like well-defined layered crystal structure (Figure 1.10). They are sandwich-like 2D clay materials composed of positively charged metal hydroxide sheets containing divalent and trivalent cations separated by charge-balancing hydrated anion-filled interlayer galleries.

The intercalated anions/interlayer spacing/metal-cation compositions of LDHs are tunable and are described by the following formula (Długosz and Banach 2022):

$$\left[M^{2+}_{1-x} M^{3+}_{x} \left(OH \right)_2 \right]^{x+} \left[\left(A^{n-} \right)_{x/n} . m H_2 O \right]^{x-} \tag{1.2}$$

where M^{2+} are divalent metal cations, M^{3+} are trivalent metal cations, and A^{n-} are anions. M^{2+} and M^{3+} cations are spread inside the hydroxide layer whereas A^{n-} anions lie in the interlayer space. Examples of M^{2+} cations are: Mg^{2+}, Co^{2+}, Ca^{2+}, Cu^{2+}, Zn^{2+}, and Ni^{2+}. Main M^{3+} cations are: Al^{3+}, Fe^{3+}, Co^{3+}, Ni^{3+}, and Cr^{3+}. The A^{n-} anions are: NO_3^-, CO_3^{2-}, and Cl^-, an oxoanion, $RCOO^-$, or a complex ion. The hydroxide layer containing the metal cations is bound together by bonding between the metal cations of the hydroxide layers and the anions in the interlayer spaces.

1.8.2 Hydrotalcite

A naturally occurring LDH is hydrotalcite [magnesium aluminum hydroxycarbonate: $Mg_6Al_2(CO_3)\,(OH)_{16} \bullet 4(H_2O)$]. In this material, the positively charged divalent ions are partially substituted by positively charged trivalent ions to produce a positive sheet charge which is compensated by negative ions lying in the interlayer galleries (Velasco et al. 2012).

1.8.3 Conductivity

Extremely high in-plane conductivities up to 10^{-1} $S\text{-cm}^{-1}$ have been shown in single-layer LDH nanosheets. These values are the highest among anion conductors (Sun et al. 2017).

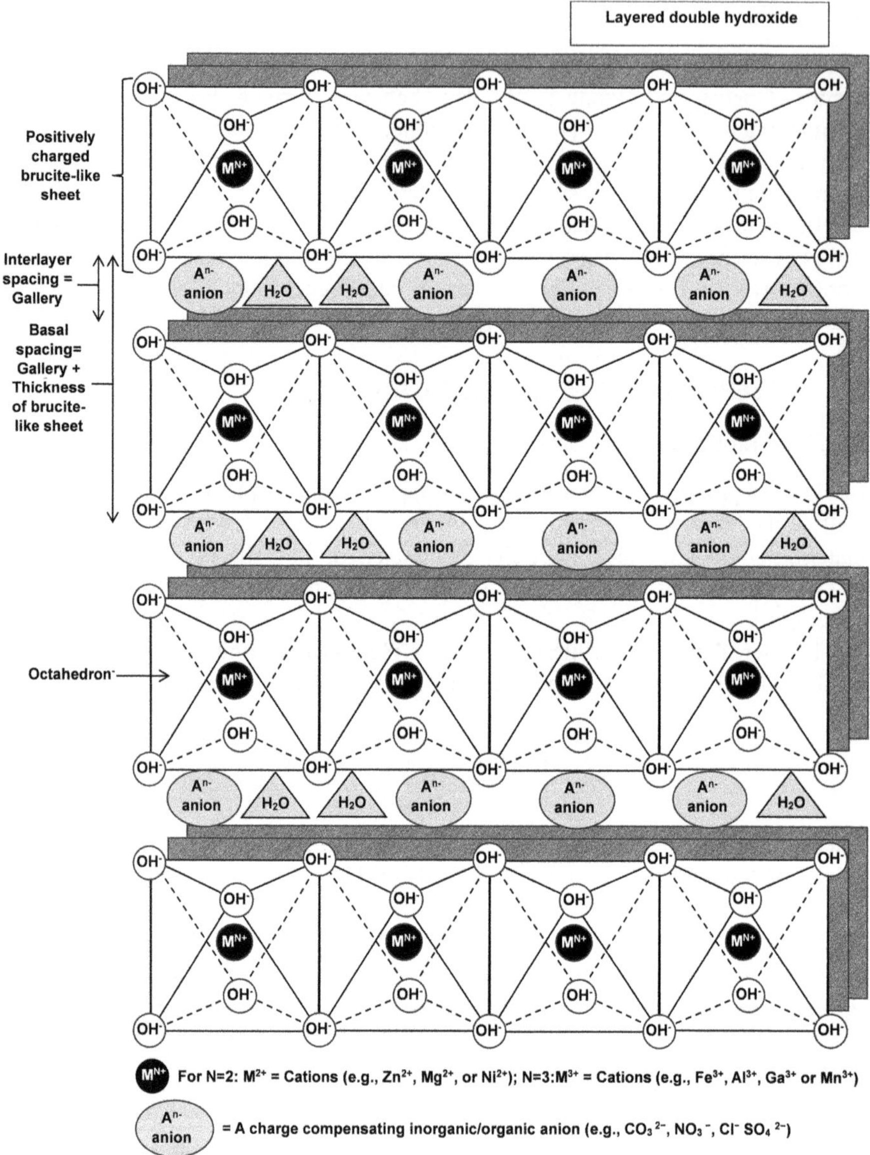

FIGURE 1.10 The layered double hydroxide (LDH) material. The diagram shows the structure of LDH. It is an ionic lamellar solid consisting of positively charged brucite-like sheets separated by an interlayer gallery. Brucite is a mineral made of magnesium hydroxide $Mg(OH)_2$. In the interlayer gallery are charge-compensating anions A^{n-} (n=1 or 2) and water molecules. At the vertices of the octahedra lie hydroxide ions while at its center are divalent or trivalent metal cations M^{N+} (N=2 or 3). Generic formula of LDH is $[(M^{2+})_{1-x}(M^{3+})_x(OH)_2]^{x+}$ $[(A^{n-})_{x/n} \cdot mH_2O]^{x-}$, x = mole fraction of M^{3+}, m is the number of water molecules. Examples of divalent metal ions M^{2+} are Mg^{2+}, Fe^{2+}, Zn^{2+}, Co^{2+}, and Cu^{2+}; trivalent metal ions M^{3+} are Al^{3+}, Fe^{3+}, Ga^{3+}, and Mn^{3+}; and anions A^{n-} are OH^-, Cl^-, NO_3^-, and CO_3^{2-}.

1.9   METAL-ORGANIC FRAMEWORKS

1.9.1   Nature and Composition

MOFs are highly porous, hybrid, crystalline materials. They are formed by metal–ligand coordination (Figure 1.11). They consist of metal-based nodes bonded to organic linkers through coordination bonding in various supramolecular architectures (Wang et al. 2022).

1.9.2   Properties

These materials offer a flexible porosity. Their porosity can be varied to form microporous to mesoporous layers by changing the length of the organic linkers. A large surface area and lateral dimension allow the availability of numerous exposed metal sites on their surfaces for attachment of molecules, making them useful for catalysis, high sensitivity bio- and gas sensors, and energy storage (Song et al. 2023). By

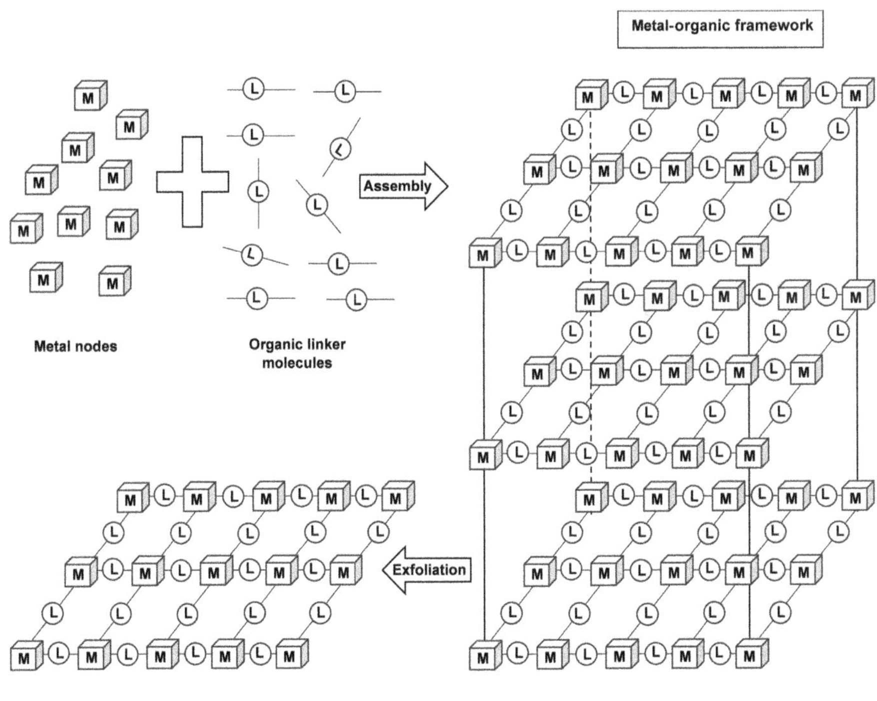

FIGURE 1.11   Metal-organic framework (MOF), also called porous coordination polymer (PCP), is an organic–inorganic hybrid crystalline ultrahigh porosity material consisting of metal ions or clusters acting as joints, coordinated by multidirectional organic ligands, acting as struts or linkers. The diagram shows the assembly of metal nodes with organic linker molecules to form 2D MOF bulk material, from which single-layer 2D MOF is obtained by exfoliation.

choosing the proper metal or organic moiety, their properties can be tuned (Ghosh et al. 2023).

1.10 2D METAL OXIDES

1.10.1 LAYERED AND NON-LAYERED OXIDES

According to their crystalline structures, 2D MOXs are of two types (Xie et al. 2022): layered, e.g., MoO_3 and V_2O_5 and non-layered, e.g., WO_3, CeO_2, In_2O_3, and SnO_2. In the layered MOXs, the in-plane atoms are joined by strong chemical bonds while the stacking layers are adhered to each other by weak vdW forces. Non-layered MOXs are formed by 3D chemical bonding.

1.10.2 PROPERTIES

Depending on their band structures and doping concentrations, 2D MOXs show conducting, semiconducting, or insulating behavior. Supercapacitors and batteries utilize their high theoretical capacitance. These devices are benefited from the large surface areas provided by MOXs along with their capabilities for oxidation–reduction reactions.

Molybdenum trioxide (Figure 1.12) is a wide bandgap semiconductor. Its bandgap is >2.7 eV. It is quasi-transparent in the visible wavelength range and also electrically conducting. It finds applications in gas sensing, optoelectronics, and flexible electronics (Puebla et al. 2021).

1.11 ATTRACTIVE PROPERTIES OF 2D MATERIALS VIS-À-VIS BULK MATERIALS FOR SENSOR APPLICATIONS

A paradigm shift in sensor technology is fostered by the wide spectrum of outstanding properties shown by 2D materials. A few illustrative examples are given in the subsections below. A vast gamut of sensing devices utilizing these properties will be presented in the book.

1.11.1 HIGH SURFACE-AREA-TO-VOLUME RATIO

Provision of innumerable reactive sites by 2D materials for interaction between the material and the analyte drastically enhances the sensitivity of sensors and lowers their limits of detection as compared to their bulk counterparts. The 2D geometry makes a large surface available for gas adsorption, which is immensely useful for chemical and gas sensors. The more the reactant from the surroundings comes in contact with the sensor surface, the stronger and faster the response is. The attachment of nanoparticles further ameliorates sensitivity. Electrocatalytic properties endowed by the active sites improve the current output. Selectivity is enhanced by the immobilization of biomolecules for the recognition of specific species by biosensors.

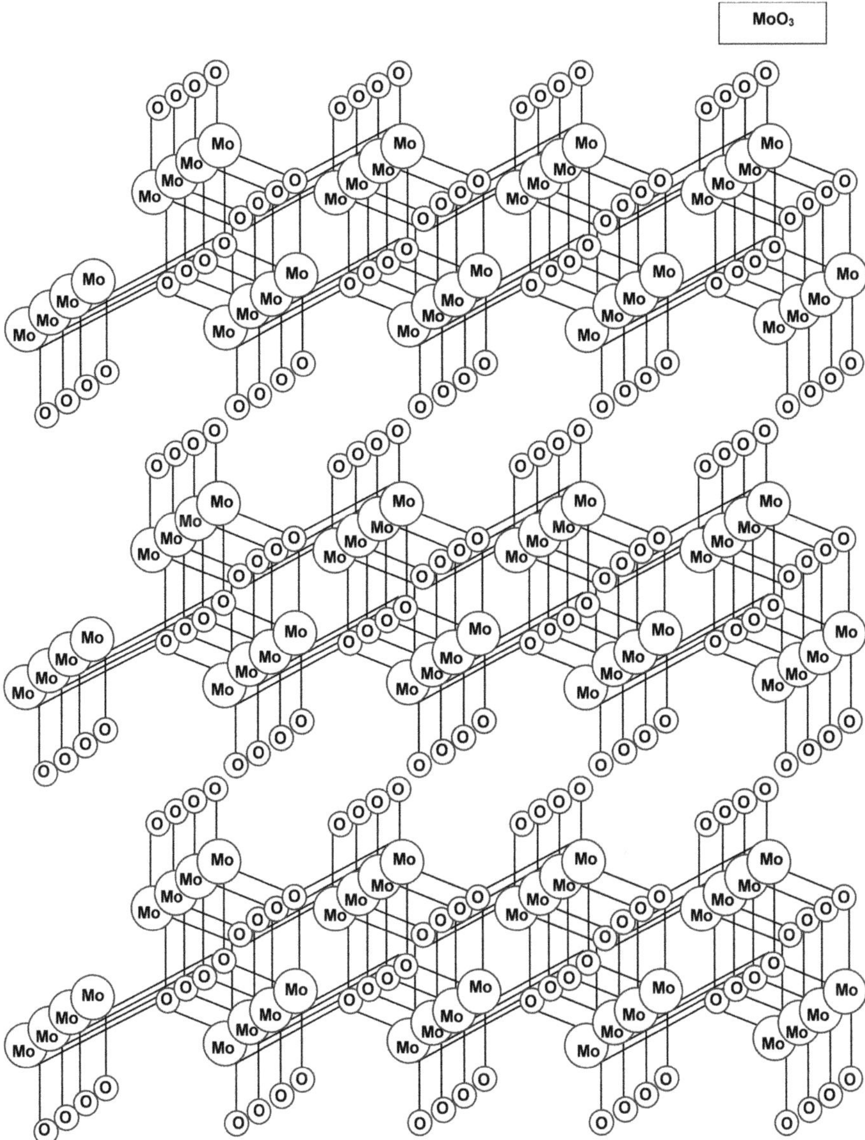

FIGURE 1.12 MoO_3 geometry consisting of layers of distorted MoO_6 octahedra stacked together in an orthorhombic crystal structure. The diagram shows the layered crystal structure of MoO_3 in which the layers are made of atomically thin sheets composed of double layers of MoO_6 octahedra oriented in the (0, 0, 1) direction. In the vertical (0,1, 0) direction, the MoO_6 octahedra are tethered by van der Waals forces causing stratification, while internally, the interactions in the octahedra take place by strong covalent and ionic bonds with Mo–O bond distances ~ 1.66–2.39 Å.

1.11.2 High Mechanical Strength and Flexibility

A 2D material is a monolayer of atoms bound together by strong covalent bonds, as opposed to a bulk material, which is made of several planes of covalently bonded atoms with the planes held together by weak vdW forces. The absence of vdW interactions in 2D materials imparts them extra-ordinarily high mechanical strength. The superb mechanical strength in combination with the extreme flexibility of 2D materials acquired by virtue of their atomic-scale thickness makes these materials excellent platforms for fabricating wearable and flexible sensors.

1.11.3 Quantum Confinement Effect and Bandgap Engineering

The quantum confinement perpendicular to the 2D plane causes discretization of energy levels and an increase of bandgap. The bandgap of a material determines its electronic and optical behavior. Quantum confinement weakens the dielectric screening between charge carriers in semiconductors. As a consequence, the Coulomb interaction increases producing more tightly bound electron–hole pairs called excitons. When the 2D material layer thickness is smaller than the Bohr radius, the quantum confinement increases the energy and hence the frequency of light absorbed or emitted by the 2D material as compared to the bulk form.

Bulk MoS_2 has an indirect bandgap of 1.2 eV which increases to a direct bandgap of 1.8 eV for single-layer MoS_2. Besides, a direct bandgap monolayer MoS_2 exhibits a high light absorption rate (11%) making it suitable for making a photodetector (Li et al. 2022b). A maximum external photoresponsivity of 880 A W^{-1} is observed with MoS_2 phototransistors at a wavelength of 561 nm (Lopez-Sanchez et al. 2013). The photodetector made with graphene/MoS_2 heterostructure achieves a photogain >10^8 (Zhang et al. 2014).

1.11.4 Tunable Electrical Conductivities

2D materials offer several opportunities to alter the conductivity, as desired. Methods to change conductivity include control of structural defects, increasing the number of constituent layers, impurity doping, or post surface modification by functional groups. These options pave the way toward the realization of various physical, chemical, and biological sensors.

1.11.5 Unique Strain-Sensing Properties of Layered Materials

Exfoliated 2D materials with sub-nanometer gaps between the layers display special piezoelectric properties such as high gauge factors for making high-sensitivity strain gauges, rendering possible the measurement of exceedingly weak forces (Sulleiro et al. 2022). A high gauge factor of 3933 is demonstrated for SnS_2 vdW-layered material-based strain sensor by controlling the density and mobility of charge carriers with the help of photoactivation with 365 nm illumination. The sensor is made by CVD of SnS_2 on fully covered Si substrate, transferring SnS_2 nanoflakes on a polydimethylsiloxane (PDMS) bar and forming Ti/Au electrodes for contacts. This phenomenon is also noticed for GaSe, GeSe, monolayer WSe_2, and monolayer MoSe (Yan et al. 2021).

1.11.6 High-Efficiency Fluorescence Quenching

Graphene and its derivatives TMDCs and MXenes act as fluorescent dye quenchers making them useful for making fluorescent probes. These probes are more efficient for detecting various analytes than their bulk forms, e.g., layered WS_2 nanosheet readily adsorbs single-strand DNA (ssDNA) chains, rapidly quenching a fluorescent dye tagged to the DNA chain. Upon interaction with other biomolecules, the adsorbed DNA chain is unfastened, giving back the fluorescence. Thus, the WS_2 nanosheet is useful as a bio-probe assembling platform (Yuan et al. 2014).

Further, WS_2 nanosheet shows differential affinity toward a short oligonucleotide fragment as opposed to ssDNA probe, efficiently quenching the adsorbed fluorescent probes. Applying this quenching ability, a method for simple, sensitive, and selective detection of miRNA is developed. In this method, the WS_2 nanosheet-based fluorescence quenching property is reinforced with duplex-specific nuclease signal amplification (DSNSA), thereby enabling highly sensitive and selective detection with a limit of 300 fM (Xi et al. 2014).

Duplex-specific nuclease (DSN) is an enzyme (Zhong et al. 2022). It is found in the hepatopancreas of Kamchatka crab. DSN has the specialty of a strong preferential cleaving of double-stranded DNA or DNA in DNA–RNA hybrid duplexes. However, it shows no activity toward single-stranded DNA or single or double-stranded RNA. These features of DSN are exploited to design highly sensitive microRNA (miRNA) detection approaches. In these approaches, the target miRNA is hybridized with the DNA probe. As a result, a DNA/miRNA heteroduplex is formed. The DSN cleaves the DNA probe in the hybrid. The released miRNA hybridizes with the remaining DNA probe to form a DNA/miRNA heteroduplex. This heteroduplex is cleaved by DSN again. Thus, a hybridization-cleavage cycle is created for isothermal signal amplification. This process is known as DSNSA.

1.11.7 Surface Plasmon Resonance-Sensitivity Enhancing Properties

1.11.7.1 SPR Principle

Surface plasmon resonance (SPR) biosensors are optical biosensors working on the measurement of refractive index changes occurring during the binding of analyte molecules in a specimen, thereby enabling bio-recognition of the molecules immobilized on the sensor surface (Janith et al. 2023). In this biosensing technique, surface polariton waves are used to probe the interactions between the biomolecules and the sensor surface. The surface polariton waves are perpendicularly confined evanescent electromagnetic waves propagating at the metal/dielectric interface. The dielectric is the sensing medium in which a change in concentration of biomolecules leads to a local change in the refractive index of the medium adjacent to the metal surface. The refractive index change causes a variation in the propagation constant of the surface polariton wave, which is measured by the attenuated reflectance (ATR) method, a contact sampling technique using a high-refractive index crystal with good IR transmission property.

1.11.7.2 The ATR Method

Figure 1.13 explains the principle of the ATR method. Generally, an IR beam traveling from a high refractive index medium, e.g., a zinc selenide or germanium crystal to a low refractive medium, the air, is partially reflected back into the low refractive

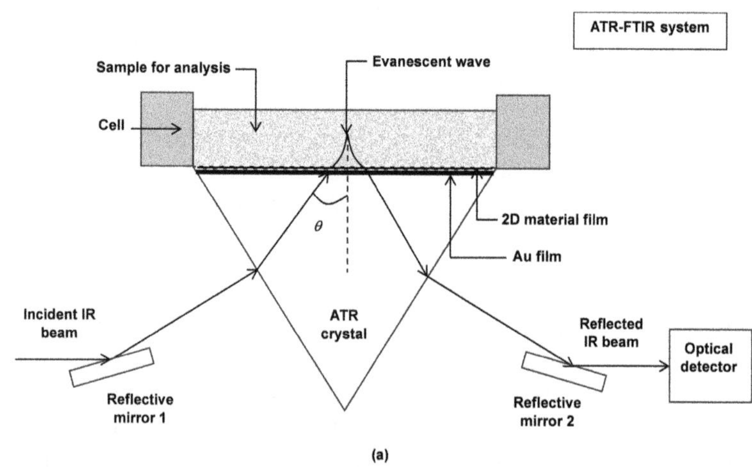

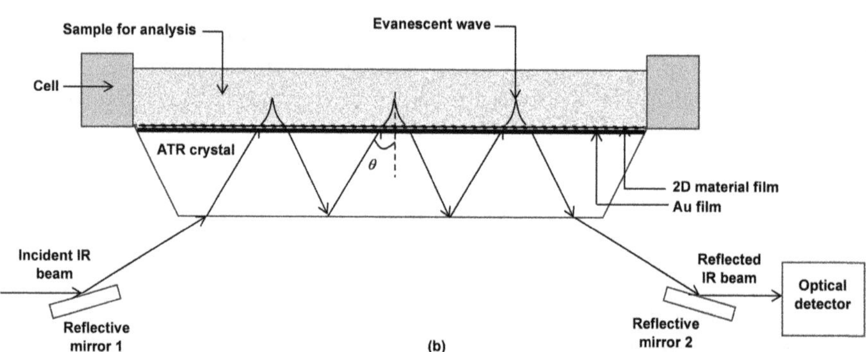

FIGURE 1.13 ATR–FTIR spectroscopy sampling system: (a) single reflection and (b) multiple reflection configurations. Part (a) shows the triangular-shaped (polyhedron) ATR crystal on the top Au-film-coated surface of which the 2D material film is deposited. The 2D material film is the floor of an analytical cell. The sample is filled with solution in this cell. IR light is reflected by mirror 1 toward the crystal. It enters the crystal and undergoes total internal reflection from the 2D material-covered surface when the angle of incidence θ of IR light is greater than the critical angle. This reflection forms an evanescent wave which penetrates into the sample. Coming out of the crystal, it falls on mirror 2 from which it is reflected to the optical detector. Part (b) shows a trapezium prism-shaped (trapezoidal) ATR crystal with the longer side covered with an Au film followed by 2D material film. The difference from part (a) is that the incident IR beam suffers several reflections from the crystal/sample interface and the opposite face of the crystal before falling on the detector. At each reflection, an evanescent wave is produced penetrating into the sample at the reflection point. With respect to a single-reflection ATR, a multiple-reflection ATR offers the advantage of amplification of the absorption spectrum. The amplification is determined by the number of additional reflections.

medium but at a particular angle of incidence greater than the critical angle, the light is almost fully reflected back. During this phenomenon, called total internal reflection, some fraction of optical energy escapes from the crystal to a short distance ~ 0.1–5 µm in the form of evanescent waves. At this point, the intensity of reflected light decreases. So, this is known as attenuated total reflectance.

When the sample is placed on the crystal, some portion of light penetrating through the crystal reaches the sample and is absorbed by it. An IR spectrum of the sample is created from this absorbance. For recording the background spectrum, a clean crystal without the sample is taken.

ATR is often used with Fourier transform infrared (FTIR) spectroscopy as ATR–FTIR spectroscopy to analyze solid and liquid samples for the identification of chemical compounds and for the examination of their molecular structures (Grdadolnik 2002).

1.11.7.3   Role of 2D Materials in SPR Sensors

Strong light absorption due to the large real part of refractive index of 2D materials together with their favorable properties of optical transparency and large specific area are greatly helpful in boosting the sensitivity of the SPR sensor, which traditionally uses a single metal film, generally an Au or Ag film. The 2D materials layer also protects the underlying metal film from oxidation.

1.11.7.3.1   Graphene-on-Au (111) SPR Sensor

Graphene stably adsorbs biomolecules with carbon-based ring structures, e.g., ssDNA resulting in a larger refractive index change near the interface between graphene and the sensing medium than without graphene. The graphene coating also modifies the propagation constant of surface plasmon polariton whereby the sensitivity to refractive index change is altered. Calculations revealed that the sensitivity increases by 25% for a graphene-on-gold SPR biosensor using 10 graphene layers (Wu et al. 2010).

1.11.7.3.2 MoS_2-on-Au SPR Biosensor

In angular modulation SPR sensor, the existence of an optimal real part of RI of 3.5–4.0 and optimal thickness is shown when selecting 2D materials (Tan et al. 2023). Furthermore, a higher sensitivity of the SPR sensor is attained for a smaller imaginary part of the refractive index of 2D material. For the highest sensitivity, the required thickness of 2D material falls with rising real and imaginary parts of the refractive index.

A 5-nm-thick MoS_2-enhanced SPR biosensor is developed for detecting sulfonamides (SAs). It is based on a group-targeting indirect competitive immunoassay. It exhibits a low sulfonamides detection limit of 0.05 µgL^{-1}. Its limit of detection is ~12-times lower than that of the undecorated Au SPR sensor (Tan et al. 2023).

1.11.7.3.3   Hybrid Nanostructures

Inadequate response of single nanomaterial coatings prompted the use of hybrid nanostructures, e.g., thin films of Si-graphene and MoS_2-graphene. The large real

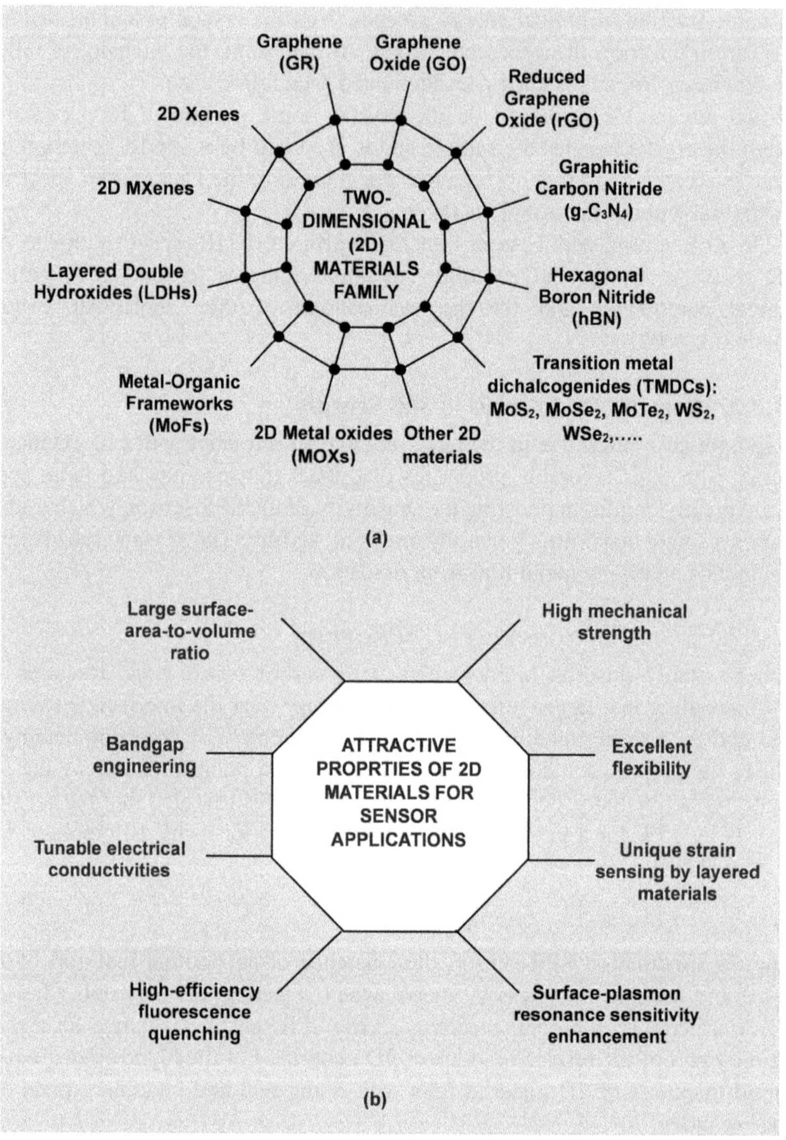

FIGURE 1.14 Pictorial summary of the (a) 2D materials family and (b) attractive properties of 2D materials utilized in sensors. Part (a) makes an exhibition of the 2D materials that have revolutionized the sensor field, outstandingly graphene, graphene oxide, reduced graphene oxide, graphitic carbon nitride, hexagonal boron nitride, transition metal dichalcogenides, 2D Xenes, 2DMxenes, LDHs, MOFs, and many more materials. Part (b) underscores the main attributes of 2D materials utilized in sensors, e.g., colossal surface area-to-volume ratio, high flexibility and mechanical strength, strain-sensing properties, tunable conductivities and bandgaps, fluorescence quenching, and surface-plasmon resonance sensitivity enhancement capability.

value of the refractive index of Si nanosheet appreciably increases the sensitivity of an SPR biosensor. The stability of the sensor is also improved by the protective action of silicon film on the metal layer underneath it. A sensitivity of 155.68°/RIU can be achieved with monolayer WS_2-coated Si nanosheet (Ouyang et al. 2016).

SPR biochemical sensors designed with heterostructures of few-layer black phosphorus (BP) and graphene/TMDCs are proposed (Wu et al. 2017). The heterostructure of BP and bilayer WSe_2 shows the maximum sensitivity ~279° RIU^{-1}, which is ~2.4 times the sensitivity of the conventional biochemical sensor. Black phosphorus, graphene, and TMDCs all have a favorable impact on SPR characteristics.

1.12 DISCUSSION AND CONCLUSIONS

Of late, the number of 2D materials has shown an exponential growth. An estimated number of 5619 layered compounds is predicted with 1825 branded as easily or potentially exfoliable. But to date, the number of experimentally synthesized or exfoliated materials is countable as meagerly a few dozens. (Mounet et al. 2018). In this chapter, we familiarized ourselves with the prominent 2D nanomaterials portfolio that has been at the focus of research in sensors (Figure 1.14). We strived to decipher and comprehend the salient features and attributes that endow them with the ability to do what bulk materials fail to achieve and so could be exploited for the development of novel high-performance sensors. These properties encompass a vast range from the semi-metallic conductivity of graphene to the semiconducting to superconducting behavior of TMDCs exhibiting spectacular band structure changes from one-to-six-layer thickness and insulating properties of hexagonal boron nitride. Conspicuously, 2D materials do not have the intrinsic properties for selective interaction with target molecules, making it possible to control the interaction between 2D nanosheets and molecules/ions by modification of their inherent properties for making sensors. In the next chapter, we will apprise ourselves with the various processes routinely followed for sensor fabrication using 2D materials.

REFERENCES

Ahn S. and S. Das Sarma 2021 Theory of anisotropic plasmons, *Physical Review B*, 103, L041303, pp. 1–5.

Bertolazzi S., J. Brivio, and A. Kis 2011 Stretching and breaking of ultrathin MoS_2, *ACS Nano*, 5(12), pp. 9703–9709.

Bhimanapati G.R., N.R. Glavin, and J.A. Robinson 2016 Chapter Three - 2D Boron Nitride: Synthesis and applications, In: Editor(s): Francesca Iacopi, John J. Boeckl, Chennupati Jagadish, *Semiconductors and Semimetals*, Academic Press/Elsevier, Cambridge, MA, Volume 95, pp. 101–147.

Bolotin K.I., K.J. Sikes, Z. Jiang, M. Klima, G. Fudenberg, J. Hone, P. Kim, and H.L. Stormer 2008 Ultrahigh electron mobility in suspended graphene, *Solid State Communications*, 146(9–10), pp. 351–355.

Butler S. Z., S. M. Hollen, L. Cao, Y. Cui, Jay A. Gupta, H. R. Gutiérrez, T. F. Heinz, S. S. Hong, et al. 2013 Progress, challenges, and opportunities in two-dimensional materials beyond graphene, *ACS Nano*, 7(4), pp. 2898–2926.

Caldwell J. D., I. Aharonovich, G. Cassabois, J. H. Edgar, B. Gil and D. N. Basov 2019 Photonics with hexagonal boron nitride, *Nature Reviews Materials*, 4, pp. 552–567.

Cheng L. and Y. Liu 2018 What limits the intrinsic mobility of electrons and holes in two-dimensional metal dichalcogenides? *Journal of the American Chemical Society*, 140(51), pp. 17895–17900.

Chen S., Q. Wu, C. Mishra, J. Kang, H. Zhang, K. Cho, W. Cai, A. A. Balandin and R. S. Ruoff 2012 Thermal conductivity of isotopically modified graphene, *Nature Materials*, 11, pp. 203–207.

Długosz O. and M. Banach 2022 Synthesis of layered zinc-aluminum double hydroxides modified with metal ions as photocatalysts with enhanced sorption properties, *Applied Physics A: Materials Science & Processing*, 128, p. 919, https://doi.org/10.1007/s00339-022-06045-3

Donga L. M., C. Yea, L. L. Zheng, Z. F. Gao and F. Xia 2020 Two-dimensional metal carbides and nitrides (MXenes): Preparation, property, and applications in cancer therapy, *Nanophotonics*, 9(8), pp. 2125–2145.

Dong J., Y. Zhang, M.I. Hussain, W. Zhou, Y. Chen, and L. N. Wang 2021 g-C_3N_4: Properties, pore modifications, and photocatalytic applications, *Nanomaterials*, 12(1), pp. 1–35.

Eda G., C. Mattevi, H. Yamaguchi, H.K. Kim, and M. Chhowalla 2009 Insulator to semi-metal transition in graphene oxide, *Journal of Physical Chemistry C*, 113(35), pp. 15768–15771.

Ghosh A., S. F. T. Kallungal, and S. Ramaprabhu 2023. 2D Metal-organic frameworks: Properties, synthesis, and applications in electrochemical and optical biosensors, *Biosensors*, 13(1), pp. 1–33.

Gogotsi Y. and B. Anasori 2019 The rise of Mxenes, *ACS Nano*, 13(8), pp. 8491–8494.

Gogotsi Y. and Q. Huang 2021 Mxenes: Two-dimensional building blocks for future materials and devices, *ACS Nano*, 15(4), pp. 5775–5780.

Graebner J.E. 1995 Thermal conductivity of diamond. In: Pan, L.S., Kania, D.R. (Eds) *Diamond: Electronic Properties and Applications*, The Kluwer International Series in Engineering and Computer Science. Springer, Boston, MA, pp. 285–318.

Grazianetti C., and C. Martella 2021 The rise of the Xenes: From the synthesis to the integration processes for electronics and photonics, *Materials (Basel)*, 14(15), pp. 1–13.

Grdadolnik J. 2002 ATR-FTIR spectroscopy: Its advantages and limitations, *Acta Chimica Slovenica*, 49(3), pp. 631–642.

Hummers W.S., Jr. and R.E. Offeman 1958 Preparation of graphitic oxide, *Journal of the American Chemical Society*, 80(6), p. 1339.

Janith G.I., H.S. Herath, N. Hendeniya, D. Attygalle, D.A.S. Amarasinghe, V. Logeeshan, P.M.T.B. Wickramasinghe and Y.S. Wijayasinghe 2023 Advances in surface plasmon resonance biosensors for medical diagnostics: An overview of recent developments and techniques, *Journal of Pharmaceutical and Biomedical Analysis Open*, 2(100019), pp. 1–12.

Jiříčková A., O. Jankovský, Z. Sofer, and D. Sedmidubský 2022 Synthesis and applications of graphene oxide, *Materials (Basel)*, 15(3), pp. 1–21.

Kang J.H., T. Kim, J. Choi, J. Park, Y. S. Kim, M. S. Chang, H. Jung, et al. 2016 Hidden second oxidation step of Hummers method, *Chemistry of Materials*, 28(3), pp. 756–764.

Lee C., X. Wei, J. W. Kysar, and J. Hone 2008 Measurement of the elastic properties and intrinsic strength of monolayer graphene, *Science*, 321(5887), pp. 385–388.

Li L., W. Han, L. Pi, P. Niu, J. Han, C. Wang, B. Su, H. Li, J. Xiong, and Y. Bando 2019, Emerging in-plane anisotropic two-dimensional materials, *InfoMat*, 1(1), pp. 54–73.

Ling X., H. Wang, S. Huang, and M. S. Dresselhaus 2015 The renaissance of black phosphorus, *Proceedings of the National Academy of Sciences (PNAS)*, 112(15), pp. 4523–4530.

Li S., G. Huang, Y. Jia, B. Wang, H. Wang, and H. Zhang 2022a Photoelectronic properties and devices of 2D Xenes, *Journal of Materials Science & Technology*, 126, pp. 44–59.

Liu K., Q. Yan, M. Chen, W. Fan, Y. Sun, J. Suh, D. Fu, et al. 2014 Elastic properties of chemical-vapor-deposited monolayer MoS_2, WS_2, and their bilayer heterostructures, *Nano Letters*, 14(9), pp. 5097–5103.

Li Y., L. Li, S. Li, J. Sun, Y. Fang, and T. Deng 2022b Highly sensitive photodetectors based on monolayer MoS_2 field-effect transistors, *ACS Omega*, 7(16), pp. 13615–13621.

Lopez-Sanchez O., D. Lembke, M. Kayci, Radenovic A. and Kis A. 2013 Ultrasensitive photodetectors based on monolayer MoS_2, *Nature Nanotechnology*, 8, pp. 497–501.

Manzeli S., D. Ovchinnikov, D. Pasquier, O. V. Yazyev, and A. Kis 2017 2D transition metal dichalcogenides, *Nature Reviews Materials*, 2, pp. 17033, https://doi.org/10.1038/natrevmats.2017.33

Mir S. H., V. K. Yadav, and J. K. Singh 2020 Recent Advances in the carrier mobility of two-dimensional materials: A theoretical perspective, *ACS Omega*, 5, pp. 14203–14211.

Mounet N., M. Gibertini, P. Schwaller, D. Campi, A. Merkys, A. Marrazzo, T. Sohier, I. E. Castelli, A. Cepellotti, G. Pizzi, and N. Marzari 2018 Two-dimensional materials from high-throughput computational exfoliation of experimentally known compounds, *Nature Nanotechnology*, 13, pp. 246–252.

Novoselov K. S., A. K. Geim, S. V. Morozov, D. Jiang, Y. Zhang, S.V. Dubonos, I. V. Grigorieva, and A. A. Firsov 2004 Electric field effect in atomically thin carbon films, *Science*, 306(5696), pp. 666–669.

Ouyang Q., S. Zeng, L. Jiang, L. Hong, G. Xu, X.-Q. Dinh, J. Qian, S. He, et al. 2016 Sensitivity enhancement of transition metal dichalcogenides/silicon nanostructure-based surface plasmon resonance biosensor, *Scientific Reports*, 6, Article number: 28190, pp. 1–13.

Prasad B.S.L. and A. R. Annamalai 2021 Effect of rhenium addition on tungsten heavy alloys processed through spark plasma sintering, *Ain Shams Engineering Journal*, 12(3), pp. 2957–2963.

Puebla S., R. D'Agosta, G. Sanchez-Santolino, R. Frisenda, C. Munuera and A. Castellanos-Gomez 2021 In-plane anisotropic optical and mechanical properties of two-dimensional MoO_3, *npj 2D Materials and Applications*, 5, pp. 1–7.

Sheehy D. E. and J. Schmalian 2009 Optical transparency of graphene as determined by the fine-structure constant, *Physical Review B*, 80(19), pp. 1–5.

Shi H., S. Fu, Y. Liu, C. Neumann, M. Wang, H. Dong, P. Kot, M. Bonn, H. I. Wang, A. Turchanin, et al. 2021 Molecularly engineered black phosphorus heterostructures with improved ambient stability and enhanced charge carrier mobility, *Advanced Materials*, 33(48), pp. 1–8.

Smith A. T., A. M. LaChance, S. Zeng, B. Liu, and L. Sun 2019 Synthesis, properties, and applications of graphene oxide/reduced graphene oxide and their nanocomposites, *Nano Materials Science*, 1(1), pp. 31–47.

Song J., H. Liu, Z. Zhao, X. Guo, C.-K. Liu, S. Griggs, A. Marks, Y. Zhu, H. K.-W. Law, I. Mcculloch, and F. Yan 2023 2D metal-organic frameworks for ultraflexible electrochemical transistors with high transconductance and fast response speeds, *Science Advances*, 9(2), pp. 1–11.

Sulleiro M.V., A. Dominguez-Alfaro, N. Alegret, A. Silvestri, I. J. Gómez 2022 2D Materials towards sensing technology: From fundamentals to applications, *Sensing and Bio-Sensing Research*, 38(100540), pp. 1–21.

Sun P., R. Ma, X. Bai, K. Wang, H. Zhu, and T. Sasaki 2017 Single-layer nanosheets with exceptionally high and anisotropic hydroxyl ion conductivity, *Science Advances*, 3(4), pp. 1–8.

Tan J., Y. Chen, J. He, L. G. Occhipinti, Z. Wang and X. Zhou 2023 Two-dimensional material-enhanced surface plasmon resonance for antibiotic sensing, *Journal of Hazardous Materials*, 455, p. 131644.

Vahidmohammadi A., J. Rosen and Y. Gogotsi 2021 The world of two-dimensional carbides and nitrides (MXenes), *Science*, 372(6547), pp. 1–14.

Velasco J.I., M. Ardanuy, and M. Antunes 2012 4-Layered double hydroxides (LDHs) as functional fillers in polymer nanocomposites, In: Fengge Gao (Ed.), Woodhead Publishing Series in *Composites Science and Engineering, Advances in Polymer Nanocomposites*, Woodhead Publishing Limited, Cambridge, UK, pp. 91–130.

Wang G., C. Robert, A. Suslu, B. Chen, S. Yang, S. Alamdari, I. C. Gerber, et al. 2015 Spin-orbit engineering in transition metal dichalcogenide alloy monolayers, *Nature Communications*, 6(10110), pp. 1–7.

Wang W., Y. Yu, Y. Jin, X. Liu, M. Shang, X. Zheng, T. Liu and Z. Xie 2022 Two-dimensional metal-organic frameworks: from synthesis to bioapplications, *Journal of Nanobiotechnology*, 20, pp. 1–17.

Wu L., H. S. Chu, W. S. Koh and E. P. Li 2010 Highly sensitive graphene biosensors based on surface plasmon resonance, *Optics Express*, 18(14), pp. 14395–14400.

Wu L., J. Guo, Q. Wang, S. Lu, X. Dai, Y. Xian and D. Fan 2017 Sensitivity enhancement by using few-layer black phosphorus-graphene/TMDCs heterostructure in surface plasmon resonance biochemical sensor, *Sensors and Actuators B: Chemical*, 249, pp. 542–548.

Xia F., H. Wang, J. C. M. Hwang, A. H. Castro Neto and L. Yang 2019 Black phosphorus and its isoelectronic materials, *Nature Reviews Physics*, 1, pp. 306–317.

Xie H., Z. Li, L. Cheng, A. A. Haidry, J. Tao, Y. Xu, K. Xu, and J. Z. Ou 2022 Recent advances in the fabrication of 2D metal oxides, *iScience*, 25(1), pp. 1–30.

Xi Q., D.-M. Zhou, Y.-Y. Kan, J. Ge, Z.-K. Wu, R.-Q. Yu, and J.-H. Jiang 2014 Highly sensitive and selective strategy for microRNA detection based on WS_2 nanosheet mediated fluorescence quenching and duplex-specific nuclease signal amplification, *Analytical Chemistry*, 86(3), pp. 1361–1365.

Yan W., H.-R. Fuh, Y. Lv, K. Q. Chen, T.-Y. Sai, Y.-R. Wu, T. H. Shieh, et al. 2021 Giant gauge factor of Van der Waals material-based strain sensors, *Nature Communications*,12, pp. 1–9.

Yuan Y., R. Li, and Z. Liu 2014 Establishing water-soluble layered WS_2 nanosheet as a platform for biosensing, *Analytical Chemistry*, 86(7), pp. 3610–3615.

Zhang T., Z. Li, S. Ma, S. Kou and H. Jing 2016 High strength steel (600–900 MPa) deposited metals: Microstructure and mechanical properties, *Science and Technology of Welding and Joining*, 21(3), pp. 186–193.

Zhang W., C.P. Chuu, J.K. Huang, C.-H. Chen, M.-L. Tsai, Y.-H. Chang, C.-T. Liang, et al. 2014 Ultrahigh-gain photodetectors based on atomically thin graphene-MoS_2 heterostructures, *Scientific Reports*, 4(3826), pp. 1–8.

Zhong Z.-T., G. Ashraf, W. Chen, L.-B. Song, S.-J. Zhang, B. Liu, and Y.-D. Zhao 2022 A new strategy based on duplex-specific nuclease and DNA aptamer with modified hairpin structure for various analytes detection, *Microchemical Journal*, 179, p. 107510.

2 Synthesis, Deposition, and Processing of 2D Materials for Sensor Fabrication

2.1 PRACTICABILITY OF 2D MATERIALS REALIZATION BY THINNING OF BULK MATERIALS

At first thinking, it might appear that a 2D material can be obtained by simply taking a bulk 3D material and reducing its thickness to the atomic scale. This simple idea is not experimentally feasible. Why? Because the material may have chemical bonds oriented in three dimensions. So, thinning it down to atomic level thickness will involve cutting through these bonds, hence, leaving a lot of dangling bonds on the surface of the thinned material. These bonds are unstable from chemical and energy viewpoints. They will bring about a rearrangement of the material to a state of lower surface energy.

However, this idea becomes practicable if the material is a layered structure in which the atoms in the constituent layers are bound together by strong chemical bonds but the layers are stacked together by weak interlayer forces. So, we have planes of chemically bonded atoms attached to each other by feeble forces. Such a situation is observed in graphite where plane sheets of single atom thickness sp^2-bonded carbon atoms are glued to each other by the frail van der Waals interaction. There are strong chemical bonds connecting the atoms in the individual planes but weak bonds in the perpendicular direction. Since the chemical bonds in graphite lie in a 2D plane, cutting a graphite slab parallel to the plane will not harm any chemical bond. Hence, the individual planes are separable without creating any dangling bonds. Each plane is called graphene.

2.2 APPROACHES TO 2D MATERIALS PREPARATION

There are two approaches to preparation of 2D materials (Alam et al. 2021):

(i)　Top-down approach: Here, the starting material is a layered bulk material. A longitudinal or transverse stress is applied on its surface to increase the distance between the layers, thereby weakening the van der Waals forces between them and peeling off a monolayer or a few-layered thick 2D materials.

(ii)　Bottom-up approach: Here, the starting materials are atomic or molecular precursors, and the required 2D material is prepared by a chemical reaction between them.

DOI: 10.1201/9781003330585-2

2.3　TOP-DOWN METHODS

2.3.1　Mechanical Exfoliation

A popular method of mechanical exfoliation uses a Scotch tape (Yi and Shen 2015). The method is used to prepare high-quality graphene flakes from highly ordered pyrolytic graphite (HOPG). The steps of this method are illustrated in Figure 2.1. The Scotch tape is pressed against the surface of bulk material and carefully pulled off to exert a force in a direction normal to its surface. After each pulling, the bulk material becomes thinner. The process of Scotch tape pressing and pulling is repeated until a single layer of the material is left. Thus, this method works by layer-by-layer peeling. Since the crystal structure is preserved and the method is straightforward and simple, it is favored for basic research. But it is time-consuming and labor-intensive and requires skilled workers. Contamination of the surface by the traces of tacky polymer from the tape is another drawback. So, it cannot be upscaled for mass-scale production.

2.3.2　Solvent-Based Ultrasonic Exfoliation

Sonication is done to propagate ultrasonic waves through the bulk material dispersed in liquid media to produce small vacuum bubbles (Figure 2.2). These bubbles collapse violently as soon as they reach a critical volume. The resulting shear forces delaminate the bulk material creating nanosheets. The choice of a suitable solvent is critical to get proper 2D material. The solvents used include (Alzakia and Tan 2021):

(i)　Organic solvents: Interfacial tension between the 2D material and the solvent must be matched to overcome the van der Waals forces and stabilize the dispersed nanosheets through solvent–nanosheet interaction. Commonly used solvents are N, N-dimethylformamide (DMF) [$HCON(CH_3)_2$] and N-methyl-2-pyrrolidone (NMP) (C_5H_9NO).

(ii)　Surfactant in aqueous media: To avoid the toxic organic solvents, environment-friendly surfactants are used. Surfactants adjust the surface tension of the liquid to facilitate exfoliation. Additionally, they adsorb on the nanosheets formed and prevent them from restacking. Hence, they act as stabilizers. Surfactants are of two types:

　　(a)　Ionic surfactants: Here, the hydrophobic tail of the surfactant adsorbs on the non-polar surface of the 2D material. Its hydrophilic head undergoes dissociation to form an electrostatic charge. The dispersion is stabilized by electrostatic repulsion between the charges. Examples of ionic surfactants are sodium deoxycholate, ($C_{24}H_{39}NaO_4$), sodium dodecylbenzene sulfonate [$CH_3(CH_2)_{11}C_6H_4SO_3Na$], sodium cholate (SC) ($C_{24}H_{39}NaO_5$), and 7,7,8,8-tetracyanoquinodimethane ($C_{12}H_4N_4$).

　　(b)　Non-ionic surfactants: Here, the hydrophobic tail is adsorbed on the nanosheets. The dispersion is stabilized by steric effects from the projecting hydrophilic tail inducting osmotic repulsion upon coming together. Examples of non-ionic surfactants include porphyrin ($C_{20}H_{14}N_4$) and P-123[203].

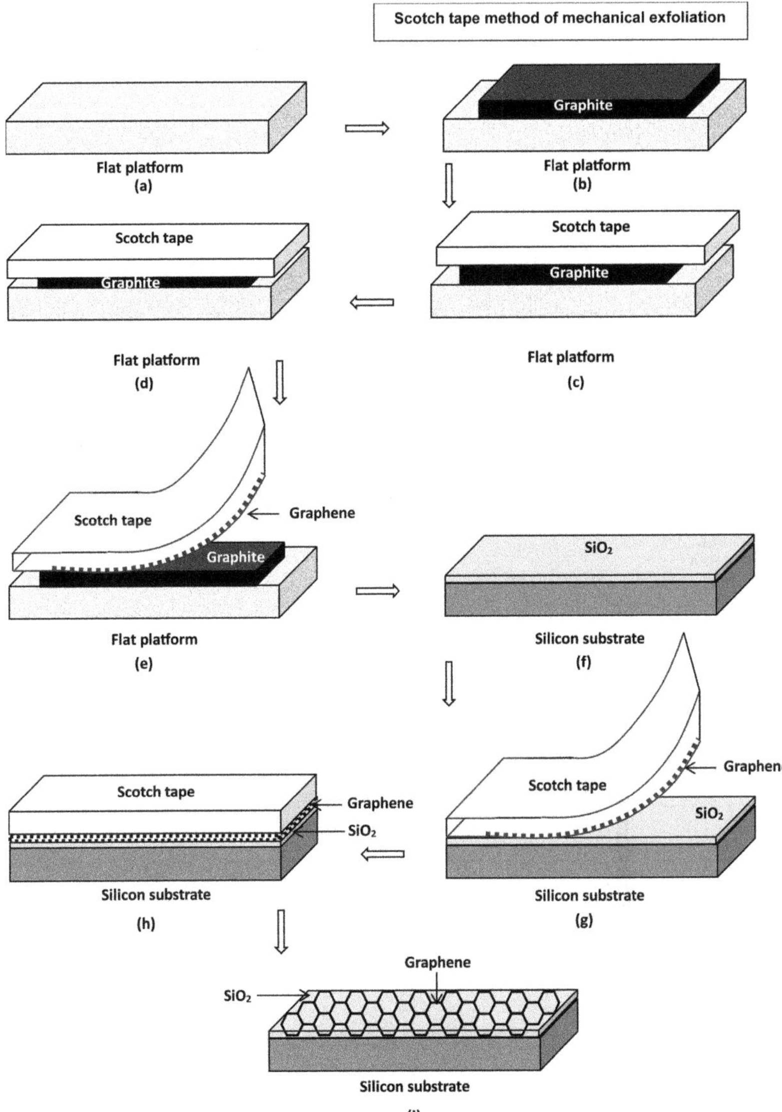

FIGURE 2.1 Scotch tape method of successive mechanical cleavage of graphite to get few-layer or single-layer graphene, and deposition of the exfoliated graphene on a silicon dioxide/silicon substrate. The parts of the diagram show the steps of exfoliation and transfer process: (a) taking a flat platform, (b) placing a graphite sheet on the flat platform, (c) applying Scotch tape on the graphite sheet, (d) pressing the Scotch tape against the graphite sheet, (e) pulling away the Scotch tape from the graphite sheet along with the exfoliated graphene layer adhered to the Scotch tape, (f) taking the SiO_2/Si substrate on which graphene is to be transferred, (g) placing the Scotch tape on the SiO_2/Si substrate with graphene layer in contact with the SiO_2 surface, (h) pressing the Scotch tape against the SiO_2 layer so that graphene is deposited on the SiO_2 surface, and (i) withdrawing the Scotch tape leaving graphene fixed on the SiO_2/Si substrate. Note that steps (c)-(e) are repeated until the monolayer graphene sheet or sheet with the desired number of graphene layers is obtained.

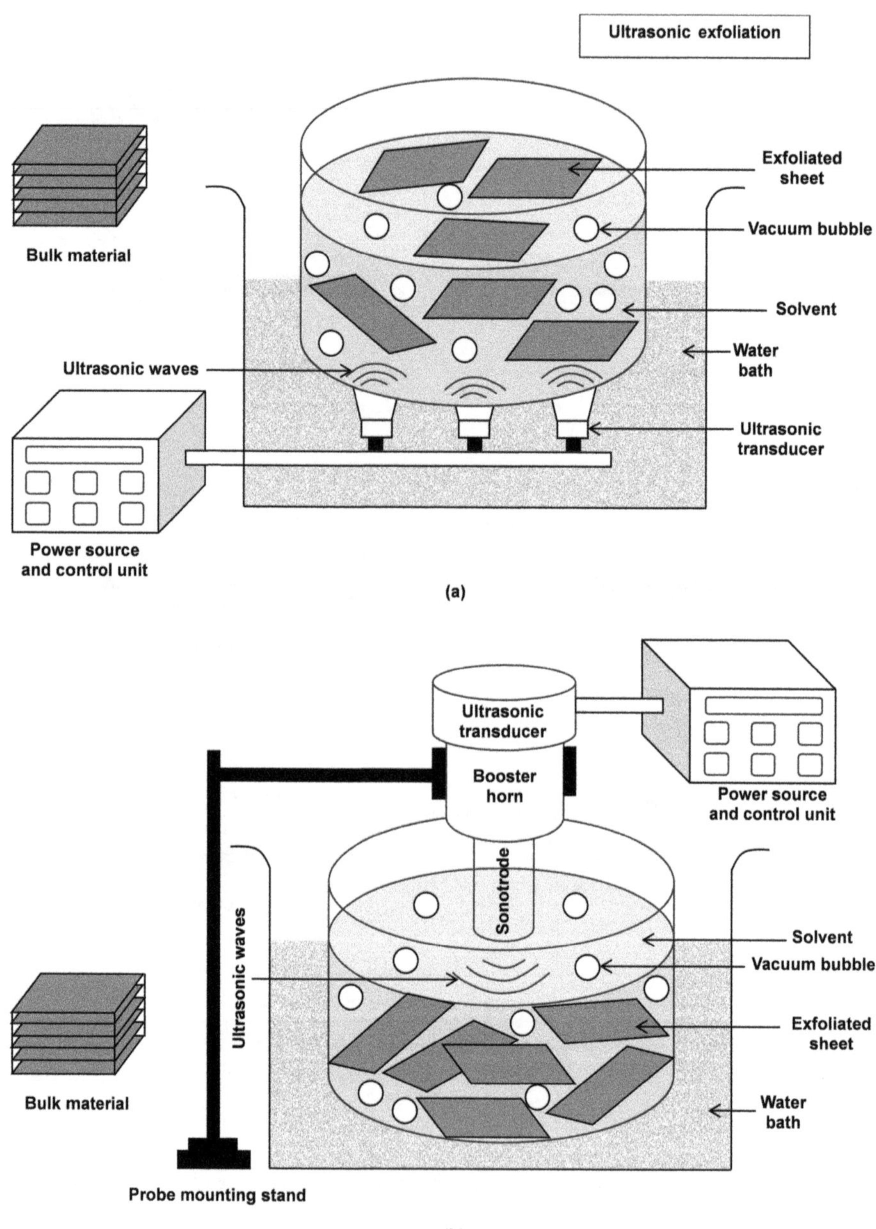

FIGURE 2.2 Ultrasonic exfoliation by (a) bath sonication and (b) probe sonication. Part (a) shows a bulk material composed of six layers on the left top side of the diagram. The bulk material is exfoliated by indirect ultrasonication in a water bath system. For this purpose, a vessel is kept in a water bath. Below the base of the vessel, three ultrasonic transducers are fitted. The transducers are activated by a power source and control unit. The power source and control unit convert the line power into a high-frequency, high-voltage signal for the conversion of electrical energy into high-frequency mechanical vibrations.

FIGURE 2.2 (Continued)   The bulk material is put inside a suitable solvent taken in the vessel and subjected to ultrasonic agitation in the solvent caused by the energy received from the transducers. Vacuum bubbles are formed in the solvent which tear apart and separate the six layers from each other. The layers are dispersed in the solvent. Part (b) shows direct ultrasonication of the bulk material using a probe sonicator. The sonicator consists of an ultrasonic transducer to produce high-frequency ultrasonic waves, a booster horn to amplify the amplitude of ultrasonic vibrations, and a tapered metal bar called the sonoprobe for supplying the amplified ultrasonic energy to the solvent. It is fixed by screwing to a stand with its free end immersed in the solvent up to a certain level. The bulk material to be exfoliated is dipped in this solvent and ultrasonication is carried out. Vacuum bubbles are formed and the exfoliated sheets are dispersed in the solvent.

(iii)   Ionic liquids: Advantages of nonflammability, thermal stability, high ionic conductivity, and low vapor pressure are offered by liquids such as 1-hexyl-3-methylimidazolium hexafluorophosphate ($C_{10}H_{19}F_6N_2P$) and 1-butyl-3-methyl-imidazolium bis(trifluoromethanesulfonyl)imide ([Bmim] [Tf2N]) ($C_{10}H_{15}F_6N_3O_4S_2$).

2.3.3   ION INTERCALATION AND ULTRASONICATION-ASSISTED ELECTROCHEMICAL EXFOLIATION

This method uses an electrochemical cell containing a liquid electrolyte (Yang et al. 2020). Three electrodes, namely, the working electrode, an Ag/AgCl reference electrode, and a counter electrode, are immersed in the electrolyte (Figure 2.3). An external power supply is connected across the cell. The cations used are the alkali metal ions, e.g., Li^+, K^+, Na^+. The anions used are the hydroxyl ion OH-, halide ions (Cl^-, Br^-, I^-), perchlorate ion ClO_4^-, sulfate ion SO_4^{2-}, etc. The electrolyte chosen not only supports ionic conduction but also has the proper surface tension to prevent the restacking of exfoliated layers.

Ion intercalation starts when a suitable voltage is applied between the working and counter electrodes. Intercalation is the incorporation of an ion or molecule into a material with a layered structure. Upon insertion of the intercalant ions, and ensuing electrolysis of intercalants, the interlayer spacings in the bulk material are enlarged. As a consequence, the interlayer interaction is weakened. Then ultrasonication can easily rupture the bonding among the layers of the material causing exfoliation. So, the bulk material is delaminated into thin layers suspended in the electrolyte.

2.3.4   AQUEOUS ACID ETCHING

Aqueous acid etching (AAE) is a common method for preparing MXenes, e.g., the MXene ($Ti_3C_2T_x$) nanosheets are produced from the parent MAX phase, titanium aluminum carbide (Ti_3AlC_2) by liquid phase exfoliation in HF at room temperature (Naguib et al. 2011). HF and fluoride-based etchants, although widely used, are risky to handle, making the process difficult for mass production. Nevertheless, the process opens a route to make other 2D materials.

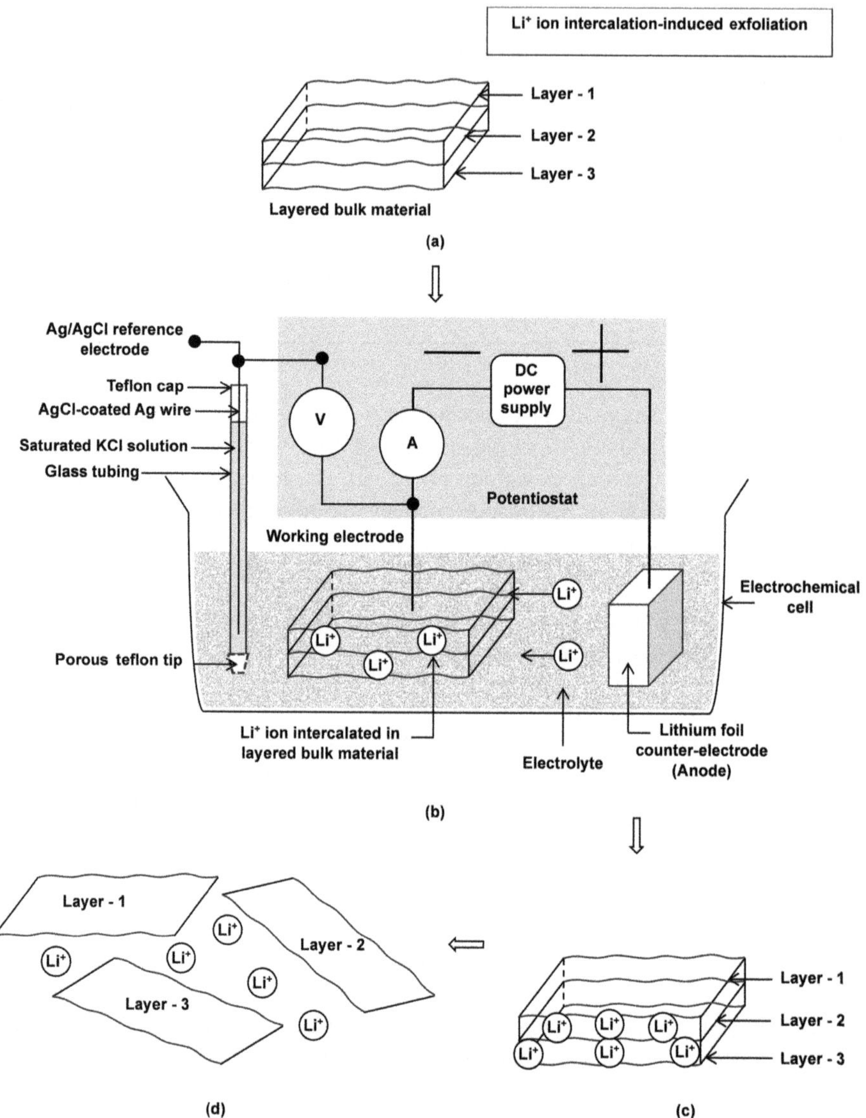

FIGURE 2.3 The process of Li⁺ ion intercalation-induced exfoliation: (a) a three-layer bulk material, (b) a three-electrode electrochemical cell for ion intercalation, (c) the bulk material with Li⁺ ions intercalated in the spaces between the layers, and (d) the three separated layers of the bulk material. Part (a) shows the three layers of the bulk-layered material numbered as layer 1, layer 2, and layer 3. Part (b) shows the glass vessel of the electrochemical cell filled with an electrolyte. The three electrodes of the cell are immersed in the electrolyte. The reference electrode consists of a silver wire coated with silver chloride. This wire is dipped in 3M potassium chloride solution in distilled water filled inside a glass tubing. On the top, the glass tubing is closed by a Teflon cap. Its bottom side is closed with a porous Teflon tip. The working electrode carries the bulk material at its slower end.

FIGURE 2.3 (Continued)   The counter electrode is a lithium foil. The reference electrode is used for controlling and measuring the potential of the working electrode without passing any current. The potential is recorded by a voltmeter in the circuit. The current is measured by an ammeter which is connected between the working and counter electrodes. The potentiostat is the electronic hardware for keeping the potential of the working electrode at a constant value relative to the reference electrode. The DC power supply applies a negative potential to the working electrode, which acts as the cathode. It applies a positive potential to the counter electrode, which acts as the anode. Lithium ions are seen migrating from the anode toward the cathode and penetrating in the intervening spaces between the layers of the bulk material. Part (c) shows the bulk material taken out from the cell after the electrochemical treatment. The intercalated Li^{3+} ions are clearly visible. Part (d) shows the layers after the bulk material subjected to the electrochemical treatment is ultrasonicated. The layers isolated into separate layers 1, 2, and 3 with the lithium ions are released from the material.

2.3.5   LASER ABLATION IN A VOLATILE SOLVENT

This process utilizes laser energy to break the carbon-to-carbon bonds of bulk materials for exfoliation into 2D form. Few-layered black phosphorous nanosheets are produced using a pulsed laser source in a volatile solvent. The number of layers in the obtained black phosphorous film and the lateral size of the film are accurately controlled by varying the energy of the laser source, the irradiation time, and the nature of the solvent used. It is a scalable process for few-layered black phosphorous nanosheet production (Zheng et al. 2020).

2.4   BOTTOM-UP METHODS

2.4.1   WET-CHEMICAL SYNTHESIS

It is a high-yield, economical, solution-based technique for bulk manufacturing of homogeneous materials with less deformation thickness than chemical vapor deposition (CVD) (Alam et al. 2021). It involves chemical reactions using correct precursors under proper experimental conditions. The reaction provides a high degree of controllability and reproducibility. Materials can be doped during synthesis. Their surfaces can be modified with ligands. Further, green chemistry criteria are fulfilled by using environment-friendly precursors.

There are various routes for 2D material synthesis, each route differs from others so that a general rule cannot be framed. Some of the routes are the following.

2.4.1.1   Hydrothermal/Solvothermal Synthesis

Hydrothermal and solvothermal syntheses are two traditional inorganic synthesis branches. Hydrothermal synthesis is done through chemical reactions in a solution made with water at a temperature higher than the boiling point of water. Solvothermal synthesis is performed in a non-aqueous solution, usually an organic solvent at a high temperature of ~300°C (Choi et al. 2022). In hydrothermal synthesis, the reaction medium is water, whereas in solvothermal synthesis, it is an organic solvent. In both cases, the reaction is carried out in a closed vessel so that a high-pressure condition exists. The controlling parameters for the synthesis are the temperature and ratio of reactants, and the time of reaction, which can be easily varied for process tuning.

2D nanosheets of metal chalcogenides and metal oxides are produced by such synthetic procedures. MoS_2 crystallites are made at a yield exceeding 90% from molybdenum trioxide (MoO_3), sulfur (S), and hydrazine hydrate ($N_2H_4.H_2O$) in pyridine (C_5H_5N). The temperature is kept at 300°C, and the time taken is 12 h (Zhan et al. 1998).

Poly(vinylpyrrolidone)-supported single-layered rhodium nanosheets are made by dissolving Rh(III) acetylacetonate, $[CH_3COCH=C(O-)CH_3]_3Rh$, and poly(vinylpyrrolidone) $[(C_6H_9NO)_n]$ in benzyl alcohol ($C_6H_5CH_2OH$) and formaldehyde (CH_2O); stirring the mixture and transferring it to an autoclave; sealing the autoclave and maintaining it at 180°C for 8 h; and cooling to room temperature. The black compound formed is precipitated with acetone, separated by centrifugation, washed with ethanol, and vacuum-dried (Duan et al. 2014).

2.4.1.2 Template Synthesis

The material is synthesized using a bulk or pre-synthesized nanostructured template. The synthesis is done either by a physical process such as through surface coating or by applying a chemical process involving addition/elimination, substitution, or isomerization reactions. After the process completion, the template is removed by dissolution or calcination. Structural, dimensional, and morphological features of the material can be finely controlled by this approach.

Ultrathin non-layered chalcopyrite-type $CuInSe_2$ nanoplatelets with a thickness of up to 2 nm are synthesized by this method for making low-cost, high-performance, flexible photodetectors (Bi et al. 2012).

2.4.1.3 Oriented Self-Assembly

In a self-assembly process, the pre-formed low-dimensional nanocrystals are spontaneously organized with each other via non-covalent interactions. Consequently, the low-dimensional nanocrystals come together to form larger crystals in two dimensions. Two-dimensional-oriented attachment is a chemical method in which non-layered structures with well-defined morphology are obtained by association and bonding of adjacent nanocrystals to form 2D nanosheets with a single-crystalline nature.

Rhenium disulfide (ReS_2) nanoflakes are made by an oriented assembly mechanism (Zhang et al. 2016). Generally, face-to-face stacking of 2D nanosheets is facilitated by the intense interlayer attraction between the constituent nanolayers. Edge-to-edge coupling is not observed for a 2D material. But ReS_2 shows a strong anisotropy and a weak van der Waals coupling. Therefore, an edge-to-edge assembly process can be induced and controlled. From these considerations, an edge-to-edge-oriented self-assembly process is realized. In this process, the aspect ratio of ReS_2 nanorolls is efficiently regulated.

2.4.1.4 Hot Injection Synthesis

The organic solvents and capping ligands are mixed together, heated to 300°C–350°C, as decided by the boiling point of the solvent. Highly reactive precursors for the desired 2D material are rapidly injected into a hot solution matrix containing the surfactant, frequently oleyl amine or oleyl acid (Tedstone et al. 2016, Kozhakhmetov et al. 2020), e.g., monodisperse, circular 1 T-WS$_2$ monolayers with diameters ~100 nm are formed by injecting oleic acid [$CH_3(CH_2)_7CH=CH(CH_2)_7COOH$]-coordinated WCl$_6$ and CS$_2$ into the surfactant oleylamine [$CH_3(CH_2)_7CH=CH(CH_2)_7CH_2NH_2$] at 320°C (Mahler et al. 2014). Reactivity of the tungsten precursor is increased by adding hexamethyldisilazane (HMDS) [$(CH_3)_3SiNHSi(CH_3)_3$] in oleylamine before injection. This allows phase tuning. The 2 H-WS$_2$ phase is stabilized over 1 T-WS$_2$. Molybdenum disulfide (MoS$_2$) (Son et al. 2016) and molybdenum (IV) ditelluride (MoTe$_2$) (Sun et al. 2016) are produced by this method.

High-purity 2D materials formed by this method display size and shape uniformity. The use of surfactants and the requirement of high reaction temperatures are the main drawbacks of this method.

2.4.1.5 Interface-Mediated Synthesis

In this method, the properties of interfaces between materials are exploited to produce various materials. High-quality monolayer hBN is synthesized on graphene by an hBN/graphene interface-mediated growth process (Wang et al. 2022). The scalable epitaxy of unidirectional monolayer hBN on graphene is rendered possible by the finding that the in-plane hBN/graphene can be accurately controlled. The method opens the path toward synthesizing ultraclean, wafer-scale 2D materials for 2D monolayer heterostructures.

2.4.2 Chemical Vapor Deposition

CVD is a versatile thermochemical process yielding uniformly thick, large-area films of a variety of 2D materials. It offers the advantage of easily controlling the quality and composition of the deposited film through temperature and pressure. It can also be scaled up for large-scale manufacturing. The CVD apparatus consists of a gas delivery set-up and a quartz reaction chamber known as the reactor, fitted with vacuum and exhaust systems. The CVD reactors come in two versions: hot-walled or cold-walled types. A hot-wall reactor is surrounded by resistive heaters. The substrate receives heat from the hot reactor walls. In the cold-wall reactor, only the substrate is heated, not the reactor walls. For 2D material deposition, volatile gaseous precursors are injected into the reactor under vacuum and heated to the reaction temperature. The precursors react and break down into the desired coating which is deposited on the substrate.

2.4.3 Magnetron Sputtering

Magnetron sputtering deposition (MSD) is a physical vapor deposition technique using a magnetic field together with an electric field to control electrically charged particles (Figure 2.4). It is called by this name because of the use of magnetic field. The magnetic field traps electrons near the surface of the target to increase the sputtering yield. After creating a high vacuum in the chamber, a plasma is produced by the ionization of a back-filled inert gas, e.g., argon. The Ar$^+$ ions are accelerated to bombard a target material. The atoms ejected from the target surface react with the relevant reaction gas incorporated in the plasma to form a film of 2D material on

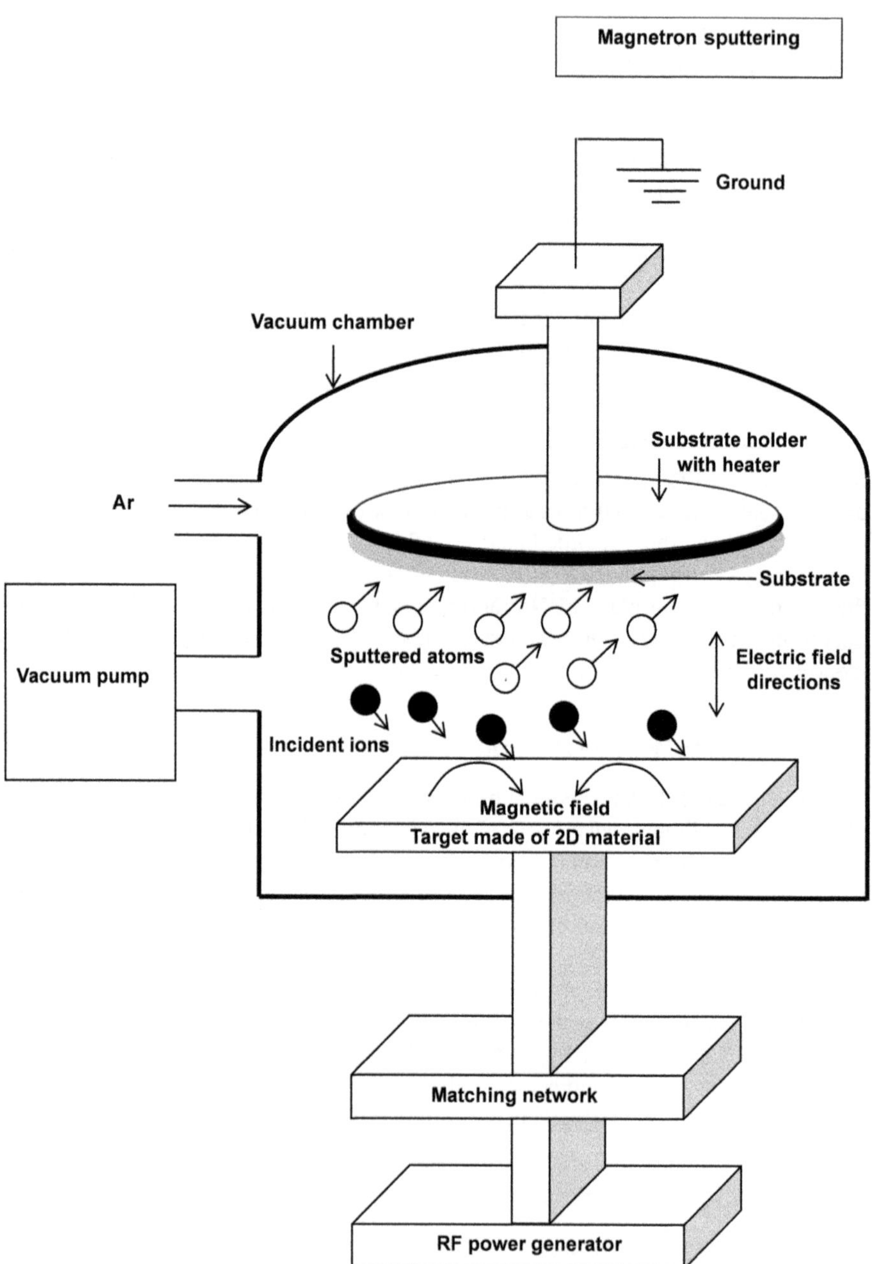

FIGURE 2.4 Magnetron sputtering system. The sputtering system consists of a vacuum chamber fitted with a vacuum pump for evacuation. The chamber has a gas inlet for introducing argon. The target made of the 2D material is mounted near the base of the chamber. It is fed by an RF power generator with a matching network. The substrate is kept in proper position by a substrate holder with the surface for 2D material deposition facing the target. The substrate holder has provision for heating the substrate during deposition. Directions of electric and magnetic fields are indicated.

the substrate kept in the sputtering chamber. Alternatively, deposition is done using a target made of the concerned 2D material. DC magnetron sputtering is employed for depositing conductive metal or alloy films. RF magnetron sputtering uses a radio-frequency electric field, typically 13.56 MHz, for Ar^+ ion acceleration. It can deposit both conducting and insulating films. It is applied for depositing In_2Se_3 films using a compound target in which the elemental ratio is In:Se=2:3. Toxic species, e.g., H_2Se or Se gas are avoided (Li et al. 2014).

2.4.4 PULSED LASER DEPOSITION

Pulsed laser ablation (PLD) is a physical vapor deposition technique. The PLD equipment is shown in Figure 2.5. The deposition is performed in an ultrahigh vacuum chamber using a high-power pulsed laser source, an optical system, and a target of the 2D material to be deposited. The target is placed inside the vacuum chamber, and a pre-heated substrate is mounted in the chamber with its surface opposite to

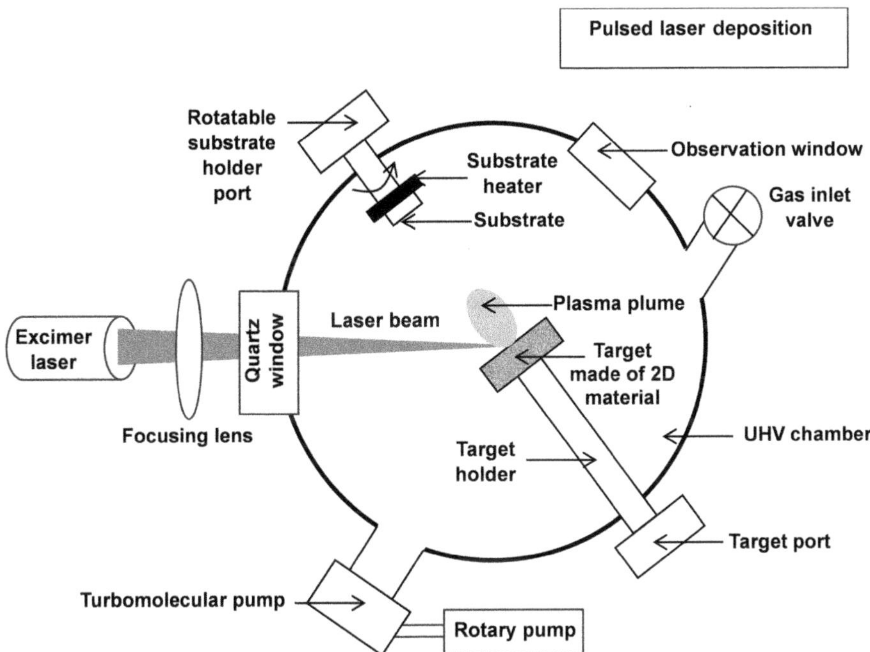

FIGURE 2.5 Pulsed laser deposition (PLD) system. The PLD system consists of an ultra-high vacuum chamber in which a vacuum is created by a turbomolecular pump backed by a rotary pump. The chamber has a target port through which a target holder is fixed. The target is firmly attached to the target holder. On the opposite side and facing the target, the chamber has a rotatable substrate holder port through which passes a substrate holder and heater assembly. The substrate surface for 2D material deposition faces the target. The deposition process can be visually monitored through an observation window aligned with the substrate position. A gas inlet valve is provided to introduce gas for breaking the vacuum and opening the system. The excimer laser beam is focused by a lens and passes through a quartz window in the vacuum chamber to strike the target for generating a plasma plume for film deposition.

the target. The laser beam is focused on the target for a short time. Laser irradiation of the target produces a plasma plume consisting of ions and atoms. As this plume enlarges, it deposits on the substrate to form a thin film of the 2D material. The quality of the film depends on the interaction between the laser source and the target, which in turn is affected by the frequency, fluence, and duration of the laser pulse. The substrate temperature influences the crystalline structure of the film.

The PLD technique is capable of producing large-area, high-quality films of 2D materials with the flexibility to control the number of layers in 2D material film by varying the number of laser pulses. The drawback of this method is the adverse effect of the droplets induced by laser ablation on the surface morphology of the film. These are avoided by methods such as using a shadow mask or a smooth, high-density target or by taking the help of an off-axis geometry (Yang and Hao 2016).

MoS_2 monolayers are synthesized on SiO_2/Si and (0001) sapphire substrates by one-step pulsed laser deposition in a vacuum at a pressure of 5×10^{-7} mbar from a MoS_2 target without any background gas using several laser pulses ranging between 10 and 50 (Miakota 2022). First, the chamber is evacuated to 5×10^{-7} mbar. Then the pressure is raised to 5×10^{-5} mbar by regulating the entry of sulfur flux, obtained from evaporation of sulfur powder kept in a reservoir at 150°C, through a valve into the chamber. The sulfur flux is aimed toward the substrate. A KrF excimer laser source of wavelength 248 nm, pulse duration 20 ns, frequency 1 Hz, and fluence 1 J/cm² ablates the substrate placed 5 cm away from the target. The MoS_2 film is grown at 700°C. The average number of pulses is 35 (Bertoldo et al. 2021).

2.4.5 Molecular Beam Epitaxy

Molecular beam epitaxy (MBE) is a clean, safe, reliable well-controlled, high-purity, layer-by-layer thin-film deposition technique. It is performed in an ultrahigh vacuum (10^{-8}–10^{-12} Torr) chamber by thermal evaporation of atomic or molecular beams of the material on the surface of a single-crystal substrate kept at a high temperature (Figure 2.6). The film is deposited at an extremely low rate ~3000 nmhr^{-1} allowing for epitaxial film growth.

The crystal growth is monitored using reflection high-energy electron diffraction (RHEED) technique. It is a powerful diffraction-based tool to investigate the near-surface structure of the film. The RHEED system consists of an electron gun, the specimen substrate, and a photoluminescent detector. High-energy electrons from the electron gun strike the surface of the substrate at a grazing angle of incidence. They undergo diffraction at the substrate surface. The diffracted electrons interfere constructively at particular angles depending on the crystal structure and spacing between atoms. Some of these electrons collide with the phosphorescent detector creating a diffraction pattern representing the surface features of the substrate. The pattern is applied for surface characterization of the specimen.

In MBE, evaporation of the chosen materials is performed from effusion or Knudsen cells situated directly below the substrate. The effusion cell consists of a crucible of pyrolytic boron nitride, quartz, or graphite with a small orifice and an orifice shutter. It has a heat shield and water-cooling system. The effusion cell serves as a highly efficient and controlled evaporation source.

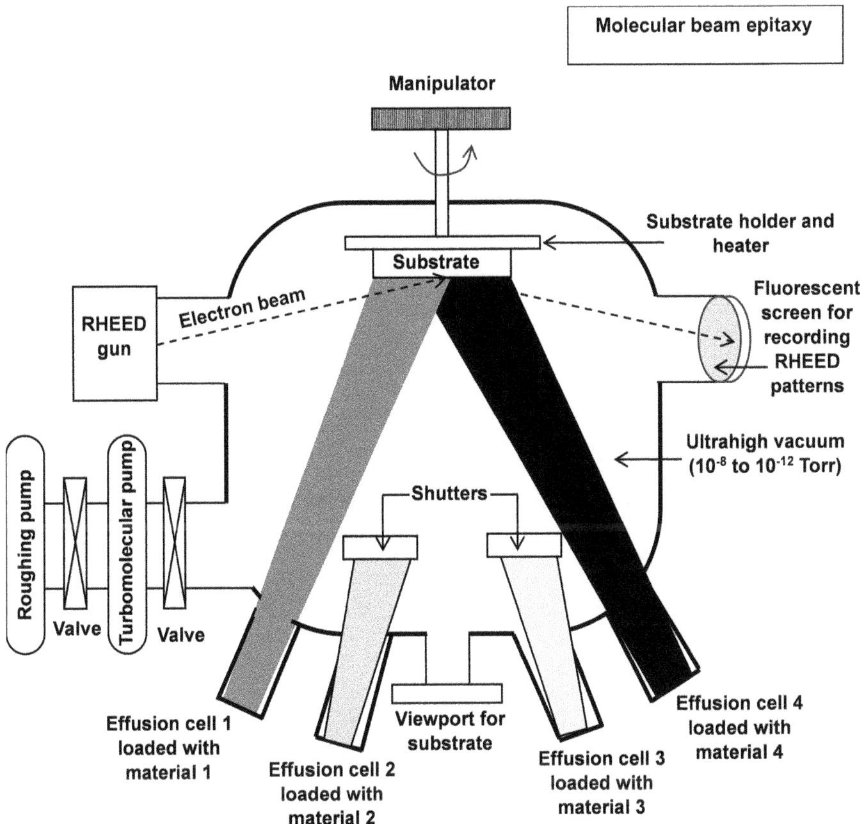

FIGURE 2.6 Molecular beam epitaxy system. The main parts of the MBE system are shown. The UHV chamber is maintained at 10^{-8}–10^{-12} Torr vacuum by a turbomolecular pump assisted by a roughing pump with a valve for interconnection and another valve for connection to the chamber. Projecting outward from the bottom surface of the chamber are effusion cells 1, 2, 3, and 4 filled with source materials 1, 2, 3, and 4, respectively. The effusion cells 2 and 3 from which deposition is not required for the operation being performed are closed by shutters. The evaporated materials 1 and 4 are emitted in the form of divergent gray and black beams. There is also a viewport from which the process is visually inspected and supervised. The material beams are incident on the substrate surface. The substrate is kept in place by a substrate holder connected to a manipulator. The substrate holder has an in-built heater. On the left side of the chamber is a reflective high-energy electron diffraction (RHEED) gun which showers an electron beam on the substrate surface at a small angle so that they do not penetrate deeply. Diffracted electrons from the top few atomic layers of the substrate fall on the photoluminescent RHEED screen on the right side of the chamber creating a diffraction pattern that gives vital information about the quality of the surface and the in-plane lattice spacing.

Heating arrangements are based on either resistive or electron beam heating. The flux of the material is monitored by temperature regulation. Resistive heating by a filament of a refractory element, e.g., tantalum (melting point 3020°C), wound around the crucible, is sufficient for a low melting point element such as selenium (220.8°C). For a high melting point metal, e.g., tungsten (3422°C), electron beam heating is employed in which thermionic electrons are accelerated to bombard the source material by applying a high voltage of a few kV.

Scalable WSe_2 thin films are epitaxially grown in layer-by-layer fashion on insulating Al_2O_3 (sapphire) (001) substrates by MBE. The epitaxial growth is performed in an ultrahigh vacuum chamber at a base pressure below $\sim 1 \times 10^{-7}$ Pa using an electron-beam evaporator cell for evaporating W and a resistive-heating evaporator cell for evaporating Se. The main growth is preceded by forming a partly covered, less-than-a-monolayer thick amorphous buffer layer consisting of W and Se. This buffer layer, formed at room temperature, is annealed at 900°C for 1 h followed by cooling the substrate to 450°C. Then the main WSe_2 layer is grown. Finally, the crystallinity of the grown film is improved by annealing the substrate at 900°C for ½ h (Nakano et al. 2017).

2.5   GRAPHENE PREPARATION

As the wonder material graphene has great significance in sensor fabrication, let us look into the different methods of preparing graphene films for device work.

2.5.1   GRAPHENE BY LASER ABLATION

Figure 2.7 shows the production of graphene by exfoliation of graphite using laser ablation. Isostatic graphite is made from graphite powder by pressing up to 400 MPa

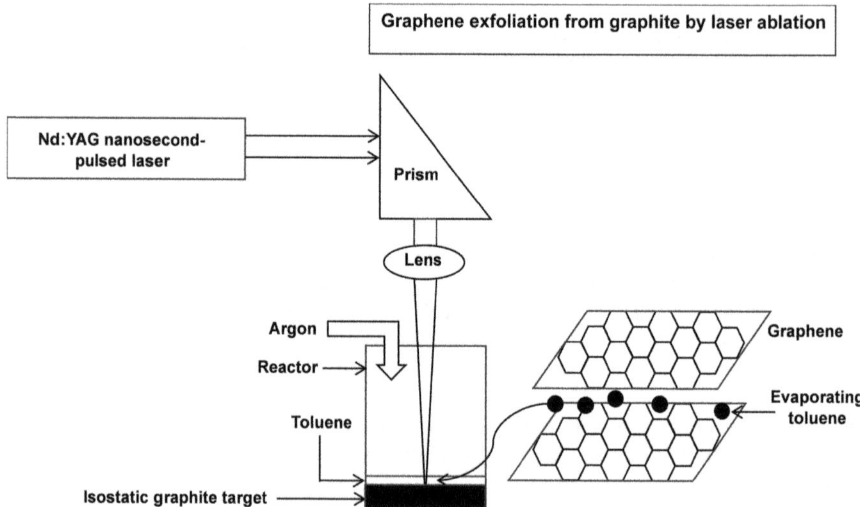

FIGURE 2.7   Exfoliation of graphene from graphite by laser ablation. The diagram shows the laser ablation system for the production of exfoliated graphene. Pulses of light from an Nd:YAG nanosecond-pulsed laser are deviated by a prism and focused by a lens toward an isostatic graphite target covered with a toluene film of sub-nanometer thickness and placed at the bottom of a reactor. The reactor is purged and filled with argon gas and sealed. The graphene layers produced by exfoliation of graphite and the flash-evaporated toluene causing this exfoliation are shown on the right-hand side of the diagram.

at temperatures up to 2000°C (Sierra-Trillo et al. 2022). The graphite thus obtained is placed at the bottom of a sealed reactor and toluene is added to form a 0.3-nm thick liquid medium above graphite. The oxidation of graphite is avoided by flooding the reactor with argon. An Nd:YAG nanosecond-pulsed laser of wavelength 1064 nm, repetition rate 15 Hz, pulse length 5 ns, and repeating time 6×10^7 ns is used for ablating graphite. The high pressure and temperature created by laser pulses cause intercalation of toluene into the layers of graphite. The intercalated toluene undergoes flash evaporation. The evaporated toluene blows up graphite causing exfoliation and producing micron-sized graphene as exfoliated graphite on the ablated graphite surface.

2.5.2 COPPER-CATALYZED CVD OF GRAPHENE

High-end electronics need high-quality, defect-free CVD graphene. The graphene film is deposited on a copper foil in a hot wall furnace (Figure 2.8). Copper acts a catalyst for graphene growth. This catalytic process works through exposure of planar copper surface at a high temperature to a gaseous precursor (Braeuninger-Weimer et al. 2016). First, the Cu foil is annealed in a hydrogen atmosphere at 1000°C.

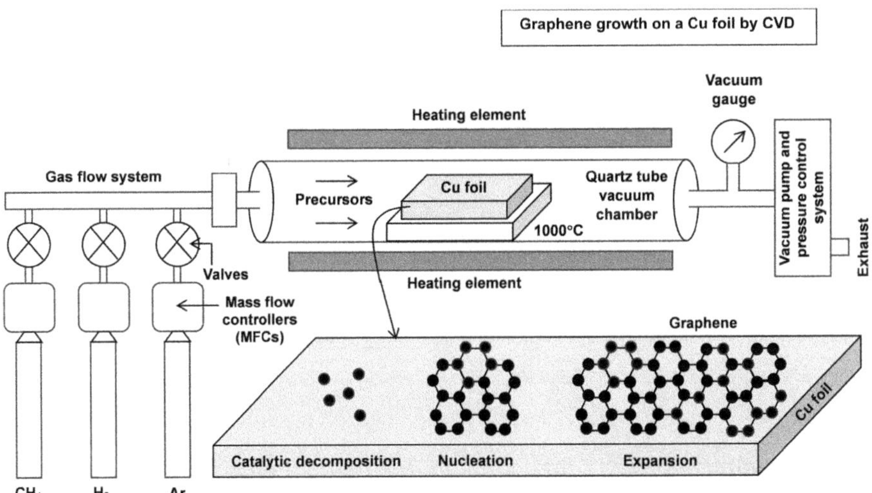

FIGURE 2.8 Growth of graphene film on a copper foil by chemical vapor deposition. The diagram shows the complete arrangement used for carrying out the graphene deposition process. Graphene CVD is done in a quartz tube vacuum chamber equipped with heating elements. The chamber constitutes a tubular vacuum furnace. On the inlet side, this furnace has a gas flow system comprising CH_4, H_2, and Ar cylinders fitted with mass flow controllers and valves. On the outlet side, the furnace is equipped with a vacuum pump, a vacuum gauge, a pressure control system, and an exhaust. A CH_4/H_2 gaseous precursor is passed over a copper foil kept in the heating zone of the furnace at 1000°C. Exposure of the planar Cu surface to gaseous precursor aids in its dissociation and formation of graphitic lattice through nucleation of graphene domains, and their evolution and merger. The three stages of graphene growth are shown: catalytic decomposition of methane, graphene nucleation, and its expansion.

Then graphene growth is started by introducing the precursor gas CH_4 (methane)/ H_2(hydrogen) into the furnace. Graphene film formation is a two-step process. In the first step, the precursor CH_4 is dissociated giving carbon atoms. In the second step, the dissociated carbon atoms are organized into graphene lattice. After the required graphene film has been deposited on the Cu foil, the furnace is cooled down to room temperature.

For electronic device fabrication, the copper substrate underlying the graphene film must be removed (Zhang et al. 2013). For substrate removal, the graphene film is covered with a polymethyl methacrylate (PMMA) layer by spin coating (Figure 2.9). The solvent in PMMA is evaporated by baking at 120°C. The Cu substrate is etched in ferric chloride copper etchant. The PMMA/graphene left behind is rinsed with DI water and placed on the targeted substrate with graphene film contacting the new substrate. The PMMA is got rid of in acetone.

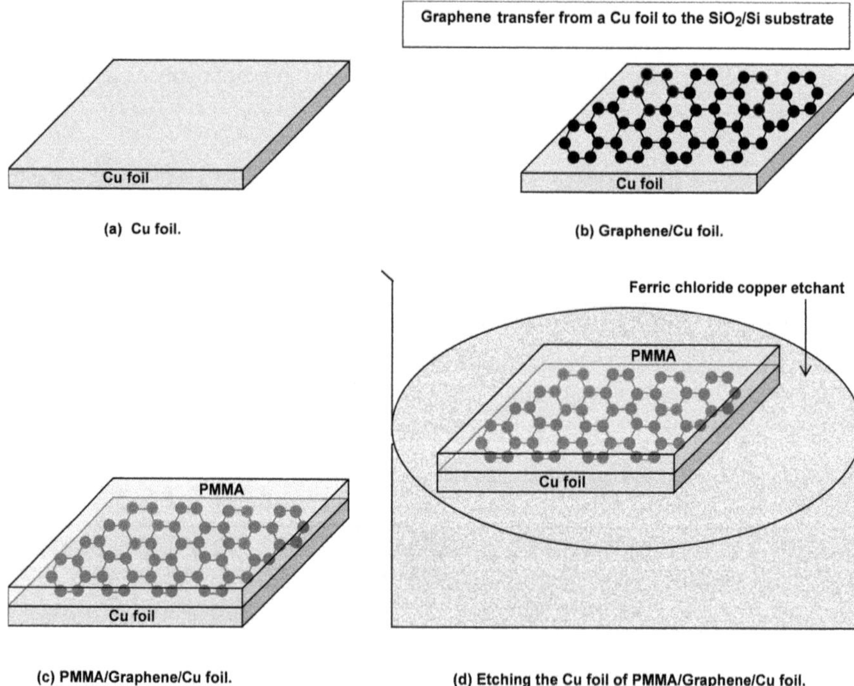

FIGURE 2.9 Graphene growth on a copper foil and transfer of the grown graphene film grown on copper foil to the target substrate. The diagram shows all the steps from graphene growth to graphene transfer: (a) taking a Cu foil, (b) catalytically synthesizing graphene on the surface of Cu foil to get: graphene/Cu foil; (c) coating the graphene-covered surface of the Cu foil with PMMA to form the structure: PMMA/graphene/Cu foil, in which the PMMA layer is used for protection of graphene from Cu etchant in the next step; (d) immersing the PMMA/graphene/ Cu foil in $FeCl_3$ solution until all copper underneath graphene is removed by etching, leaving behind PMMA/graphene.

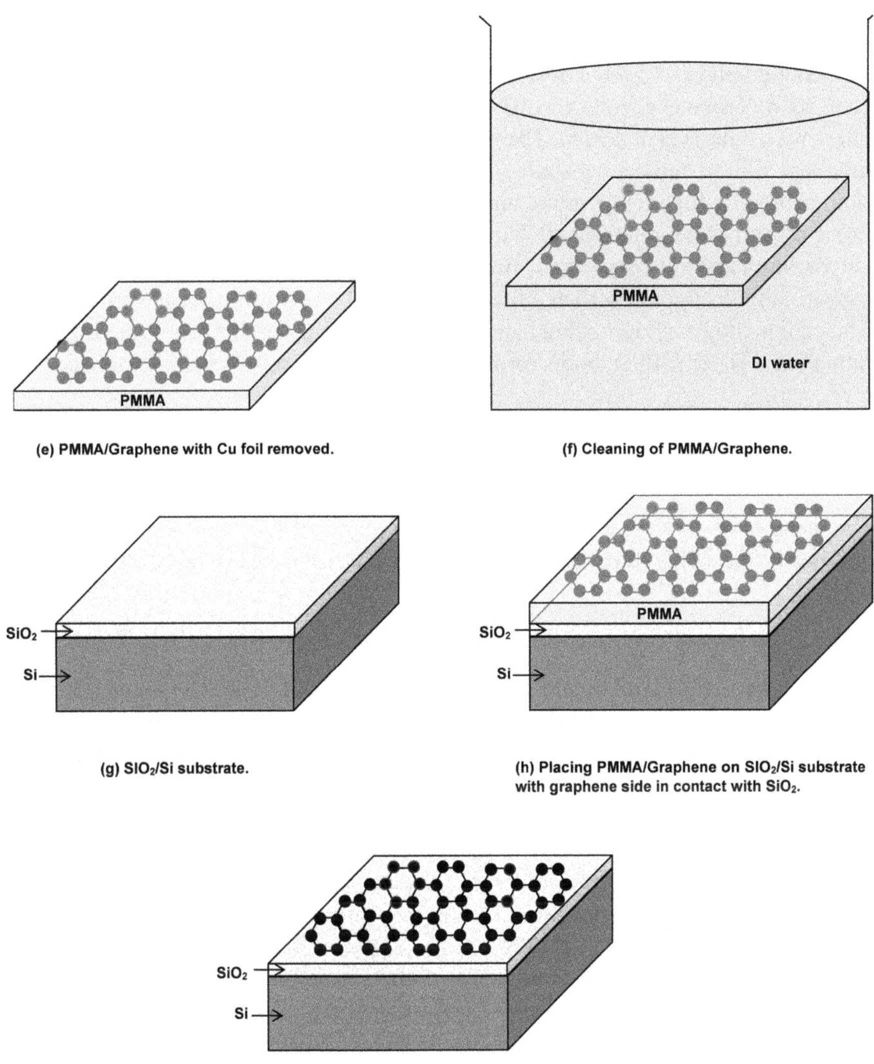

FIGURE 2.9 (Continued) (e) taking out PMMA/graphene from the copper etching solution; (f) rinsing and cleaning the PMMA/graphene in DI water; (g) taking an SiO₂/substrate; (h) placing and pressing the PMMA/graphene over the SiO₂/Si substrate with the graphene film contacting the SiO₂ surface; and (i) removing PMMA in acetone to get graphene/SiO₂/Si.

2.5.3 Epitaxial Growth of Graphene on SiC or SiC/Si Substrates

Epitaxial graphene is grown over commercially available bulk SiC substrates (Yazdi et al. 2016). Graphene growth on 3C SiC (1 1 1)/Si (1 1 1) substrates is a cost-effective alternative (Gupta et al. 2014). The main advantage of this method is that transfer of graphene film to another substrate is not necessary. The growth is done by sublimation method by high-temperature annealing of SiC at 1125°C to 1375°C in vacuum or Ar overpressure atmosphere. The growth mechanism is the depletion of surface silicon atoms of silicon carbide. Because silicon has a higher vapor pressure than carbon, silicon atoms desorb at a faster rate from the SiC surface than carbon atoms. The carbon atoms are left behind and gradually the SiC surface becomes rich in carbon atoms. These carbon atoms form a large-area, uniform-thickness graphene film.

2.6 GRAPHENE OXIDE AND REDUCED GRAPHENE OXIDE PREPARATION

Graphene oxide (GO) and reduced graphene oxide (rGO) are extensively used in sensing devices. Therefore, methods of their preparation are most relevant for sensors.

2.6.1 Hummers' Method of Graphene Oxide Synthesis

GO is commonly synthesized by Hummers' method (Hummers and Offeman 1958, Eigler and Hirsch 2014). A picturesque depiction of the method is given in Figure 2.10. It is a fast method of GO synthesis. The reaction time is 8–12 h. It consists of a reaction between graphite, concentrated sulfuric acid (H_2SO_4), sodium nitrate ($NaNO_3$), and potassium permanganate ($KMnO_4$). The reaction mixture consists of an excess of H_2SO_4 and $KMnO_4$ and a small quantity of $NaNO_3$. At the end of the reaction, the excess potassium permanganate left after the reaction is neutralized with dilute H_2O_2. In this reaction, sulfuric acid and potassium permanganate act as oxidizing agents. Graphite is peeled off to yield GO because the tightly bound graphite layers are loosened by the introduction of oxygen atoms. Thus, GO is produced by chemical oxidation and exfoliation of pristine graphite. The method is safe because no explosive chlorine dioxide (ClO_2) is released. But it is environment-unfriendly due to the evolution of NO_x.

2.6.2 Reduced Graphene Oxide or Graphene by Graphene Oxide Reduction

rGO or graphene is made by chemically or thermally reducing GO. The rGO produced in the two cases is called chemically reduced graphene (CRG) and thermally reduced graphene (TRG) respectively.

(i) Chemical reduction: CRG offers an easier and cost-effective approach for large-scale production of graphene than TRG. Reducing agents used are hydrazine or its derivatives, e.g., hydrazine hydrate [$NH_2NH_2.xH_2O$] and dimethylhydrazine [$(CH_3)_2NNH_2$]. Ascorbic acid (vitamin C) ($C_6H_8O_6$)

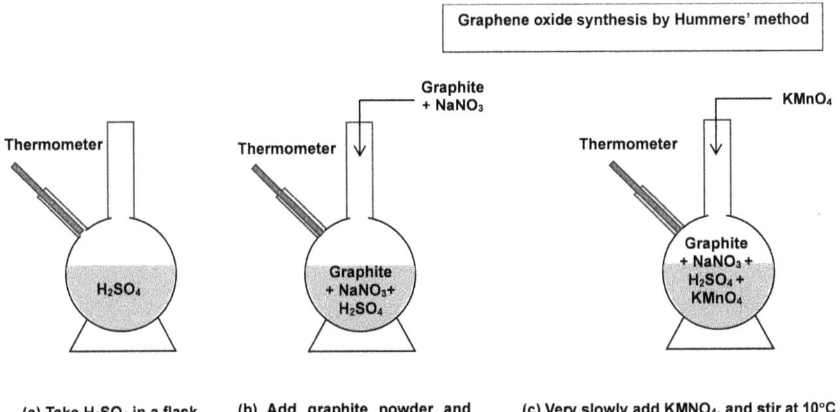

(a) Take H_2SO_4 in a flask and cool in ice bath.

(b) Add graphite powder and $NaNO_3$, stir vigorously at 50°C for 2h, and cool to temperature below 10°C in ice bath.

(c) Very slowly add $KMNO_4$, and stir at 10°C for 2h.

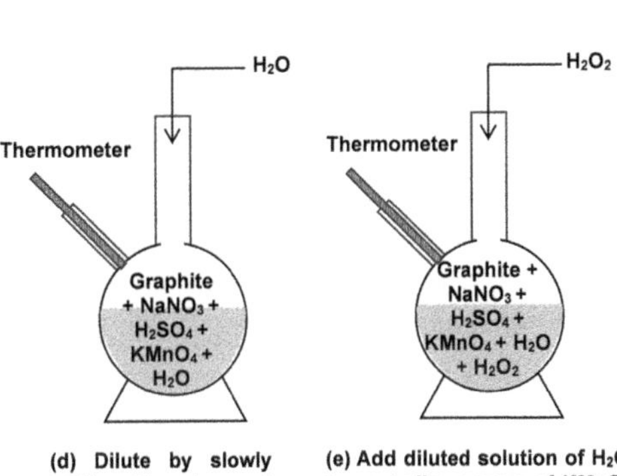

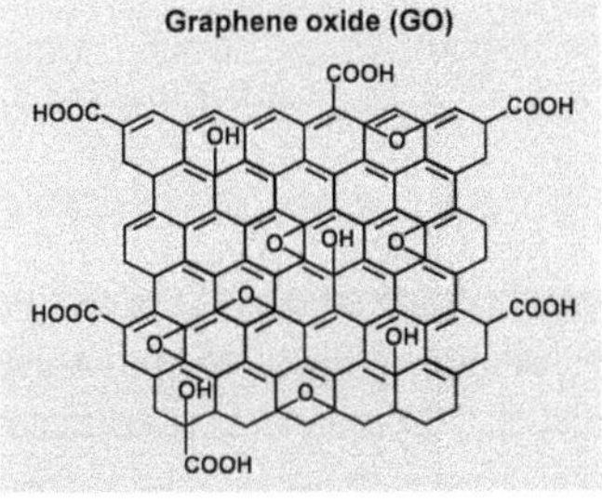

(d) Dilute by slowly adding DI water, and stir vigorously at 98°C for 2h.

(e) Add diluted solution of H_2O_2 to neutralize excess of $KMnO_4$; solution color changes from brown to bright yellow indicating that GO is formed.

(f) GO separation and purification: (i) Sieve and filter, (ii) Centrifuge the filtrate and decant away the supernatant, (iii) Wash the residue with dilute HCl, and then DI water until pH=7, (iv) Wash the residue with ethanol and decant away the supernatant, and (v) Dry to get graphene oxide.

FIGURE 2.10 Synthesis of graphene oxide by Hummers' method. The diagram shows the various consecutive steps followed during GO synthesis: (a) taking H_2SO_4 in a flask and cooling, (b) adding (graphite+$NaNO_3$), stirring and cooling, (c) adding $KMnO_4$ and stirring, (d) diluting with DI water and stirring, (e) adding dilute H_2O_2, and (f) separating and purifying graphene oxide.

is a good substitute for hydrazine (N_2H_4). For chemical reduction, the chosen liquid reagent is added to a suspension of GO in water. The reaction results in agglomeration of graphene-based nanosheets because the hydrophobicity rises.

(ii) Thermal reduction: TRG is produced by thermally annealing GO at a high temperature. Rapid heating to high temperature decomposes oxygen-containing functional groups accompanied by the liberation of CO, CO_2, and H_2O. The released gases create enormous pressure between the layers of GO which induces exfoliation of GO. At 300°C, the pressure is 40 MPa which increases to 130 MPa at 1000°C. A small pressure ~ 2.5 MPa is sufficient to separate two layers of GO. Toxic hydrocarbons are emitted during the TRG process. Complete deoxygenation of GO requires temperatures exceeding 1500°C (Rozada et al. 2013).

2.7 SYNTHESIS OF g-C$_3$N$_4$ BY THERMAL POLYMERIZATION

The synthesis is done by thermal polymerization of different precursors in air or nitrogen flow:

(i) Melamine precursor: 1 g melamine (C$_3$H$_6$N$_6$) powder placed in an alumina crucible in a muffle furnace is covered with a lid and heated up to 550°C for 3 h in a semi-closed system by raising the temperature at 20°C min^{-1} (Al Marzouqi et al. 2019).

(ii) Urea precursor: 5 g urea [CO(NH$_2$)$_2$] is placed in a crucible with a cover and heated up to 550°C in air ambiance for 4 h by raising the temperature at 10°C min^{-1} followed by natural cooling to room temperature (Yang et al. 2018).

2.8 CVD OF COMMONLY USED 2D MATERIALS IN SENSORS

For 2D materials, CVD is a scalable, cost-effective method serving as the main route for obtaining scalable, large-area films. So, this section is devoted to the CVD of selected 2D materials used in sensor technology.

2.8.1 CVD OF HEXAGONAL BORON NITRIDE

The CVD process for hBN production is shown in Figure 2.11. Copper foils are polished with dilute copper etchant, rinsed with 20% HF, and loaded in the furnace heating zone kept at 1050°C under a flowing (Ar+H$_2$) mixture (Jang et al. 2016). The Cu foils are heated at this temperature for 2 h. CVD of hBN is done for 1 h with borazine (B$_3$N$_3$H$_6$) gas as a precursor supplied by bubbling nitrogen through liquid borazine at −15°C temperature. After the CVD process, the furnace is cooled down in Ar/H$_2$ flow.

The mechanism of hBN growth consists of three steps (Liu et al. 2021):

(i) Precursor decomposition into B- and N-containing species.

(ii) Deposition and clustering of B- and N-containing species on the metallic substrate.

(iii) Growth of clusters into larger islands, and their merger to form a continuous hBN film.

Besides borazine, another precursor for hBN is ammonia borane or borazane (BH$_6$N).

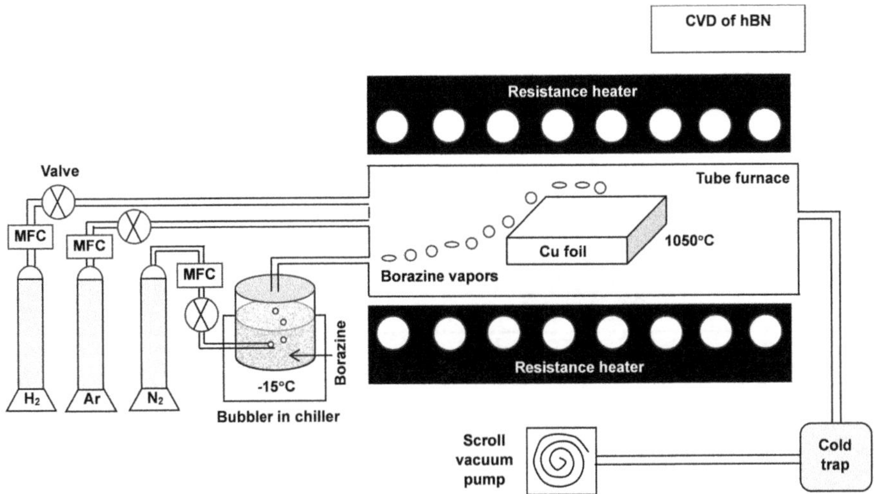

FIGURE 2.11 Growth of hexagonal boron nitride on a copper foil by chemical vapor deposition. The diagram shows the furnace used for CVD and its supporting gas flow and pumping facilities. It is a tube furnace with resistive heating. Cu foil is placed in the hot zone at 1050°C. The gas system and the control panel contain three cylinders filled with H_2, Ar, and N_2 gases, each with its own MFC and valve. Nitrogen is bubbled through a flask containing liquid borazine. Borazine vapors are delivered through nitrogen carrier gas to the furnace tube. The borazine flask is cooled by keeping it in a chiller at -15°C. The outlet of the furnace tube is connected to a cold trap to prevent any contaminants from reaching the scroll pump used for the evacuation of the furnace tube.

2.8.2 CVD OF MoS₂ FLAKES

Large-size MoS_2 flakes of size > 500 μm have been synthesized by CVD (Seravalli and Bosi 2021). An SiO_2/Si substrate is preferred because of its high melting point and compatibility with silicon process technology. The substrate is placed on a graphite susceptor in zone 2 of a quartz tube furnace with resistance heating (Figure 2.12). MoO_3 powder is commonly used as the precursor for molybdenum. The deposition temperature for MoS_2 is 600°C–800°C. The precursor for sulfur is sulfur powder which is placed in a quartz crucible upstream at a distance of 10–15 cm from the substrate and heated to 100°C–200°C in a heating zone 1. Sulfur vapors are transported to the substrate through a carrier gas (N_2, Ar or H_2). The deposition of MoS_2 involves the delivery of Mo and S precursors to the substrate maintained at a high temperature. Chemical reactions take place and MoS_2 is deposited over the substrate.

Alternative liquid precursors for Mo are sodium molybdate (VI) dihydrate ($Na_2MoO_4.2H_2O$) and ammonium heptamolybdate [$(NH_4)_6Mo_7O_{24}$], which are diluted with water, mixed with a promoter (NaOH) and the mixture is deposited over the substrate by spin coating. A liquid precursor for sulfur is dodecane-1-thiol ($C_{12}H_{25}SH$), through which a carrier gas is bubbled for delivering its vapors to the CVD reactor. A gaseous precursor for sulfur is H_2S.

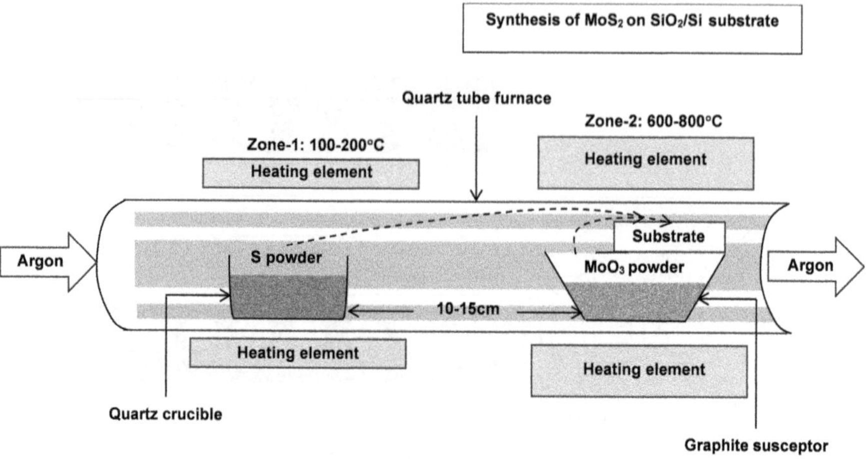

FIGURE 2.12 Synthesis of MoS$_2$ flakes on an SiO$_2$/Si substrate by CVD. The diagram shows a quartz tube furnace with heating elements for two heating zones: a low-temperature zone 1 at 100°C–200°C and a high-temperature zone 2 at 600°C–800°C. The zones are at a distance of 10–15 cm apart. In zone 1, sulfur powder is kept in a quartz crucible. In zone 2, MoO$_3$ powder is kept in a graphite susceptor. The substrate is loaded on the top of the graphite susceptor. Argon gas flows through the furnace tube. Dashed lines with arrows show the transport of sulfur and MoO$_3$ vapors from their individual containers to the substrate upon which MoS$_2$ deposition takes place.

For device fabrication, it is often necessary to transfer the MoS$_2$ film to a different substrate. Then the MoS$_2$ film is spin coated with poly (methyl methacrylate) (PMMA), the underlying SiO$_2$ is etched in HF to separate the MoS$_2$ film from Si, the leftover MoS$_2$/PMMA is placed on the substrate of interest with MoS$_2$ in contact with the substrate, and the PMMA coating is removed in acetone followed by cleaning with isopropanol.

2.8.3 CVD of MoSe$_2$ on Sapphire Substrate

Crystalline monolayers of molybdenum diselenide (MoSe$_2$) are synthesized over sapphire substrates in a hot-wall CVD system by the reaction between MoO$_3$ and Se powder precursors for 15 min. The MoO$_3$ powder is placed in the center of heating zone in a ceramic boat and the temperature is raised to 700°C–900°C while the Se powder is kept in a quartz boat upstream at 270°C. The MoO$_3$ and Se vapors are transported in Ar/H$_2$ flowing gas on sapphire substrates kept on the downstream side near MoO$_3$ powder (Chang et al. 2014).

2.8.4 CVD of MoSe$_2$ on Mo Foil

Phase-pure 2H-MoSe$_2$ is grown on a dilute-HCl etched Mo foil in an ambient-pressure chemical vapor deposition (AP-CVD) system (Konar et al. 2022). The system comprises two furnaces F1 and F2: furnace F1 for elemental Se and furnace F2 for Mo foil (Figure 2.13). The two furnaces F1 and F2 are placed in a single fume hood. The

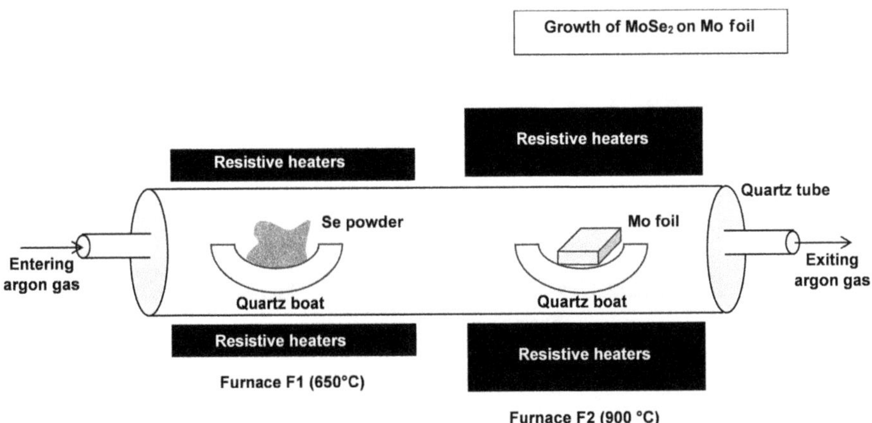

FIGURE 2.13 Growing MoSe$_2$ flakes on Mo foil in an ambient-pressure CVD system. The diagram shows the components of the CVD system. The system consists of a resistively heated quartz tube in which argon gas is continuously flowing from one end to the other. There are two furnaces: F1 at 650°C for selenium and F2 at 900°C for Mo. Selenium powder and Mo foil are placed in quartz boats at their respective temperature zones, selenium in furnace F1 and Mo foil in F2. MoSe$_2$ flakes are formed on the Mo foil.

furnaces are fitted with a single common quartz tube and built-in thermocouples. Elemental Se and Mo foils are placed in separate quartz boats in the quartz tube. The temperature of F1 is set at 650°C and that of F2 at 900°C. The MoSe$_2$ growth is carried out under a steady flow of argon. The nucleation of MoSe$_2$ begins within a span of 15 min of the reaction between Se and Mo. In 1 h, uniform-size MoSe$_2$ flakes are formed.

2.8.5 CVD OF WSE$_2$ FLAKES

A three-zone quartz tube furnace is used (Liu et al. 2015). The CVD setup is shown in Figure 2.14. Selenium powder is placed in the temperature zone at 540°C. The WO$_3$ powder is placed in the temperature zone at 950°C. The SiO$_2$/Si substrate is placed at a small height above the WO$_3$ powder with the SiO$_2$ side facing down toward WO$_3$. The distance between the zones of placement of selenium and WO$_3$ powders is kept 55 cm to ensure uniformity of Se along the substrate length. To start the process, the quartz tube is flushed with argon. Then the furnace is ramped to the required temperatures at the respective zones. After the reaction is completed, the furnace is cooled down to 200°C under Ar/H$_2$ flow, and the substrate is unloaded.

2.8.6 CVD OF BLACK PHOSPHOROUS

An *in situ* CVD method is employed for growing 2D black phosphorous from red phosphorous film directly on a silicon substrate (Smith et al. 2016). In the first step, red phosphorous film is vapor deposited on the silicon substrate. In the second step, this red phosphorous film is used to grow 2D black phosphorous film on the same substrate.

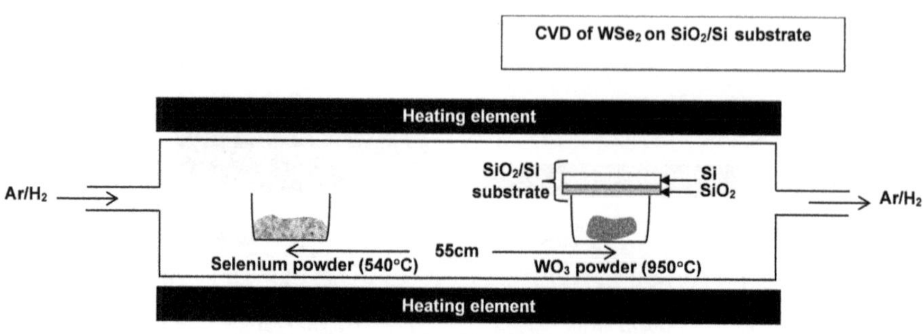

FIGURE 2.14 Synthesizing WSe$_2$ flakes on an SiO$_2$/Si substrate by chemical vapor deposition. The diagram shows a three-zone quartz tube furnace enveloped with heating elements on two sides. The Se powder is placed in a carrier at 540°C temperature zone. WO$_3$ powder is placed in another carrier in the 950°C temperature zone. The SiO$_2$/Si substrate is placed on the top surface of the WO$_3$ powder with the SiO$_2$ surface facing downward toward the WO$_3$ powder side. The WO$_3$ powder carrier is 55 cm farther from the Ar/H$_2$ gas inlet than the Se powder carrier. Ar/H$_2$ gas mixture flows from one end of the furnace tube to the opposite end. Due to sequential placement of Se and WO$_3$ powders, the incoming Ar/H$_2$ gas interacts with the selenium powder coming first on its way and then moves on to interact with the WO$_3$ powder, resulting in the deposition of WSe$_2$ flakes on the SiO$_2$/Si substrate in the flowing Ar/H$_2$.

Step (i): Red phosphorous powder or a bulk BP piece is heated in a tubular furnace at 600°C for ½ h (Figure 2.15(a)). The heating is done under vacuum in the presence of a silicon substrate. Then the furnace is switched off and cooled down. The silicon substrate with the amorphous red phosphorous film is unloaded.

(ii) Step (ii): The red phosphorous-coated Si substrate is transferred to a glass centrifuge tube containing mineralizing agents (Sn and SnI$_4$), as shown in Figure 2.15(b). After placing the centrifuge tube in a pressure vessel reactor, the vessel is sealed, evacuated, and filled with Ar at 27.2 atm. The vessel is placed in a tube furnace and its temperature is increased to 950°C. After attaining this temperature, the furnace is cooled at the rate of 100°C per h. The furnace is kept at 700°C for 1 h, then at 600°C for 1 h, and turned off. 2D BP films with areas larger than 3 µm² and thickness ~4 layers are obtained. Larger-area 2D BP specimens ~9000 µm² of various thicknesses in the typical range of 3.4–600 nm are also observed.

2.8.7 CVD OF MXENES

Scalable synthesis of MXenes has been done by a direct synthetic route (Wang et al. 2023). CVD growth of directly synthesized-Ti$_2$CCl$_2$ (ds-Ti$_2$CCl$_2$) MXene is accomplished by the reaction between Ti, graphite, and TiCl$_4$ at 950°C (Figure 2.16(a)). The ds-Ti$_2$CCl$_2$ formation commences at 850°C. The maximum yield is achieved at 950°C. At temperatures >1000°C, the main product of the reaction changes to titanium carbide (TiC$_x$). The growth of ds-Ti$_2$CCl$_2$ is a self-limiting process because the gaseous reagents have to diffuse through the already-grown film to reach the reaction

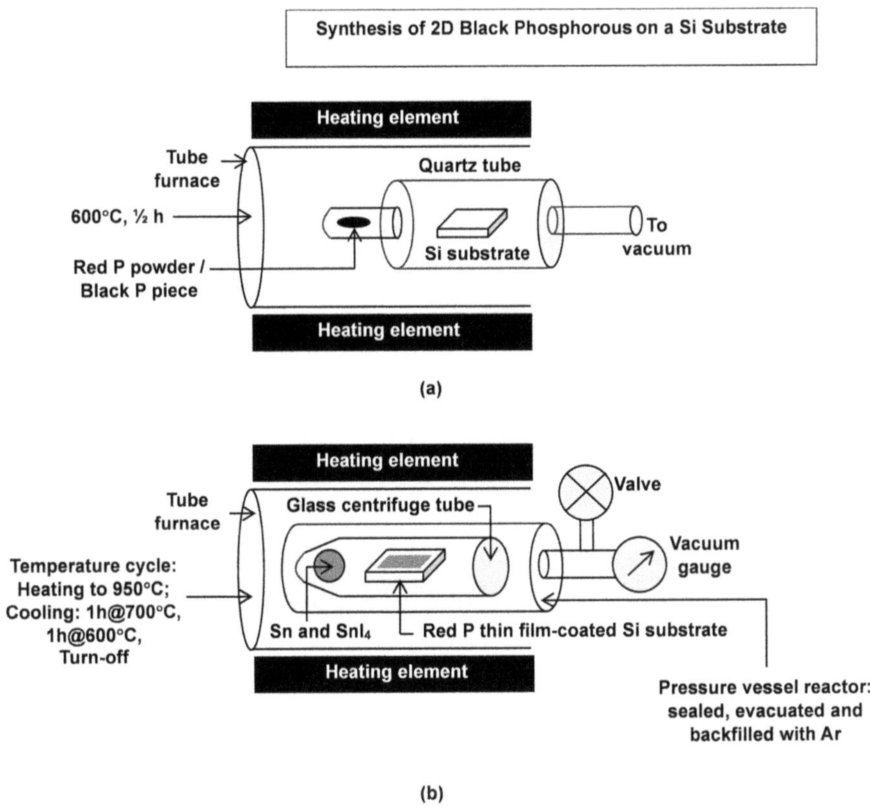

FIGURE 2.15 2D black phosphorous synthesis by CVD: (a) vapor deposition of amorphous red phosphorous powder/black phosphorous piece on a silicon substrate under vacuum and (b) synthesis of 2D black phosphorous film from red phosphorous directly on an Si substrate. Part (a) shows a tube furnace with heating elements. A quartz tube is loaded inside the furnace. At one end of the quartz tube, red phosphorous powder or a piece of black phosphorous is placed. In the quartz tube, a silicon substrate is kept. The opposite end of the quartz tube is connected to a vacuum system. Part (b) shows the same tube furnace with heating elements. A glass centrifuge tube containing Sn, SnI_4, and amorphous red phosphorous thin film-coated silicon substrate is placed in a pressure vessel reactor, which is sealed, evacuated, and backfilled with argon. The reactor has a vacuum gauge and valve. It is loaded in the tube furnace and heated to 950°C, cooled to 700°C, kept at 700°C for 1 h, cooled to 600°C, kept at 600°C for 1 h, and switched off. Red phosphorous is evaporated at 600°C for ½ h.

site. As this film becomes thicker, the gaseous diffusion through it becomes slower, and so also the $ds\text{-}Ti_2CCl_2$ growth reaction.

To get separate 2D $ds\text{-}Ti_2CCl_2$ monolayers from the layered $ds\text{-}Ti_2CCl_2$ MXene prepared by synthesis, the layered $ds\text{-}Ti_2CCl_2$ MXene is delaminated by subjecting it to solution processing (Figure 2.16(b)). For delamination, it is intercalated with Li^+ ions by treating with an organolithium reagent, n-butyllithium (C_4H_9Li) in hexane (C_6H_{14}) solution. This treatment yields a stable suspension of 2D $ds\text{-}Ti_2CCl_2$ sheets

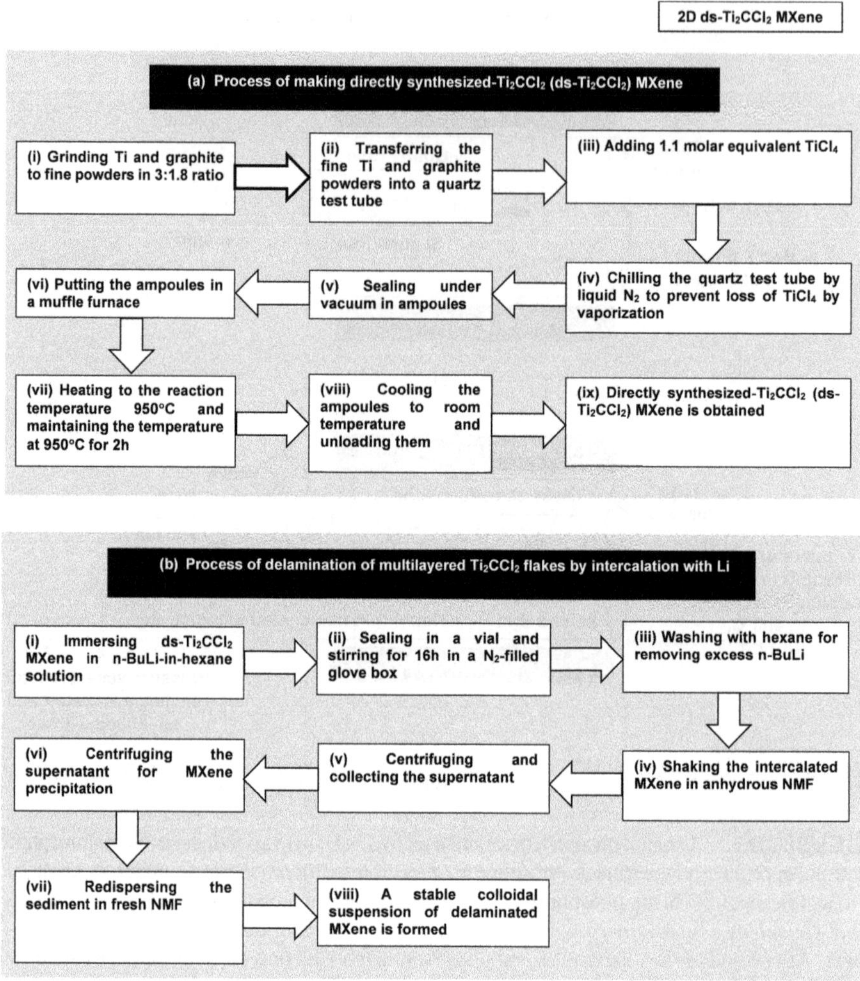

FIGURE 2.16 Making 2D ds-Ti$_2$CCl$_2$ sheets by CVD: (a) making ds-Ti$_2$CCl$_2$ MXene and (b) delaminating multilayered Ti$_2$CCl$_2$ flakes. Part (a) shows the steps for making ds-Ti$_2$CCl$_2$. The steps are: grinding Ti and graphite to fine powders, transferring the fine powders to a quartz tube, adding TiCl$_4$, chilling the quartz tube using liquid N$_2$, sealing the ampoules under vacuum, loading the ampoules in a furnace, heating to and maintaining the reaction temperature for the required period, and cooling and unloading the ampoules. Part (b) shows the steps for delaminating the flakes produced in part (a). The steps are: dipping ds-Ti$_2$CCl$_2$ in n-BuLi-in-hexane solution, sealing in a vial and stirring in a glove box, washing with hexane, shaking in anhydrous NMF and centrifuging, centrifuging the supernatant, and redispersing the sediment in NMF to obtain a colloidal suspension of delaminated MXene.

on shaking with N-methylformamide (NMF) (HCONHCH$_3$). TiC$_x$ being insoluble is precipitated when centrifugation is done.

For synthesizing Zr$_2$CCl$_2$ and Zr$_2$CBr$_2$ MXenes, Zr foil is exposed to methane gas. Along with CH$_4$ gas, ZrCl$_4$ vapors are used for Zr$_2$CCl$_2$ MXenes, and ZrBr$_4$ vapors for Zr$_2$CBr$_2$ MXenes. The zirconium MXenes have a vertically aligned structure that looks like a carpet. Their morphology resembles that of the titanium MXenes.

2.9 DISCUSSION AND CONCLUSIONS

Extensive studies have been carried out by chemists and physicists on the synthesis and fabrication of laminar 2D nano-materials, as well as the integration of the 2D materials with precisely controlled properties in the fabrication process sequences of various sensors. Numerous techniques and procedures for orchestrating high-quality and ultrathin nanosheets have emerged.

A résumé of the top-down and bottom-up techniques discussed in this chapter is shown in Figure 2.17. Each technique has its own set of benefits and drawbacks in producing 2D materials of specified thicknesses and lateral dimensions. The most suitable technique for the fabrication of a particular sensor must be selected from the wide choices and options available.

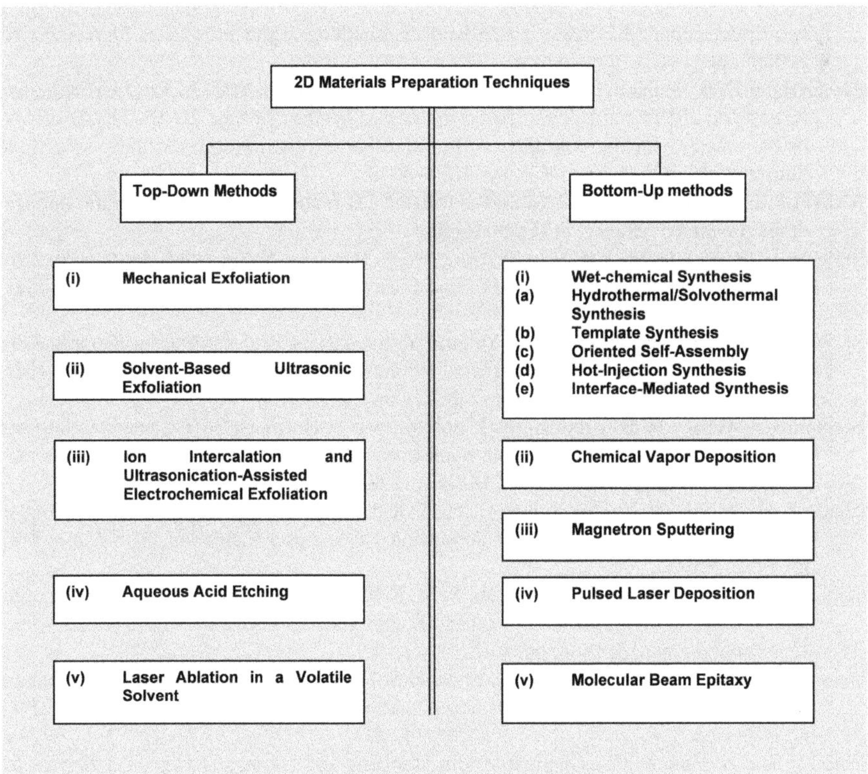

FIGURE 2.17 Résumé of the synthesis techniques of 2D materials. The diagram shows the subdivision of the 2D material synthesis techniques into top-down and bottom-up classes. The top-down methods are: mechanical, ultrasonic and electrochemical exfoliation; aqueous acid etching and laser ablation in a solvent. The bottom-up approaches include: wet-chemical synthesis (hydro-/solvothermal, template, self-assembly, hot-injection, and interface-mediated), chemical vapor deposition, magnetron sputtering, laser ablation, and molecular beam epitaxy.

Furthermore, the same 2D material can be produced by a multiplicity of routes. Hence, after the generalized description of different synthesis methods, important example cases were deliberated, namely:

(i)   Graphene production by laser ablation, CVD, epitaxial growth on SiC or SiC/Si substrates, reduction of Hummers'-method synthesized GO.

(ii) g-C_3N_4 formation by thermal polymerization,.

(iii) CVD of hexagonal boron nitride, MoS_2 flakes, $MoSe_2$ on Mo foil, WSe_2 flakes, black phosphorous, and MXenes.

Equipped with the knowledge of prime 2D materials in Chapter 1 and the principal processes of their synthesis in Chapter 2, we proceed to the study of mechanical 2D materials-based sensors in the next chapter.

REFERENCES

Alam S., M. A. Chowdhury, A. Shahid, R. Alam, and A. Rahim 2021 Synthesis of emerging two-dimensional (2D) materials–Advances, challenges and prospects, *FlatChem*, 30, 100305, pp. 1–16.

Al Marzouqi F., B. Al Farsi, A. T. Kuvarega, H. A. J. Al Lawati, S. M.Z. Al Kindy, Y. Kim, and R. Selvaraj 2019 Controlled microwave-assisted synthesis of the 2D-BiOCl/2D-g-C_3N_4 heterostructure for the degradation of amine-based pharmaceuticals under solar light illumination, *ACS Omega*, 4(3), pp. 4671–4678.

Alzakia F. I. and S. C. Tan 2021 Liquid-exfoliated 2D materials for optoelectronic applications, *Advanced Science*, 8(11), pp. 1–54.

Bertoldo F., R. R. Unocic, Y.-C. Lin, X. Sang, A. A. Puretzky, Y. Yu, D. Miakota, et al. 2021 Intrinsic defects in MoS_2 grown by pulsed laser deposition: From monolayers to bilayers, *ACS Nano*, 15, pp. 2858–2868.

Bi W., M. Zhou, Z. Ma, H. Zhang, J. Yu, and Y. Xie 2012 $CuInSe_2$ ultrathin nanoplatelets: Novel self-sacrificial template-directed synthesis and application for flexible photodetectors, *Chemical Communications*, 48(73), pp. 9162–9164.

Braeuninger-Weimer P., B. Brennan, A. J. Pollard, and S. Hofmann 2016 Understanding and controlling Cu-catalyzed graphene nucleation: The role of impurities, roughness, and oxygen scavenging, *Chemistry of Materials*, 28(24), pp. 8905–8915.

Chang Y.-H., W. Zhang, Y. Zhu, Y. Han, J. Pu, J.-K. Chang, W.-T. Hsu, et al 2014 Monolayer $MoSe_2$ grown by chemical vapor deposition for fast photodetection, *ACS Nano*, 8(8), pp. 8582–8590.

Choi S. H., S. J. Yun, Y. S. Won, C. S. Oh, S. M. Kim, K. K. Kim and Y. H. Lee 2022 Large-scale synthesis of graphene and other 2D materials towards industrialization, *Nature Communications*, 13, 1484, pp. 1–5.

Duan H., N. Yan, R. Yu, C. R. Chang, G. Zhou, H.S. Hu, H. Rong, Z. Niu, J. Mao, H. Asakura and T. Tanaka 2014 Ultrathin rhodium nanosheets, *Nature Communications*, 5(1), pp. 3093, pp. 1–8.

Eigler S. and A. Hirsch 2014 Chemistry with graphene and graphene oxide - Challenges for synthetic chemists, *Angewandte Chemie*, 53(30), pp. 7720–7738.

Gupta B., M. Notarianni, N. Mishra, M. Shafiei, F. Iacopi, and N. Motta 2014 Evolution of epitaxial graphene layers on 3C SiC/Si (111) as a function of annealing temperature in UHV, *Carbon*, 68, pp. 563–572.

Hummers W.S., Jr. and R.E. Offeman 1958 Preparation of graphitic oxide, *Journal of the American Chemical Society*, 80, p. 1339.

Jang S. K., J. Youn, Y. J. Song and S. Lee 2016 Synthesis and characterization of hexagonal boron nitride as a gate dielectric, *Scientific Reports*, 6, 30449, pp. 1–9.

Kozhakhmetov A., R. Torsi, C. Y. Chen and J. A. Robinson 2020 Scalable low-temperature synthesis of two-dimensional materials beyond graphene, *Journal of Physics: Materials*, 4(1), pp. 1–16.

Li S., Y. Yan, Y. Zhang, Y. Ou, Y. Ji, L. Liu, C. Yan, Y. Zhao, and Z. Yu 2014 Monophase γ-In_2Se_3 thin film deposited by magnetron radio-frequency sputtering, *Vacuum*, 99, pp. 228–232.

Liu B., M. Fathi, L. Chen, A. Abbas, Y. Ma, and C. Zhou 2015 Chemical vapor deposition growth of monolayer WSe_2 with tunable device characteristics and growth mechanism study, *ACS Nano*, 9(6), pp. 6119–6127.

Liu H., C. Y. You, J. Li, P. R. Galligan, J. You, Z. Liu, Y. Cai, and Z. Luo 2021 Synthesis of hexagonal boron nitrides by chemical vapor deposition and their use as single photon emitters, *Nano Materials Science*, 3, pp. 291–312.

Mahler B., V. Hoepfner, K. Liao and G.A. Ozin 2014 Colloidal synthesis of 1T-WS_2 and 2H-WS_2 nanosheets: Applications for photocatalytic hydrogen evolution, *Journal of the American Chemical Society*, 136(40), pp. 14121–14127.

Miakota D. I. 2022 Pulsed laser deposition of 2D and quasi1D transition metal dichalcogenides, Doctoral thesis in Electrical and Photonics Engineering, Technical University of Denmark, p. 30.

Konar R., B. Rajeswaran, A. Paul, E. Teblum, H. Aviv, I. Perelshtein, I. Grinberg, Y. Raphael Tischler, and G. D. Nessim 2022 CVD-assisted synthesis of 2D layered $MoSe_2$ on Mo foil and low frequency Raman scattering of its exfoliated few-layer nanosheets on CaF_2 substrates, *ACS Omega*, 7(5), pp. 4121–4134.

Naguib M., M. Kurtoglu, V. Presser, J. Lu, J. Niu, M. Heon, L. Hultman, Y. Gogotsi, and M. W. Barsoum 2011 Two-dimensional nanocrystals produced by exfoliation of Ti_3AlC_2, *Advanced Materials*, 23, pp. 4248–4253.

Nakano M., Y. Wang, Y. Kashiwabara, H. Matsuoka, and Y. Iwasa 2017 Layer-by-layer epitaxial growth of scalable WSe_2 on sapphire by molecular-beam epitaxy, *Nano Letters*, 17(9), pp. 5595–5599.

Rozada R., J. I. Paredes, S. Villar-Rodil, A. Martínez-Alonso and J. M. D. Tascón 2013 Towards full repair of defects in reduced graphene oxide films by two-step graphitization, *Nano Research*, 6, pp. 216–233.

Seravalli L, and M. A. Bosi 2021 Review on chemical vapour deposition of two-dimensional MoS_2 flakes, *Materials (Basel)*, 14(24), pp. 1–28.

Sierra-Trillo M. I., R. Thomann, I. Krossing, R. Hanselmann, R. Mülhaupt and Y. Thomann 2022 Laser ablation on isostatic graphite—A new way to create exfoliated graphite, *Materials*, 15, 5474, pp. 1–13.

Son D., S. I. Chae, M. Kim, M. K. Choi, J. Yang, K. Park, V. S. Kale, J. H. Koo, C. Choi, M. Lee and J. H. Kim 2016 Colloidal synthesis of uniform-sized molybdenum disulfide nanosheets for wafer-scale flexible nonvolatile memory, *Advanced Materials*, 28(42), pp. 9326–9332.

Sun Y., Y. Wang, D. Sun, B. R. Carvalho, C.G. Read, C. H. Lee, Z. Lin, K. Fujisawa, J. A. Robinson, V. H. Crespi and M. Terrones 2016 Low-temperature solution synthesis of few-layer 1T'-$MoTe_2$ nanostructures exhibiting lattice compression, *Angewandte Chemie*, 128(8), pp. 2880–2884.

Smith J. B., D. Hagaman and H.-F. Ji 2016 Growth of 2D black phosphorus film from chemical vapor deposition, *Nanotechnology*, 27, 215602, pp. 1–8.

Tedstone A. A., D. J. Lewis and P. O'Brien 2016 Synthesis, properties, and applications of transition metal-doped layered transition metal dichalcogenides, *Chemistry of Materials*, 28(7), pp. 1965–1974.

Wang D., C. Zhou, A. S. Filatov, W. Cho, F. Lagunas, M. Wang, S. Vaikuntanathan, C. Liu, R. F. Klie, and D. V. Talapin 2023 Direct synthesis and chemical vapor deposition of 2D carbide and nitride Mxenes, *Science*, 379(6638), pp. 1242–1247.

Wang P., W. Lee, J. P. Corbett, W. H. Koll, N. M. Vu, D. A. Laleyan, Q. Wen, Y. Wu, A. Pandey, et al 2022 Scalable synthesis of monolayer hexagonal boron nitride on graphene with giant bandgap renormalization, *Advanced Materials*, 34(21), pp. 1–9.

Yang L., X. Liu, Z. Liu, C. Wang, G. Liu, Q. Li, and X. Feng 2018 Enhanced photocatalytic activity of g-C_3N_4 2D nanosheets through thermal exfoliation using dicyandiamide as precursor, *Ceramics International*, 44(17), pp. 20613–20619.

Yang S., P. Zhang, A. S. Nia, and X. Feng 2020 Emerging 2D materials produced via electrochemistry, *Advanced Materials*, 32(10), pp. 1–19.

Yang Z. and J. Hao 2016 Progress in pulsed laser deposited two-dimensional layered materials for device applications, *Journal of Materials Chemistry C*, 4, pp. 8859–8878.

Yazdi G. R., T. Iakimov and R. Yakimova 2016 Epitaxial graphene on SiC: A review of growth and characterization, *Crystals*, 6, 53, pp. 1–45.

Yi M. and Z. Shen 2015 A review on mechanical exfoliation for the scalable production of graphene, *Journal of Materials Chemistry A*, 3, pp. 11700–11715.

Zhang Y., L. Zhang, and C. Zhou 2013 Review of chemical vapor deposition of graphene and related applications, *Accounts of Chemical Research*, 46(10), pp. 2329–2339.

Zhan J. H., Z. D. Zhang, X. F. Qian, C. Wang, Y. Xie and Y. T. Qian 1998 Solvothermal synthesis of nanocrystalline MoS_2 from MoO_3 and elemental sulfur, *Journal of Solid State Chemistry*, 141(1), pp. 270–273.

Zhang Q., W. Wang, X. Kong, R. G. Mendes, L. Fang, Y. Xue, Y. Xiao, M. H. Rümmeli, S. Chen, and L. Fu 2016 Edge-to-edge oriented self-assembly of ReS_2 nanoflakes, *Journal of the American Chemical Society*, 138(35), pp. 11101–11104.

Zheng W., J. Lee, Z.-W. Gao, Y. Li, S. Lin, S. P. Lau, and L. Y. S. Lee 2020 Laser-assisted ultrafast exfoliation of black phosphorus in liquid with tunable thickness for Li-ion batteries, *Advanced Energy Materials*, 10(31), 1903490. https://doi.org/10.1002/aenm.201903490

3 2D Materials-Based Mechanical Sensors

After an introduction to 2D materials and their amazing properties in Chapter 1, followed by a description of a wide variety of synthesis techniques for these materials in Chapter 2, we have sufficient knowledge to start the discussion of different types of sensors that have been developed from these materials. We begin with the sensors for mechanical signals. This chapter deals with mechanical sensors which form a class of sensors for detecting physical variables related to mechanical phenomena such as pressure, flow, acceleration, and ultrasound (Yang et al. 2022). Among the several mechanical transduction techniques available, the focus will be on those using piezoresistive, capacitive, squeeze film, and resonant effects. Since 2D materials are essentially a class of nanomaterials, these sensors are referred to as nanoelectromechanical sensors.

3.1 PRESSURE SENSORS

3.1.1 Piezoresistive Pressure Sensors

Deformation of a diaphragm by the application of pressure produces a high level of mechanical stress in the diaphragm, particularly at the edges of the diaphragm. Either semiconductor resistors correctly positioned on the diaphragm or the resistance of the diaphragm film itself transforms the stress variations into changes in resistance by piezoresistive effect. Resistance of a bar of piezoresistive material of length L and cross-sectional area A is given by the standard formula:

$$R = \frac{\rho_{\text{Piezo}} L}{A} \tag{3.1}$$

where the piezoresistivity ρ_{Piezo} of the material is a function of pressure. The length L and cross-section A too are pressure-dependent variables.

3.1.1.1 Pressure Sensor Using Graphene Piezoresistors on an SiN_x Membrane

The graphene-based piezoresistive sensor (Figure 3.1) consists of graphene resistors located at maximum strain areas on an SiN_x membrane suspended on a cavity in a silicon wafer (Zhu et al. 2013). It works on the piezoresistive effect in graphene. The graphene resistors are connected to a Wheatstone bridge. When a differential pressure is applied to the SiN_x membrane, it undergoes deformation into a concave shape, and the output voltage of the bridge changes. The sensor is calibrated by measuring the output voltage of the bridge at different applied pressures.

DOI: 10.1201/9781003330585-3

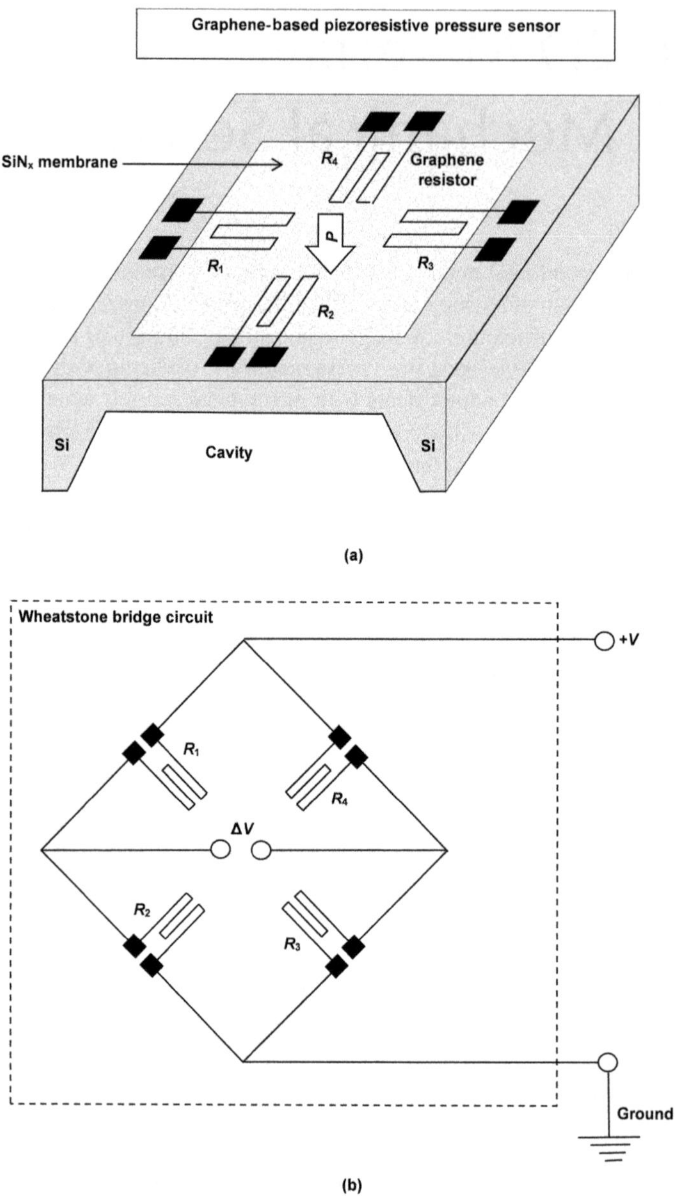

FIGURE 3.1 2D material pressure sensor: (a) graphene-based piezoresistive pressure sensor and (b) the Wheatstone bridge circuit for pressure measurement. Part (a) shows a silicon substrate with a trapezoidal-shaped cavity over which an SiN_x membrane is suspended. Meander-shaped graphene resistors R_1, R_2, R_3, and R_4 are patterned on the surface of the SiN_x membrane with their metallic contact pads lying over silicon. A pressure P is acting vertically downward on the SiN_x membrane in the direction of the arrow. Part (b) shows the four graphene resistors R_1, R_2, R_3, and R_4 connected to form a Wheatstone bridge. One pair of diagonally opposite points of the Wheatstone bridge is connected to a power source, +V and ground. Across the other pair of diagonally opposite points of the Wheatstone bridge, the output voltage ΔV is measured.

The sensor is made by bulk micromachining of silicon using microelectromechanical systems (MEMS) technology. On a <100> silicon wafer with both sides polished, SiN$_x$ film of thickness 100 nm is formed by low-pressure chemical vapor deposition (LPCVD). A square window is opened through SiN$_x$ on the backside of the Si wafer by reactive ion etching (RIE) using CHF$_3$/Ar (trifluoromethane/argon). Then anisotropic etching of silicon is done by KOH until SiN$_x$ film is reached. After etching, an SiN$_x$ membrane of dimensions 280 μm × 280 μm ×100 nm is obtained.

Separately centimeter-scale graphene is grown by CVD over nickel on a Ni (300 nm)/SiO$_2$(300 nm)/Si substrate. The Ni film below graphene is etched in 1 M aqueous FeCl$_3$ solution. The graphene film floats on the solution. The floating graphene film is transferred over the window in SiN$_x$ film by direct contact (Kim et al. 2009).

Contact electrode and alignment marker are patterned. Metallization is done by electron-beam evaporation. Graphene meanders are patterned by electron beam lithography and O$_2$ RIE. The winding or curving course of graphene is necessary to increase the length of graphene wire and, hence, its resistance. Wire bonding completes the sensor fabrication.

The output voltage vs. applied pressure characteristic of the sensor is linear in the range 0–500 mbar, and the sensitivity is 8.5 mVmbar^{-1}. The dynamic range of the pressure sensor is 0–700 mbar. The gauge factor of the graphene piezoresistor is (Zhu et al. 2013)

$$G = \frac{\text{Change in resistance}\left(\Delta R\right) / \text{Original resistance}\left(R\right)}{\text{Change in length}\left(\Delta L\right) / \text{Original length}\left(L\right)} \approx 1.6 \tag{3.2}$$

3.1.1.2 Pressure Sensor Using Large-Area–Layered PtSe$_2$ Film

The piezoresistive PtSe$_2$ pressure sensor consists of a PtSe$_2$ membrane suspended over a cavity (Wagner et al. 2018). Figure 3.2 shows the sensor. It utilizes the piezoresistive effect in PtSe$_2$. Air from the environment is trapped in the cavity below the PtSe$_2$ membrane. So, the pressure in the cavity is environmental pressure. The pressure above the PtSe$_2$ membrane is the variable pressure to be measured. Depending on this pressure, a differential pressure exists across the PtSe$_2$ membrane. This pressure causes a change in the shape of the membrane. When the membrane shape changes, the PtSe$_2$ resistance is altered giving an estimate of the pressure.

Fabrication of the sensor begins with cavity formation in an SiO$_2$/Si substrate using RIE. After cavity etching, metal contacts are deposited.

Separately PtSe$_2$ film is synthesized by thermally assisted conversion (TAC) of pre-deposited platinum through vaporization of solid selenium precursor (Figure 3.3). For this synthesis, Pt is deposited on the SiO$_2$ layer of an SiO$_2$/Si substrate, Pt thickness = 1 nm. We get:

$$\text{Pt} / \text{SiO}_2 / \text{Si.}$$

which is loaded in a quartz tube furnace in zone 2. Selenium is vaporized in zone 1 of the furnace kept at 220°C and transported to zone 2 of the furnace maintained at

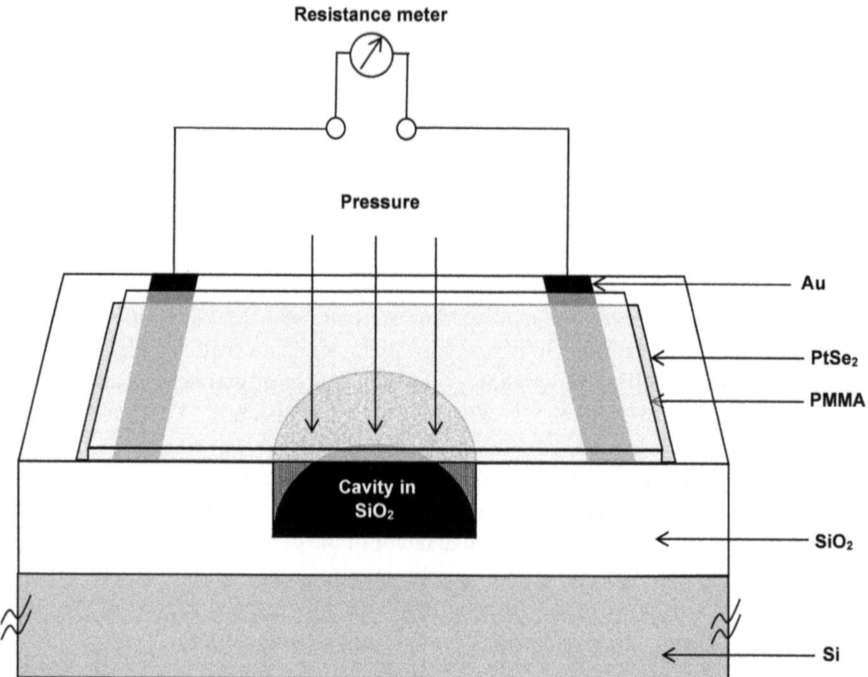

FIGURE 3.2 Piezoresistive PtSe$_2$ pressure sensor showing a PMMA-coated PtSe$_2$ film suspended over a cavity in the silicon dioxide film of an SiO$_2$/Si substrate. The PtSe$_2$ film is in contact with Au electrodes deposited on the SiO$_2$ film before PtSe$_2$ transfer. Pressure is applied on the PMMA-coated PtSe$_2$ film. A resistance meter is connected across the film contacts for electrical measurements.

400°C by flowing Ar/H$_2$. PtSe$_2$ film is formed on the SiO$_2$ layer of SiO$_2$/Si substrate. This synthesis yields:

$$PtSe_2 \,/\, SiO_2 \,/\, Si$$

The thickness of PtSe$_2$ film is 4.5 nm for 1-nm thick Pt. Polymethyl methacrylate (PMMA) is spin coated on PtSe$_2$ to get:

$$PMMA \,/\, PtSe_2 \,/\, SiO_2 \,/\, Si$$

The PtSe$_2$ film covered with PMMA is transferred over the cavity in an SiO$_2$/Si substrate made at the beginning of the process resulting in a PtSe$_2$ film bridging over the cavity with metal contacts on its two sides. The overlying PMMA film provides environmental protection to PtSe$_2$.

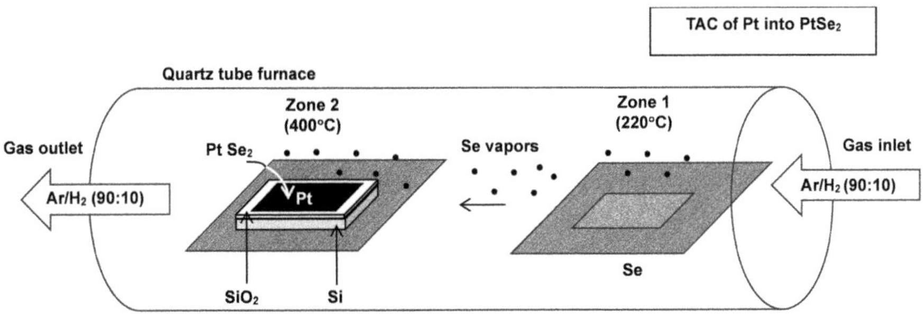

FIGURE 3.3 Thermally assisted conversion of platinum into platinum selenide. The diagram shows a quartz tube furnace in which there are two zones: zone 1 at 220°C and zone 2 at 400°C. In zone 1, a selenium source is placed while in zone 2, a platinum thin film-coated SiO_2/Si wafer is kept. The Ar/H_2 gas flows through the tube over the Pt/SiO_2/Si structure from the gas inlet to the gas outlet. Selenium is vaporized to form Se vapors which are carried by the Ar/H_2 (90/10) gas toward the Pt/SiO_2/Si substrate. The reaction of selenium vapors with platinum transforms the Pt film on the surface of SiO_2 into $PtSe_2$.

Pressure–response parameters of the sensor are: pressure sensitivity = {change in resistance (ΔR)/original resistance (R)}/change in pressure (ΔP) is 5.5×10^{-4} mbar^{-1}. The sensitivity per unit membrane area is 8.85×10^{-7} mbar^{-1} μm^{-2}, which is orders of magnitude higher than that of several nanoelectronic devices, e.g., SWCNT (1.01×10^{-8}), suspended graphene (5.05×10^{-9}), and Si nanowires (8.33×10^{-10}). The gauge factor = relative change in resistance (ΔR)/mechanical strain (ε) = -84.8 (Wagner et al. 2018).

3.1.2 CAPACITIVE PRESSURE SENSORS

In a capacitive pressure sensor, the upper plate is the pressure-sensitive diaphragm covering a cavity. The lower plate is a metal layer coated on the base of the cavity. Generally, the lower plate is fixed while the upper plate is deflected by pressure changes. When the external pressure is the same as the pressure in the cavity, the device is a parallel-plate capacitor because the diaphragm is planar in shape, not curved. Its capacitance is given by the well-known parallel plate capacitance formula for the capacitance C of a capacitor having plates of area A located at a distance d apart with intervening space filled with a medium of relative permittivity ε_r:

$$C = \frac{\varepsilon_0 \varepsilon_r A}{d} \tag{3.3}$$

where ε_0 is the permittivity of free space.

When the external pressure is less than the pressure in the cavity, the diaphragm is curved upward. By upward curving of the diaphragm, the top and bottom plates of the capacitor separate farther apart from each other. The distance between them increases and the capacitance decreases.

When external pressure is greater than cavity pressure, the diaphragm is curved downward. By downward curving of the diaphragm, the top and bottom plates of the capacitor come closer to each other. The distance between them decreases and the capacitance increases.

3.1.2.1   Pressure Sensor Using a Single-Graphene Drum

This sensor (Figure 3.4) consists of a capacitor formed between a graphene drum suspended over a cavity and a bottom metal electrode (Davidovikj et al. 2017). The device is made on a quartz substrate. The substrate is coated with two layers of PMMA resist. E-beam lithography is done to define the pattern for the bottom electrode consisting of a circular disk with an arm. After developing the photoresist, $Ti/Au_{0.6}Pd_{0.4}$ evaporation is done. The photoresist is removed. During photoresist removal, the metal film is detached in areas in which there was a photoresist layer between the metal and the substrate but is retained in other regions. The method is known as lift-off lithography. A circular bottom electrode with an arm is formed.

Spin-on-glass (SOG) is spin coated over the metal pattern formed on quartz substrate, baked at 150°C (3 min), 250°C (3 min), and placed at 500°C. Then two PMMA resist layers are coated as before. The pattern for the top electrode is defined using E-beam lithography. This pattern consists of a circular hole inside a rectangle. Photoresist development is followed by $Ti/Au_{0.6}Pd_{0.4}/Cr$ evaporation on SOG. By lift-off process, the top electrode is delineated. Subsequently, RIE is done in CHF_3: Ar to

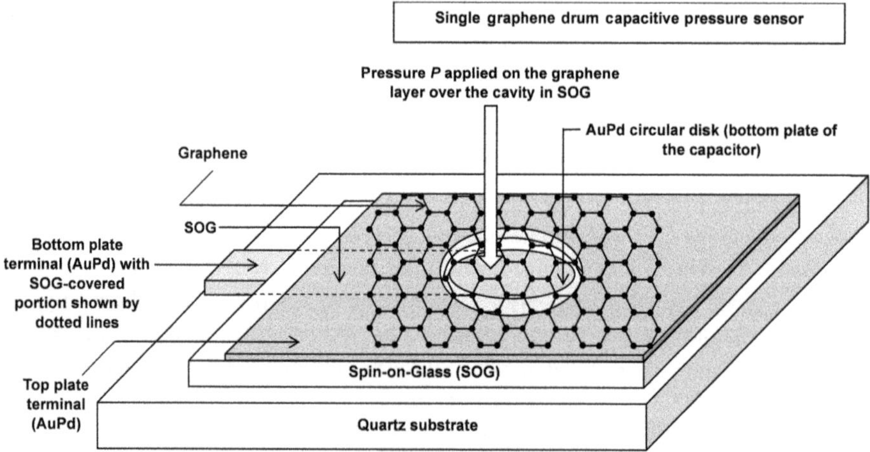

FIGURE 3.4   The single-graphene drum capacitive pressure sensor showing from the bottom upward: the quartz substrate and the AuPd bottom disk electrode with the arm hidden under spin-on-glass (SOG) layer acting as the bottom plate of the capacitor. The hidden arm is shown by dotted lines. The SOG is coated with AuPd film forming the top plate of the capacitor. A circular cavity is opened through the top AuPd film and SOG layer, in alignment with the circular disk of the bottom electrode. Over this cavity, the graphene film is laid down. The pressure P is applied over the graphene layer hanging on the cavity in SOG. Its direction is indicated by the arrow.

remove SOG surrounding the rectangle but the rectangular portion is protected by the Cr layer acting as a hard mask. After RIE, the remnant Cr is removed in Cr etchant.

Dry transfer process for graphene is used to produce a graphene layer over the cavity (Kim et al. 2009). This graphene layer is the diaphragm of the pressure sensor. In the dry-transfer process for graphene, a PDMS stamp is attached to the graphene film. The nickel layer below graphene is removed by etching in $FeCl_3$ solution. Then graphene is transferred using the PDMS stamp to its position over the cavity.

The final structure consists of a circular AuPd electrode at the bottom. There is a circular cavity over the electrode. Cavity walls are made of SOG. The AuPd film covers the upper surface of SOG around the cavity. Graphene is suspended over the cavity and rests on AuPd film which provides mechanical support to graphene and also acts as the top electrode. The device looks like a drum and is therefore referred to as a graphene drum.

Capacitance measurements are performed with an LCR meter at a frequency of 1 MHz. Capacitance changes as low as 50aF are revealed so that pressure steps down to 25 mbar are detectable by this sensor (Davidovikj et al. 2017).

3.1.2.2 Pressure Sensor with Graphene Membrane Arrays

Since the sensitivity for pressure measurements obtained by using a single-graphene drum is insufficient for practical utilization, a commercially acceptable pressure sensor is formed by connecting a large number of graphene drums in parallel (Šiškins et al. 2020). Device design and simulations have shown that the number of drums of diameter 5 μm required to achieve a sensitivity that can compete with the commercially available devices is ~10,000. Further calculations indicate that placement of these drums on a hexagonal grid having a pitch of 10 μm between centers of the drums results in a square chip of size 1 mm × 1 mm.

Figure 3.5 shows a single capacitor of the array. To realize this design, a silicon substrate with a 285-nm thick silicon dioxide layer is taken. Ti/Au electrodes for contacting graphene are defined on the surface of SiO_2 layer. A pattern of circular holes is defined on the silicon dioxide layer by RIE. The depth of etching is restricted to 240 nm leaving a 45-nm thick layer at the bottom of the circular holes. The purpose of this bottom SiO_2 layer is to prevent electrical shorting between the top and bottom electrodes of the capacitor when graphene is placed over the cavity.

Apart from the thin bottom SiO_2 layer, another precautionary technique adopted in this sensor for improving reliability involves using two layers of graphene instead of a single-layer graphene. In double-layer graphene (DLG), any defect in the layers will impair performance only when the defects are aligned with each other. As the defects occur at random locations, possibilities of such defect alignment are extremely remote. So DLG is more reliable than single-layer graphene. To make DLG, two layers of graphene are stacked together after synthesizing graphene.

DLG is transferred over the pre-patterned substrate containing circular holes in SiO_2 layer and Ti/Au electrodes. PMMA is used as a supporting layer during this transfer.

Readout electronics for this device is made on a battery-powered circuit board. The responsivity of the sensor is found to be 47.8 aFPa^{-1}mm^{-1} (Šiškins et al. 2020).

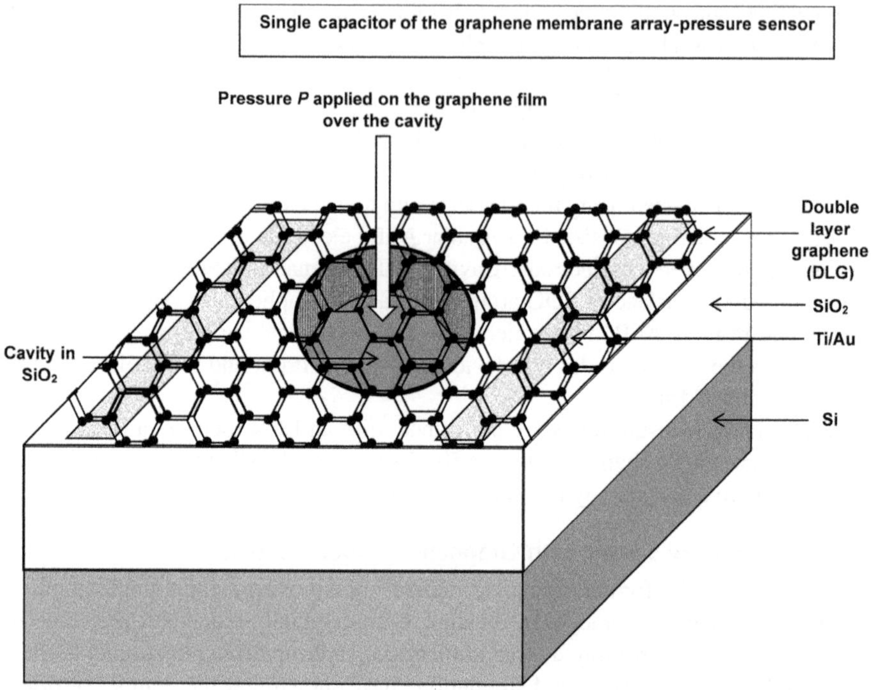

FIGURE 3.5 Cross-sectional diagram of a single capacitor of the array of capacitors showing from bottom upward: the silicon substrate, the silicon dioxide film over Si, the cavity formed by etching silicon dioxide leaving a thin layer of silicon dioxide at the bottom of the cavity for insulation, the Ti/Au electrodes on both sides of the cavity, and the double-layer graphene covering the cavity and touching the Ti/Au electrodes for making contact. The direction of applied pressure P on the graphene film hanging over the cavity in SiO_2 is indicated by an arrow.

3.1.3 Iontronic Pressure Sensors

3.1.3.1 Principle

Iontronics are concerned with electronic–ionic coupled properties and functions (Li and Xiao 2022). The concept of iontronics combines the advantages of electrons and ions. An iontronic pressure sensor is one which utilizes the advantages offered by iontronics. It is an interfacial capacitive sensor working on the ultra-high capacitance of an electrical double layer formed at an electrode/electrolyte interface.

3.1.3.2 Iontronic Pressure Sensor Using Monolayer MoS₂/Au Composite Electrode

Metal electrodes have a small specific surface area. They also have low electrical double-layer capacitance. MoS_2 has been widely used in supercapacitors and batteries. Its outstanding intercalating pseudo-capacitive and electrical double-layer properties favor its application as an electrode material for iontronic pressure sensor together with a gold layer to make an MOS_2/Au electrode for a flexible pressure

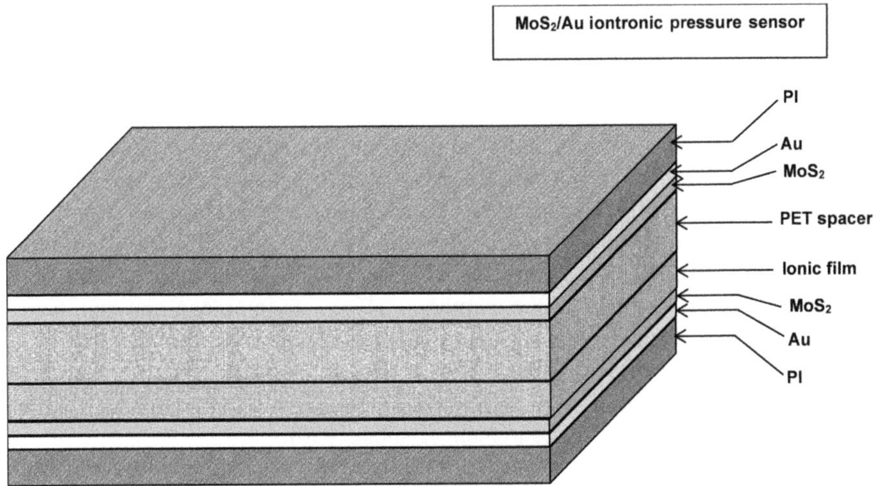

FIGURE 3.6 Sequence of layers in the iontronic pressure sensor. On the upper side, there is a stack of PI/Au/MoS$_2$. On the lower side also, there is a stack MoS$_2$/Au/PI. Between these two stacks lie two layers: the PET spacer and the ionic film.

sensor (Xu et al. 2022). Various layers in the structure of this pressure sensor are shown in Figure 3.6. Fabrication of the sensor involves:

(i) preparation of MoS$_2$/Au electrode on a PI (polyimide) substrate giving: MoS$_2$/Au electrode/PI substrate, and

(ii) preparation of PVA/H$_3$PO$_4$ (polyvinyl alcohol/phosphoric acid) ionic film.

Then the different structures are stacked together with a polyethylene terephthalate (PET) spacer layer in the sequence from bottom upward: (i) PI substrate/Au/MoS$_2$ electrode, (ii) PET spacer layer, (iii) PVA/H$_3$PO$_4$ ionic film, and (iv) MoS$_2$/Au electrode/PI substrate. They are securely bound with 3M tape.

The capacitance variation of the sensor is measured with respect to pressure. Performance parameters of the pressure sensor are: the maximum sensitivity S_{max} is 89.75 kPa^{-1} up to a pressure of 55.6 kPa, the minimum sensitivity S_{Min} is 10.24 kPa^{-1} in the pressure range of 55.6–722.2 kPa, the sensing range is 722.2 kPa, the response time is 3 ms, and the relaxation time is also 3 ms. The relative capacitance change $\Delta C/C_0$ over the sensing range is 10^4 times. The sensor shows stable behavior upon cycling over 5000 cycles at 139.8 kPa (Xu et al. 2022).

3.1.4 Squeezed -Film Pressure Sensors

Pressure sensors described so far in this chapter relied on a sealed cavity for their operation. The limitations imposed by the requirement of a sealed cavity will be discussed below and a viable solution for pressure measurement without a reference sealed cavity will be proposed. The device working without a sealed cavity is called a squeezed-film pressure sensor.

3.1.4.1   Squeeze Film Effect as a Means of Overcoming the Limitation of a Sealed Cavity Pressure Difference Pressure Sensor

The principle of operation of a squeezed film pressure sensor can be understood with reference to Figure 3.7. Imagine a vibrating graphene membrane. There is a channel below the membrane through which gas is introduced at a very high frequency. In this arrangement, gas is trapped between the bottom plate of the channel and the membrane because of viscous forces. The trapping forms a squeezed film of gas. This squeezed film stiffens the membrane and the degree of stiffness of the membrane depends on the pressure of the gas. The higher the gas pressure, the stiffer the membrane is. The stiffness of the membrane affects its force constant and, hence, resonance frequency. So, the resonance frequency of the membrane depends on its stiffness. In turn, stiffness is controlled by the pressure of gas. Thus, the resonance frequency of membrane becomes a parameter varying with the pressure of gas and can be applied for pressure measurement (Dantan 2020).

What is the advantage of squeezed film pressure sensor relative to the pressure difference sensors in which there is a sealed cavity covered by a membrane? In the pressure difference sensors, the membrane is distended by an externally applied pressure and the deformation of the membrane is correlated with pressure. The obvious shortcoming of this type of pressure sensor is the problem of maintaining a constant pressure in the sealed cavity, which is particularly demanding when leakage through a graphene film is considered over the long operational life of the sensor extending to several years. To avoid this difficulty, the squeeze film effect is applied to build a reliable pressure sensor.

3.1.4.2   Equation for Pressure Dependence of Resonance Frequency of a Graphene Membrane

The dependence of resonance frequency of the graphene membrane on pressure is expressed by the equation (Dolleman et al. 2016):

$$f^2_{\text{Resonance}} - f_0^2 = \frac{P_{\text{Ambient}}}{4\pi^2 g_0 \rho h} \tag{3.4}$$

where

$f_{\text{Resonance}}$ is the resonance frequency of the membrane at ambient pressure P_{Ambient}.

f_0 is the resonance frequency of the membrane in a vacuum.

P_{Ambient} is the pressure of ambient gas.

g_0 is the distance between the membrane and the stationary substrate.

ρh is the mass per unit area of the membrane.

As the 'ρh' parameter has an infinitesimally small value for graphene, the right-hand side of eq. (3.4) is extremely large which implies that a miniscule pressure change will lead to a broad shift in resonance frequency.

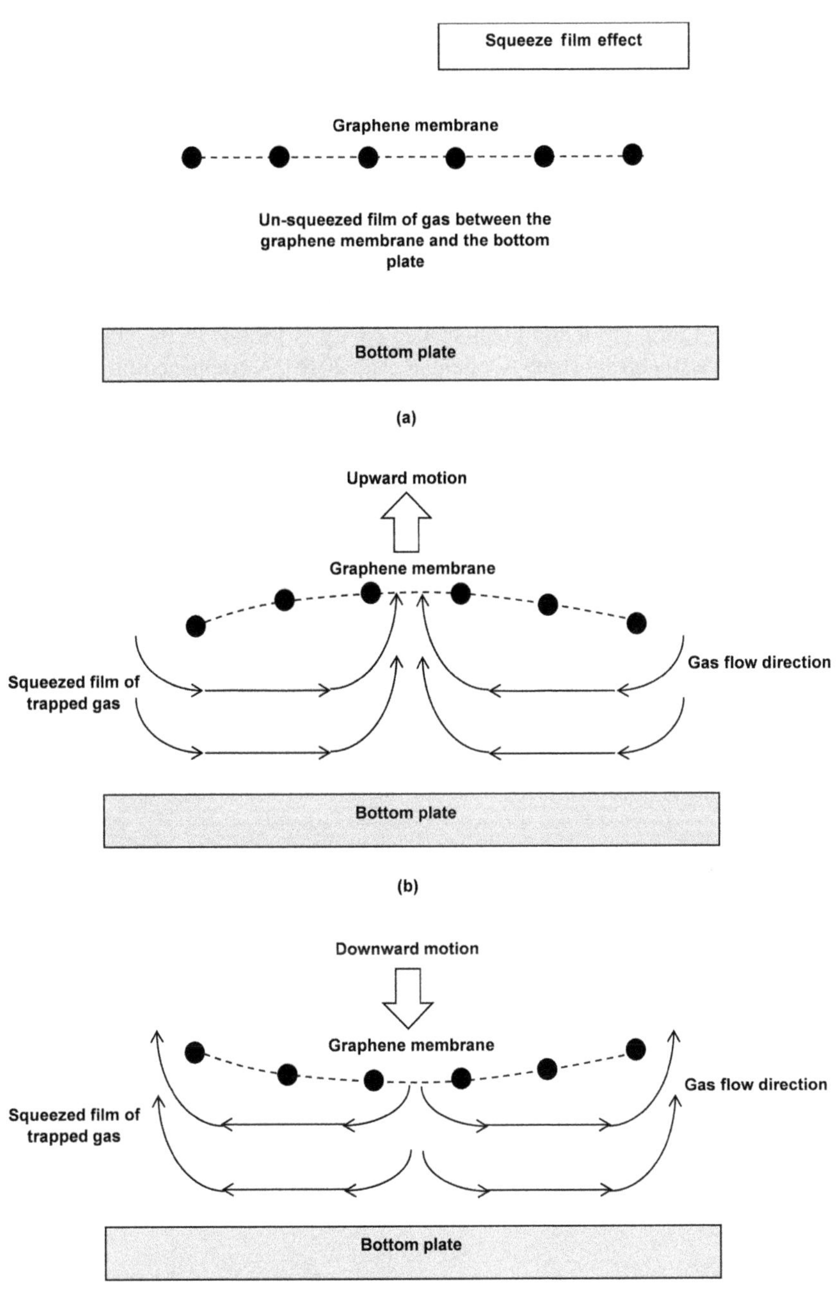

FIGURE 3.7 Squeeze film effect: (a) a graphene membrane and the bottom plate, (b) upward motion of the graphene membrane, and (c) downward motion of graphene membrane. In both cases (b) and (c), the gas flow directions are marked. Gas flows into the gap between the membrane and bottom plate in (b) and flows out of the gap in (c). The squeezed film of gas trapped between the graphene membrane and the bottom plate is indicated.

The equation is valid under the conditions:

(i)   the frequency of membrane vibration is very high, and
(ii) the compression of gas into the cavity of the sensor takes place isothermally.

The pressure dependence of resonance frequency is described well by this equation at low pressures less than 100 mbar. Above a pressure of 100 mbar, an appreciable deviation from this equation is noticed.

3.1.4.3   Structure of Graphene-Squeezed Film Pressure Sensor

Referring to Figure 3.8, a dumbbell-shaped cavity is formed in the silicon dioxide film covering a silicon substrate (Dolleman et al. 2016). A graphene film is laid down

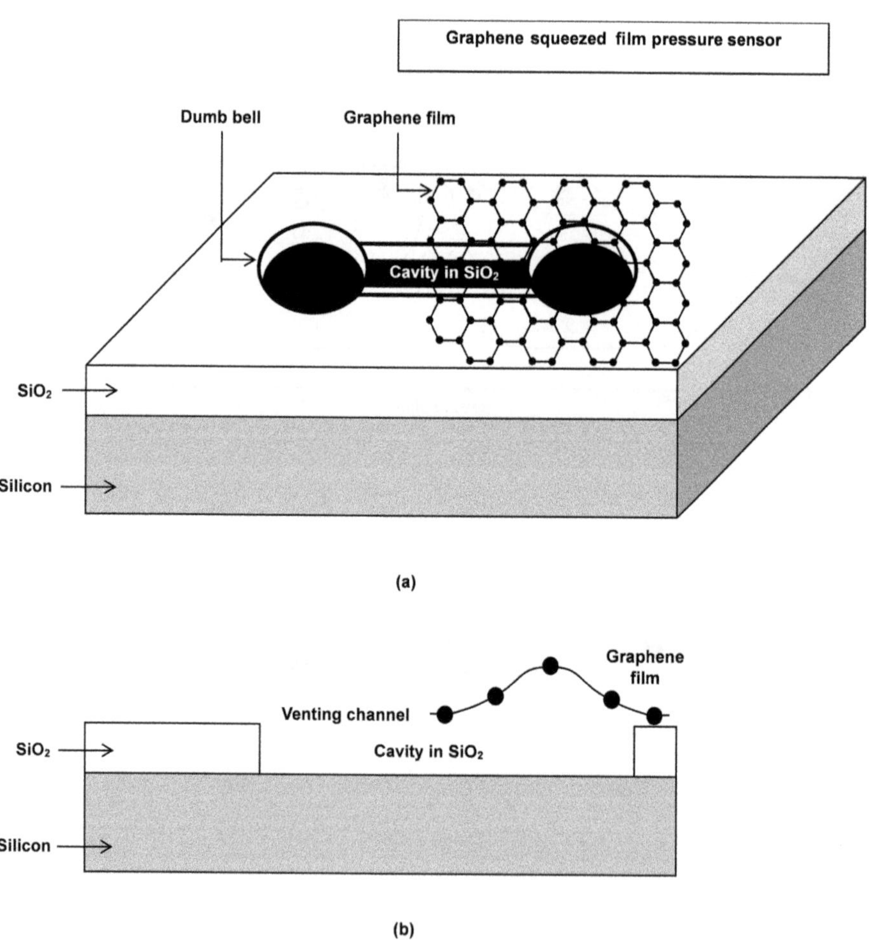

FIGURE 3.8   Graphene squeezed-film pressure sensor: (a) 3D view showing a dumbbell-shaped cavity formed in the silicon dioxide film grown over a silicon substrate, and a part of the dumbbell covered with a graphene film; (b) cross-sectional diagram of the 3D view of sensor shown in Figure 3.8(a).

over the cavity in such a manner that it covers one circular cavity and a part of linear cavity connecting the two circular cavities. In this way, a venting channel is formed. This channel is directed toward the environment. It prevents the building up of any pressure difference across the surface of the membrane and maintains the pressure in the cavity, i.e., ambient pressure. By this mechanism, it is ensured that during the operation of the sensor, the gas is being compressed in the cavity at ambient pressure, so that the effect of any contribution to tension of the membrane from pressure difference across the membrane is completely eliminated, and the squeeze film effect becomes solely responsible for frequency shift of the membrane.

3.1.4.4 Excitation of Vibrations in the Graphene Membrane

The operation of the sensor requires a vibrating graphene membrane. So, the membrane must be set into vibration for pressure measurements using this device. An advantage of suspended 2D materials is their high thermal conductivity and low heat capacitance (Steeneken et al. 2021). So, the optical power can be used for quickly heating them in an efficient manner. The method is called optothermal actuation (Figure 3.9). When the graphene membrane is heated, it contracts moving downward. Note that graphene has a negative coefficient of thermal expansion, which for the monolayer graphene is $-3.2 \times 10^{-6} K^{-1}$ (McQuade et al. 2021). Upon cooling, it expands moving upward. A blue power-modulated laser supplies the necessary time-varying heating power.

3.1.4.5 Optical Setup for Resonance Frequency Detection

The setup (Figure 3.10) works on interferometry principle. A 405-nm blue laser actuates the graphene membrane. Simultaneously a 632-nm red laser is directed on the graphene drum and the bottom of the cavity. The vibration of the graphene drum modulates the light intensity. A photodiode is used for detecting the interference.

The setup also contains a vector network analyzer. For the determination of the frequency spectrum of the graphene membrane, the vector network analyzer modulates the blue-light intensity and detects the red-light intensity on the photodiode.

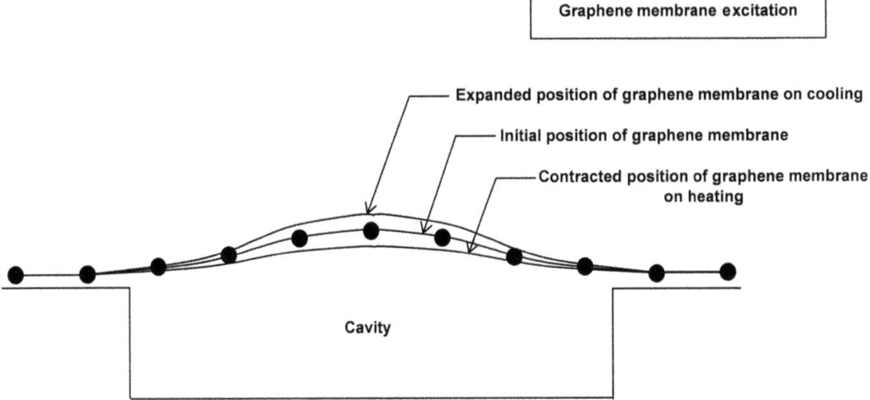

FIGURE 3.9 Method of optothermal excitation of graphene membrane vibration showing a graphene membrane suspended over a cavity; the initial, expanded, and contracted positions of the membrane are indicated.

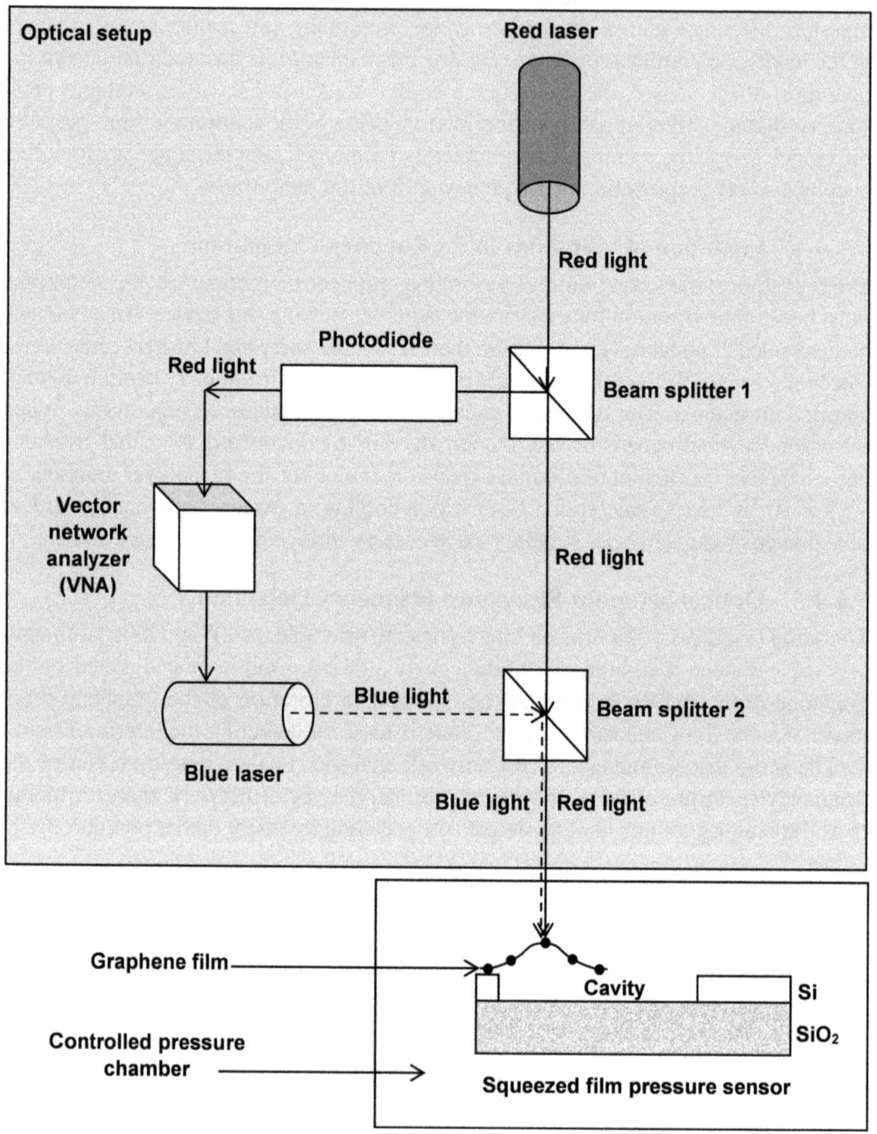

FIGURE 3.10 Interferometry setup for pressure measurement consisting of red and blue lasers, beam splitters 1 and 2, a vector network analyzer, a photodiode, and a pressure sensor with its graphene film hanging over a cavity in SiO_2/Si substrate placed inside a controlled pressure chamber.

3.1.4.6 Pressure Measurement

The pressure is ramped up/down at a fixed rate, e.g., 3.3 mbar-s^{-1} using a pressure controller. During the pressure ramping, the vector network analyzer records the frequency spectra from 5 to 30 MHz. The variation of resonant frequency with pressure is extracted from the measurement data by fitting the data with a damped harmonic oscillator model. The resonant frequency is found to shift by 4 MHz for a pressure change from 8 mbar to 1000 mbar (Dolleman et al. 2016, 2021).

3.2 ACCELEROMETERS

3.2.1 WORKING PRINCIPLE

The accelerometer is an electromechanical device. It is used to measure forces arising from acceleration of a body, either static forces such as the force of gravity or dynamic forces exerted on a moving body. Its operation follows Newton's second law of motion. The acceleration a produced in a body of mass m by a force F is given by

$$a = F / m \qquad (3.5)$$

Piezoresistive accelerometer is one subgroup of the three types of accelerometers, namely, piezoelectric, piezoresistive, and capacitive. In a piezoresistive accelerometer, the resistance of a piezoresistive material changes when a force originating from acceleration produces a mechanical strain in the material.

3.2.2 SUSPENDED GRAPHENE MEMBRANE PIEZORESISTIVE ACCELEROMETER

The graphene accelerometer (Figure 3.11) consists of a clamped graphene membrane from which a silicon proof mass is suspended over a trench etched in the oxidized silicon device layer of a silicon-on-insulator (SOI) wafer (Fan et al. 2019). What happens when the device is subjected to an acceleration? The proof mass is deflected from its position by the force acting on it due to the acceleration. Consequent upon the deflection of the proof mass, a strain is produced in the suspended section of graphene membrane. The resistance of the suspended section is altered because of the piezoresistive effect in graphene. The change in resistance of the suspended part of graphene membrane helps in the derivation of the acceleration. An electrical equivalent circuit model of the resistance variation produced in the membrane is applied to obtain the acceleration.

Accelerometer fabrication consists of the following stages.

3.2.2.1 Part A: Preprocessing of SOI Substrate

(i)   Starting SOI wafer: Thickness of silicon device layer = 15 μm, buried oxide (BOX) = 2 μm, and handle layer = 400 μm.

(ii) Thermal oxidation (front and back): 1.4 μm.

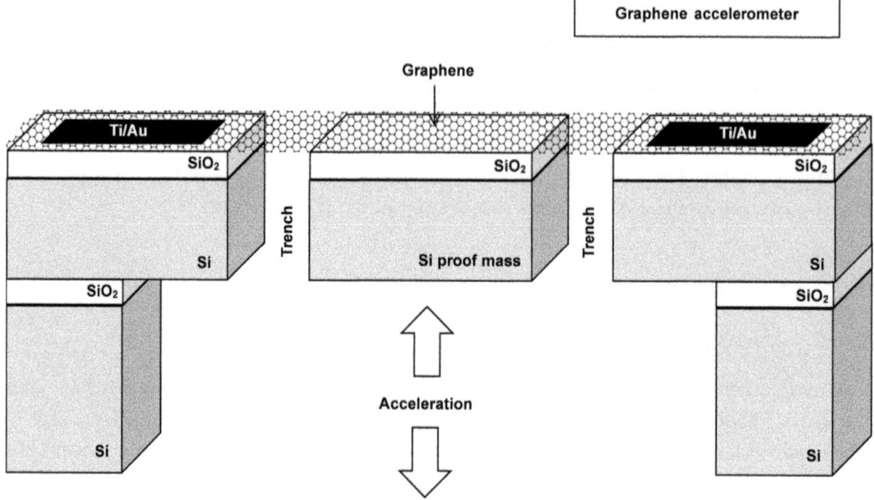

FIGURE 3.11 Graphene accelerometer consisting of a silicon proof mass formed by etching trenches through the silicon device layer of an SOI wafer. The proof mass hangs from a double-layer graphene film over a cavity in silicon. The graphene film is placed over an SiO_2 layer with pre-deposited and patterned Ti/Au pads to form electrical contacts on its two sides. Any acceleration along the vertical direction bends the graphene film at the portions over the trenches. Consequent change in resistance of the suspended part of graphene film gives the acceleration.

(iii) Metal patterning by lift-off process: The steps are: spin-coating photoresist, metal pattern masking, exposure, developing and etching 300 nm deep cavities by RIE, Ti/Au evaporation, and photoresist removal. The metal electrodes formed protrude 20 nm outside the oxide surface.

(iv) Creation of trenches on all sides of the proof mass: After photoresist coating and lithography for defining the trenches surrounding the proof mass, first 1.4-μm thick SiO_2 is removed by RIE, and then 5-μm thick Si below the oxide layer is removed by deep reactive ion etching (DRIE).

(v) Backside process: The steps include: photoresist coating, lithography to define squares in the same areas as the proof mass on the front side, etching SiO_2 by RIE, and etching silicon device layer by DRIE until the BOX layer is reached. O_2 plasma etching is performed for the removal of residues of photoresist.

(vi) Dicing into 8 mm × 8 mm chips: each chip contains 64 devices.

3.2.2.2 Part B: Double-Layer Graphene Preparation

DLG is made from commercial single-layer graphene on a copper foil. PMMA solution is spin coated on the graphene surface of single-layer graphene/copper foil and cured. Carbon residues on the copper foil surface are removed by O_2 plasma treatment. We have now:

PMMA/graphene layer/copper foil,

which is floated on iron chloride solution with the copper foil side immersed in the solution for copper etching. After 2 h, copper is removed and the PMMA/graphene layer stack is cleaned in DI water and dilute HCl, followed by DI water rinsing.

Another graphene layer/copper foil stack is taken, and layered with the previous one to form the stack:

PMMA/graphene layer/graphene layer/copper foil

or

PMMA/DLG/copper foil

The copper foil at the bottom is removed by etching in iron (III) chloride solution as done before, leaving:

PMMA/DLG.

3.2.2.3 Part C: Double-Layer Graphene Transfer Over Diced Chip

PMMA/DLG is transferred over the diced chip obtained in part A. Baking is done at 45°C for 10 min to reinforce the bonding between the graphene and the chip. PMMA is removed in acetone, and acetone residues are removed with isopropanol.

3.2.2.4 Part D: Graphene Patterning and Release of Proof Mass

After spin-coating photoresist and lithography to define the graphene pattern, graphene is etched in O_2 plasma. Photoresist is removed. The SiO_2 layer below the proof mass is partially etched with RIE and then completely etched by HF. The two-step etching decreases the risk of damaging graphene.

3.2.2.5 Part E: Packaging

A ceramic package with an open cavity is taken. The chip is glued in the package. Electrical connections between the electrode pads on the chip and the bond pads of the package are made through gold wires fixed by wire bonding.

3.2.2.6 Utility as a NEMS Accelerometer

Device with a proof mass of size 100 μm × 100 μm × 16.4 μm and trench width of 3 μm is found to be capable of sustaining an indentation force of 4070 nN at the center of proof mass, which is comparable to the force of 0.039 nN resulting from an acceleration of 1 g on a proof mass of above size. This shows that the device can be used as a NEMS accelerometer (Fan et al. 2019).

3.3 FLOW SENSORS

Flow sensors or flow meters constitute a class of devices used to measure the flow rate of a fluid, a liquid or gas, through a pipe or tube.

3.3.1 Self-Powered Single-Graphene Microelectrode Device for Liquid Flow Measurement

Suppose a liquid is flowing over monolayer graphene. The flowing liquid conveys a charge to graphene. This charge handover from liquid to graphene is a consequence of the rearrangement of electrical double layer at the graphene–liquid interface. It gives rise to a charge-transfer current. Thus, the liquid flow is transduced in real time into a charge transfer current. The charge transfer current thus produced is used for the measurement of liquid flow rate. The sensor utilizing such contact electrification of graphene needs no external power supply. It is a self-powered sensor.

In the device configuration shown in Figure 3.12, a graphene film extends over a microfluidic channel engraved in an acrylic sheet (Zhang et al. 2021). The graphene film overlaps a Cr/Au layer from which it is connected to the inverting input of the amplifier of an instrument used to measure the charge or quantity of electricity. The instrument is called a Coulombmeter. When an aqueous solution flows through the channel, a charge is harvested by the graphene film, which is stored in the feedback capacitor of the amplifier for quantification by the Coulombmeter. The charge transferred is measured as a function of time for different velocities of the liquid. It is found that the charge-transfer current is linearly related to the flow velocity. Hence, the current response (defined as the flow-induced current change with respect to the current at zero flow velocity) is proportional to the flow velocity. The resolution of the device is 0.49 μms^{-1}, which is suitable for the application to investigate extremely low flow rate phenomena <10 μms^{-1}(Zhang et al. 2021).

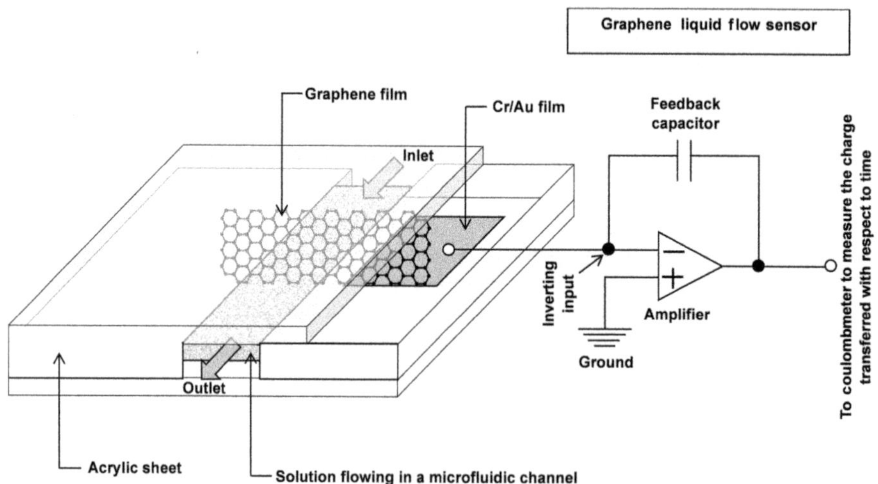

FIGURE 3.12 Single-graphene microelectrode device consisting of a graphene film stretched over a microfluidic channel made in an acrylic sheet. An aqueous solution is flowing through the channel. Inlet and outlet sides of the channel are shown. The graphene film embracing the flowing liquid is in contact with a Cr/Au pad. The metal pad is connected to the inverting input of the amplifier of a Coulombmeter. A feedback capacitor is connected between the output terminal of the amplifier and its inverting input terminal for storing the charge acquired by graphene due to liquid flow.

3.3.2 Airflow Sensor Using Si₃N₄–Graphene Composite Diaphragm with Microholes

This sensor (Figure 3.13) takes the help of piezoresistive effect like the pressure sensor (Wang et al. 2018). The airflow distends a microhole-riddled diaphragm carrying a graphene film, and the resistance of graphene film varies in proportion to airflow rate.

To fabricate this sensor, a silicon nitride film of thickness 200 nm is formed over a silicon substrate by chemical vapor deposition. An array of microholes (diameter 2.5 μm, period 5 μm) is made in the membrane in a triangular lattice using photolithography and RIE. A window is opened on the backside of the silicon substrate by lithography. Through this window, silicon is etched resulting in a perforated silicon nitride membrane suspended over a cavity in silicon. A few-layered graphene of thickness ~1 nm is transferred over the perforated silicon nitride film. It is patterned to form a resistor. Gold contacts for the graphene layer are deposited by electron-beam evaporation. The fabricated structure is flipped over and attached to a silicon wafer with photoresist. Graphene is etched in an oxygen plasma using silicon nitride as a shadow mask. Microholes are produced in graphene at the locations at which these holes exist in silicon nitride Thus a composite membrane of size (490 μm × 490 μm) is realized. This membrane comprises a graphene film over a silicon nitride layer with a periodic array of microholes across graphene and silicon nitride.

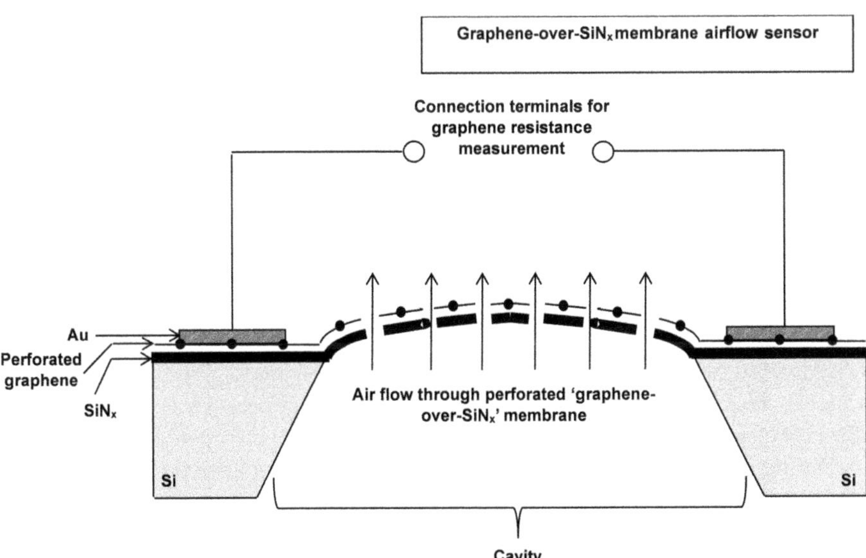

FIGURE 3.13 Airflow sensor consisting of a graphene-over-silicon nitride membrane having microscale through-holes and gold contacts on its sides. The membrane is suspended over a cavity formed by anisotropic etching in the silicon substrate. When air flows through the membrane, it is deformed in accordance with airflow rate. The resulting change in graphene resistance is an indicator of the extent of membrane deformation and hence the airflow rate.

When air flows through the membrane, it bends with the degree of bending determined by the airflow velocity. The graphene film experiences a stress and its resistance changes depending on the airflow rate. Hence, the deformation of the membrane induced by airflow velocity is transformed into electrical resistance variation. The sensor provides airflow velocity measurements in a wide dynamic range of $5\ ms^{-1}$–$56\ ms^{-1}$. The sensitivity is $7 \times 10^{-8}\ (ms^{-1})^{-1}$ (Wang et al. 2018).

3.4 ULTRASOUND DETECTORS

Ultrasound detection provides non-invasive medical diagnostic imaging and non-destructive testing of materials for defect analysis.

3.4.1 GRAPHENE RESONATOR AS AN ULTRASONIC DETECTOR

A graphene resonator is used as an ultrasonic detector (Verbiest et al. 2018). It consists of a graphene membrane stretched across a cavity in a silicon dioxide layer over silicon (Figure 3.14). The membrane has metal contacts on its two sides.

3.4.1.1 Fabrication

Over an SiO_2 (285 nm thickness)/Si substrate, mechanically exfoliated graphene is laid down. Over graphene, Cr (5 nm)/Au (120 nm) contacts are patterned by e-beam lithography. They act as clamps for the graphene, holding it in its place. The SiO_2 underneath graphene is etched in 1% HF solution. The critical point dryer is used for drying graphene to prevent it from collapsing under the action of capillary forces. Thus, a graphene membrane is suspended over a cavity in SiO_2. The SiO_2 layer at the bottom of the cavity is the gate insulator. The silicon layer below it acts as the gate terminal. The contacts on the two sides of graphene are the source and drain terminals.

3.4.1.2 Operation

This device is essentially a parallel-plate capacitor formed between graphene membrane as the top electrode and silicon layer floor of the cavity as the bottom electrode. The air filled in the cavity together with SiO_2 layer below air in the cavity and above silicon substrate is the dielectric medium. One plate of this capacitor, the silicon layer is fixed, while the other plate, the graphene membrane is flexible. It becomes a variable capacitor when the graphene membrane vibrates because the inter-electrode distance changes.

During operation, a DC bias is applied between the graphene membrane and the silicon layer for stiffening the membrane by the attractive force between the positive and negative plates. Incident ultrasound waves cause vibrations in the graphene membrane. The vibrations vary the capacitance of the capacitor in accordance with the undulations in the ultrasound wave. Hence, an alternating current is produced in harmony with the ultrasonic-induced capacitance variations, which is the output signal of the detector.

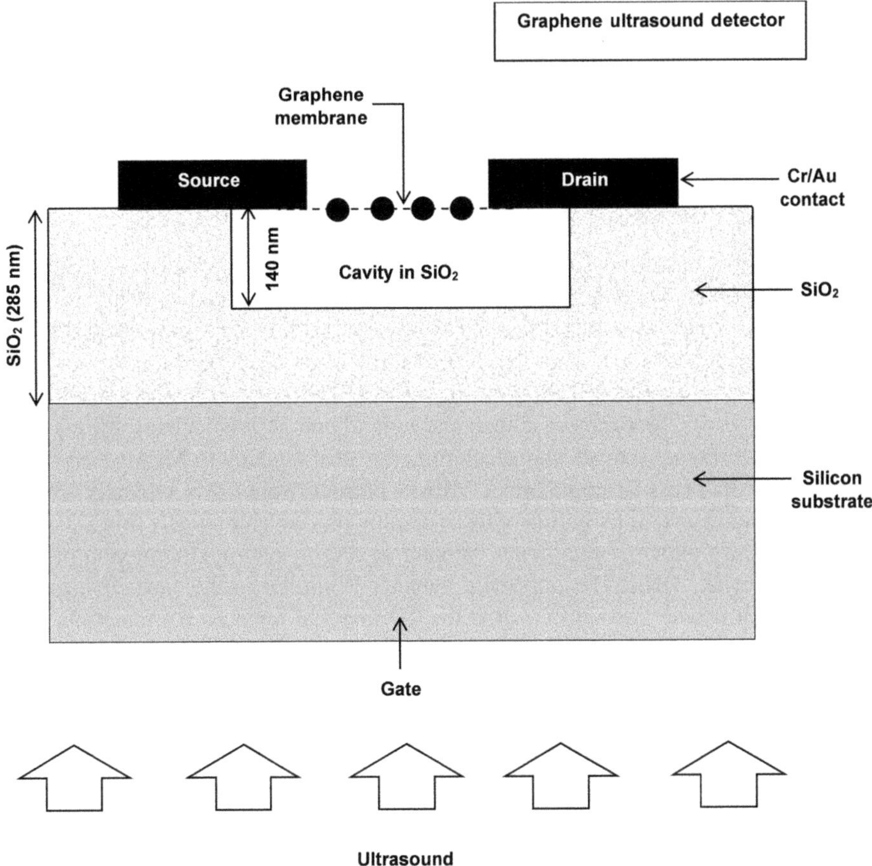

FIGURE 3.14 Graphene resonator for ultrasonic detection: The resonator is fabricated on an Si/SiO$_2$ substrate by suspending the graphene membrane over a cavity in the SiO$_2$ layer. The graphene is clamped at the two ends under Cr/Au contacts serving as the source and drain terminals of the ultrasonic detector while the silicon layer underlying SiO$_2$ is the gate terminal. The distance of SiO$_2$ layer below the graphene membrane is 140 nm. The contacts are under-etched by an approximately equal distance.

3.4.1.3 Detection Arrangement

The detection of ultrasound was shown by mechanically shaking the substrate with the help of a piezoelectric element having a resonant frequency of 4.5 MHz. As the ultrasound waves travel through the substrate to the gold contacts, the graphene resonator is actuated. Actuation of the resonator results in the production of an electric signal, which is detected by a circuit. This circuit measures the resonance frequency of the resonator. At frequencies up to 100 MHz, the resolution achieved in ultrasound amplitude is ~7 pm($\sqrt{\text{Hz}}$)$^{-1}$. Besides amplitude, the resonance frequency itself is also used for detecting ultrasound. When the vibration of amplitude 100 pm is induced with a resolution of 25 pm, a shift of 120 kHz is noted at a resonance frequency of 65 MHz. Thus, ultrasound detection is achieved with a resolution of 20–25 pm by using the vibration amplitude away from the resonance frequency as well as by using the resonance frequency itself (Verbiest et al. 2018).

3.5 DISCUSSION AND CONCLUSIONS

Salient features of various 2D material sensors described in this chapter are summarized in Table 3.1. The role played by 2D material is highlighted in each case. The 2D material is used as a membrane either alone or with a supporting film. Besides using a 2D material for making flexible diaphragms or beams of sensors, it is also used as an electrode material. As expected, the 2D material mechanical sensor technology follows the footsteps of silicon and non-silicon MEMs technology, e.g., the 2D material pressure sensors and accelerometers bear analogy to MEMs structures. Diaphragm thickness in commercial MEMS pressure sensors is typically several hundred nanometers. Substitution of these diaphragms with atomically thin 2D material membranes allows a significant increase in responsivity and reduction of area (Dolleman et al. 2016). The ultrasonic detector is similar to the capacitive micromachined ultrasonic transducer. Often the 2D material replaces the materials used in conventional MEMS structures. But there are also major differences. A notable example is the graphene flow sensor for liquids.

After perusal of mechanical sensors in this chapter, the next chapter will deal with photodetectors made with 2D materials taking advantage of their ultrabroad wavelength detection range.

TABLE 3.1

Summary of 2D Materials-Based Mechanical Sensors

S. No.	Sensor	Notable Features	Reference
1.	Pressure sensor using graphene piezoresistors on an SiN_x membrane	Piezoresistive effect in graphene is applied by forming graphene meander patterns on the maximum strain areas of an SiN_x supporting membrane	Zhu et al. (2013)
2.	Pressure sensor using large-area–layered $PtSe_2$ film	A high-sensitivity, high gauge factor sensor is demonstrated using piezoelectric effect in $PtSe_2$ film grown at a low temperature (400°C)	Wagner et al. (2018)
3.	Pressure sensor using a single-graphene drum	Pressure-induced capacitance changes of a single-graphene drum are detected up to 50aF	Davidovikj et al. (2017)
4.	Pressure sensor with graphene membrane arrays	Capacitance changes of pressure sensors comprising arrays of ~10000 graphene membranes are read out with a low-priced battery-powered circuit	Šiškins et al. (2020)

(Continued)

TABLE 3.1 (Continued)

S. No.	Sensor	Notable Features	Reference
5.	Iontronic pressure sensor using monolayer MoS_2/Au composite electrode	The favorable electrical double-layer properties and large intercalation capacitance of MoS_2 are exploited to make a composite electrode with Au for a high-sensitivity iontronic sensor	Xu et al. (2022)
6.	Graphene-squeezed film pressure sensor	Large change in resonance frequency of a low-mass, strong, and flexible graphene membrane originating from the squeezed-film effect enables its use for pressure sensing	Dolleman et al. (2016, 2021)
7.	Suspended graphene membrane piezoresistive accelerometer	A NEMS accelerometer is demonstrated using a graphene membrane fully clamped at its circumference with attached silicon proof mass	Fan et al. (2019)
8.	Self-powered single-graphene microelectrode device for liquid flow measurement	Flowrate of blood through a microchannel is measured with a single-graphene microelectrode device without an external power source, and superiority of the technique over existing approaches is shown	Zhang et al. (2021)
9.	Airflow sensor using Si_3N_4-graphene composite diaphragm with microholes	A MEMS flow sensor for gases uses a graphene-over-silicon nitride membrane with through-holes. When it is stressed by gas flow, the piezoresistance of graphene changes	Wang et al. (2018)
10.	Graphene resonator as an ultrasonic detector	Ultrasound detection is shown via vibration amplitude away from the resonance frequency of graphene as also from the resonance frequency itself.	Verbiest et al. (2018)

REFERENCES

Dantan A. 2020 Membrane sandwich squeeze film pressure sensors, *Journal of Applied Physics*, 128, 091101-1–091101-9.

Davidovikj D., P. H. Scheepers, H. S. J. van der Zant, and P. G. Steeneken 2017 Static capacitive pressure sensing using a single graphene drum, *ACS Applied Materials & Interfaces*, 9(49), pp. 43205–43210.

Dolleman R. J., D. Davidovikj, S. J. Cartamil-Bueno, H. S.J. van der Zant, and P. G. Steeneken 2016 Graphene squeeze-film pressure sensors, *Nano Letters*, 16(1), pp. 568–571.

Dolleman R. J., D. Chakraborty, D. R. Ladiges, H. S. J. van der Zant, J. E. Sader, and P. G. Steeneken 2021 Squeeze-film effect on atomically thin resonators in the high-pressure limit, *Nano Letters*, 21, pp. 7617–7624.

Fan X., F. Forsberg, A. D. Smith, S. Schröder, S. Wagner, M. Östling, M. C. Lemme, and F. Niklaus 2019 Suspended graphene membranes with attached silicon proof masses as piezoresistive nanoelectromechanical systems accelerometers, *Nano Letters*, 19, pp. 6788–6799.

Kim K. S., Y. Zhao, H. Jang, S. Y. Lee, J. M. Kim, K. S. Kim, J.-H. Ahn, P. Kim, J.-Y. Choi and B. H. Hong 2009 Large-scale pattern growth of graphene films for stretchable transparent electrodes, *Nature Letters*, 457(5), pp.706–710.

Li T. and K. Xiao 2022 Solid-state iontronic devices: Mechanisms and applications, *Advanced Materials Technologies*, 7, 2200205, pp. 1–17, https://doi.org/10.1002/admt.202200205

McQuade G. A., A. S. Plaut, A. Usher, and J. Martin 2021 The thermal expansion coefficient of monolayer, bilayer, and trilayer graphene derived from the strain induced by cooling to cryogenic temperatures, *Applied Physics Letters*, 118, 203101-1–203101-5.

Šiškins M., M. Lee, D. Wehenkel, R. van Rijn, T. W. de Jong, J. R. Renshof, B.C. Hopman, W. S. J. M. Peters, D. Davidovikj, H. S. J. van der Zant and P. G. Steeneken 2020 Sensitive capacitive pressure sensors based on graphene membrane arrays, *Microsystems & Nanoengineering*, 6(102), pp. 1–9.

Steeneken P. G., R. J. Dolleman, D. Davidovikj, F. Alijani and H. S. J. van der Zant 2021 Dynamics of 2D material membranes, *2D Materials*, 8, pp. 1–39.

Verbiest G. J., J. N. Kirchhof, J. Sonntag, M. Goldsche, T. Khodkov, and C. Stampfer 2018 Detecting ultrasound vibrations by graphene resonators, *Nano Letters*, 18(8), pp. 5132–5137.

Wagner S., C. Yim, N. McEvoy, S. Kataria, V. Yokaribas, A. Kuc, S. Pindl, C.-P. Fritzen, T. Heine, G. S. Duesberg, and M. C. Lemme 2018 Highly sensitive electromechanical piezoresistive pressure sensors based on large-area layered $PtSe_2$ films, *Nano Letters*, 18, pp. 3738–3745.

Wang Q., Y. Wang, and L. Dong 2018 MEMS flow sensor using suspended graphene diaphragm with microhole arrays, *Journal of Microelectromechanical Systems*, 27(6), pp. 951–953.

Xu D., L. Duan, S. Yan, Y. Wang, K. Cao, W. Wang, H. Xu, Y. Wang, L. Hu and L. Gao 2022 Monolayer MoS_2-based flexible and highly sensitive pressure sensor with wide sensing range, *Micromachines*, 13, 660, pp. 1–11.

Yang T., X. Jiang, Y. Huang, Q. Tian, L. Zhang, Z. Dai, and H. Zhu 2022 Mechanical sensors based on two-dimensional materials: Sensing mechanisms, structural designs and wearable applications, *iScience*, 25, pp. 1–28.

Zhang X., E. Chia, X. Fan and J. Ping 2021 Flow-sensory contact electrification of graphene, *Nature Communications*, 12, 1755, pp. 1–7, https://doi.org/10.1038/s41467-021-21974-y

Zhu S.-E., M. K. Ghatkesar, C. Zhang, and G. C. A. M. Janssen 2013 Graphene based piezoresistive pressure sensor, *Applied Physics Letters*, 102, 161904-1–161904-3.

4 2D Materials-Based Photodetectors

The photodetectors, also called photosensors, are optical-to-electrical signal converting devices which produce an electrical signal in proportion to the intensity of incident light. They are the key structural components of optoelectronic telecommunication links as well as optical imaging and sensing systems. The photodetectors fall under the class of optical sensors. In this chapter, our attention is focused on optical nanosensors working on the extraordinary response of 2D materials, which display strong electron–photon interactions when illuminated with light as compared to their bulk equivalents (Gupta et al. 2018).

4.1 FROM SILICON TO 2D MATERIAL PHOTODETECTORS

4.1.1 Advances in Silicon Photodetector Technology

Silicon photonics has been the technological workhorse for several decades. Here the information is encoded on optical signals. Waveguides carry the optical signals in place of metal wires. Monolithic optoelectronic integration is achieved between various silicon photonic devices and complementary metal–oxide semiconductor circuits. This has paved the way for remarkable progress in chip-to-chip and board-to-board short-range communication. Silicon photodetectors are widely used in the visible portion of the light spectrum spanning from 400 nm to 700 nm.

4.1.2 Shortcomings of Silicon Photodetectors

Silicon photodetectors fail to detect near-infrared (NIR) radiation. The reason for their failure is that the energy of NIR photons at the wavelengths used in telecommunication ranges between 0.78 and 0.95 eV. These low-energy photons cannot cause photogeneration of carriers in silicon with an energy gap of 1.12 eV. Obviously, insignificant photocurrent is produced.

4.1.3 Alternative Routes to Silicon Photodetectors

Germanium with a bandgap of 0.67 eV comes to the rescue of silicon. SiGe devices have worked as a valuable option. Compound III–V materials have also been used for photodetection in the telecom frequencies in association with silicon CMOS. These approaches introduce unnecessary process complexities compelling the pursuit of easier possibilities.

DOI: 10.1201/9781003330585-4

4.1.4 Recourse to 2D Materials

2D materials exhibit bandgaps in a broad range of energies. Their range encompasses from zero bandgap of graphene to a band gap of 6 eV for h-BN. Narrow bandgap 2D materials such as Nb_2SiTe_4 (0.39 eV) are good candidates for the detection of mid-infrared frequencies (Zhao et al. 2019), whereas wide bandgap 2D materials e.g., TMDs (E_G=1-2.5 eV) and h-BN are suitable for ultraviolet detection.

4.2 MECHANISMS OF PHOTOCURRENT PRODUCTION IN A SEMICONDUCTOR

Many mechanisms are utilized for fabricating photodetectors. Some of these mechanisms are briefly explained below (Zheng et al. 2016).

4.2.1 Photoconductive Effect (PCE)

Photons with energy exceeding the bandgap of an illuminated semiconductor dislodge electrons from their shells creating free electron–hole pairs. Under an applied bias, the electrons and holes drift in opposite directions toward the respective electrodes leading to flow of an electric current. As the increase in free carrier population by shining light on the semiconductor results in an increase in electrical conductance of the semiconductor, this effect is referred to as the photoconductive effect.

4.2.2 Photovoltaic Effect (PVE)

Interfaces such as P–N junction or the Schottky barrier junction set up a built-in electric field. Under the influence of this electric field, the photogenerated electrons and holes are separated from each other causing an electric current to flow in the circuit. Owing to the built-in electric field, such photodetectors can work without an externally applied bias and are said to be photovoltaic.

4.2.3 Photothermoelectric Effect (PTE)

Light has a heating effect. Localized illumination such as by a small laser spot can establish a temperature gradient in a semiconductor. This temperature gradient produces a thermo EMF by the Seebeck effect causing current flow. As the process responsible for current generation is the thermal effect of illumination of semiconductor with light, this effect is termed photothermoelectric.

4.2.4 Photogating Effect (PGE)

The free charge carriers resulting from impinging light can be localized in trapping centers in the semiconductor. These trapped carriers have a gating effect whereby the conductivity of the channel is modulated; hence, this effect is known as the photogating effect.

4.2.5 Photobolometric Effect (PBE)

Here the resistivity of a temperature-sensitive semiconductor changes when light falls on it. Unlike the photothermoelectric effect in which the photocurrent is driven by a thermo EMF, in the photobolometric effect, the photocurrent produced flows on applying an external voltage.

4.3 PERFORMANCE METRICS FOR COMPARING PHOTODETECTORS

Vital performance parameters of photodetectors are defined to enable the comparison of devices of different sizes working under different conditions on a common yardstick (Long et al. 2019).

4.3.1 Photoresponsivity (R_I)

A figure of merit for optical-to-electrical conversion efficiency of a photodetector, photoresponsivity of a photodetector, is defined as

$$R_I = \frac{\text{Photogenerated current}\left(I_P\right)\text{in mA}}{\text{Incident optical power}\left(P\right)\text{in mW}} \tag{4.1}$$

or,

$$R_I = \frac{\text{Photogenerated voltage}\left(V_P\right)\text{in mV}}{\text{Incident optical power}\left(P\right)\text{in mW}} \tag{4.2}$$

It is a wavelength-dependent parameter.

4.3.2 External Quantum Efficiency (EQE)

EQE of a photodetector at a given wavelength λ is given by the equation

$$\text{EQE}\left(\lambda\right) = \frac{\begin{array}{c}\text{Number of free charge carriers collected by}\\ \text{the photodetector electrodes per unit time}\left(N_C\right)\end{array}}{\begin{array}{c}\text{Number of incident photons of}\\ \text{wavelength }\lambda\text{ per unit time}\left(N_I\right)\end{array}} \tag{4.3}$$

4.3.3 Internal Quantum Efficiency (IQE)

It is the ratio of the number of photogenerated carriers per second to the number of photons of wavelength λ absorbed per second in the active layer of photodetector. IQE and EQE are related by the equation:

$$\text{IQE}\left(\lambda\right) = \frac{\text{EQE}\left(\lambda\right)}{1 - T - R} \tag{4.4}$$

where T is the transmittance and R is the reflectance of the photodetector. Hence,

$$\mathrm{IQE}(\lambda) > \mathrm{EQE}(\lambda) \tag{4.5}$$

This is because EQE (λ) takes into account the effects of optical losses taking place by reflection of light from and transmission of light through the photodetector while IQE (λ) does not consider these losses.

4.3.4 RISE TIME (T_R)

It is the time taken by the photocurrent to rise from 10% to 90% of the final value.

4.3.5 FALL TIME (T_F)

It is the time taken by the photocurrent to decay from 90% to 10% of the maximum value.

4.3.6 CUT-OFF FREQUENCY (F_C)

Also called 3 dB bandwidth, it is the frequency at which photoresponsivity R of the photodetector decreases to 0.707 R_0 where R_0 is its low-frequency photoresponsivity.

4.3.7 NOISE EQUIVALENT POWER (NEP)

It is the light signal power giving a signal-to-noise ratio of unity in a frequency bandwidth of 1 Hz. It is expressed in $\mathrm{WHz}^{-1/2}$.

4.3.8 PHOTOCONDUCTIVE GAIN (G)

It is defined as

$$
\begin{aligned}
G &= \frac{\text{Number of photogenerated free charge carriers collected by the photodetector electrodes}}{\text{Number of photons absorbed by the photodetector}} \\[2mm]
&= \frac{N_{Collected}}{N_{Absorbed}}
\end{aligned}
\tag{4.6}
$$

4.3.9 SPECIFIC DETECTIVITY (D^*)

It is given by

$$D^* = \frac{\sqrt{\text{Area of photodetector} \times \text{Bandwidth}}}{\text{Noise equivalent power}} = \frac{\sqrt{AB}}{NEP}\,\mathrm{cmHz}^{1/2}\mathrm{W}^{-1} \tag{4.7}$$

The unit of detectivity is $\mathrm{cmHz}^{1/2}\mathrm{W}^{-1}$ which is called Jones.

4.3.10 SPECTRAL RESPONSE

It describes the sensitivity of the photodetector to light of different frequencies.

4.3.11 NOISE SPECTRUM

It is the noise voltage or current in the photodetector as a function of frequency.

4.3.12 DARK CURRENT

It the current flowing in the photodetector in darkness when no light is falling on it.

4.3.13 LINEAR DYNAMIC RANGE (LDR)

It is defined by the equation

$$\mathrm{LDR} = 20\log\frac{\text{Photocurrent at }1\mathrm{mWcm}^{-2}\text{ optical intensity}}{\text{Dark current}} = 20\log\frac{I_\mathrm{P}^*}{I_\mathrm{Dark}} \quad (4.8)$$

4.4 COMMON PHOTODETECTOR CLASSES

Photodetectors are generally made in any one of the following three structures (Taffelli et al. 2021), as explained in sub-sections below.

4.4.1 PHOTOCONDUCTORS

A film of 2D material photoactive medium is deposited on a glass substrate and ohmic contacts are made at its two ends.

4.4.2 PHOTODIODES

There are three structures:

   (a) Homojunctions: These are P–N junctions in the same material.
   (b) Heterojunctions: These are P–N junctions between different materials.
   (c) Metal–semiconductor junctions: These are Schottky barrier diodes.

Photodiodes are fabricated either in in-plane (or horizontal form) or in out-of-plane (or vertical) form. In the former, the constituent layers are deposited side-by-side while in the latter, they are arranged as one layer on top of another.

4.4.3 PHOTOTRANSISTORS

The 2D material film is deposited on a dielectric film such as SiO_2 grown on an N-doped silicon substrate. Ohmic contacts made on the opposite ends of the 2D material film serve as source and drain electrodes. The N-doped silicon substrate acts as the gate electrode. The gate bias is controlled to deplete the semiconductor film to minimize the dark current of the photodetector. This helps to improve the signal-to-noise ratio of the device. Furthermore, the gate electrode modulates the carrier mobility to achieve a large ratio between on-state and off-state currents. In this way, the responsivity of the detector is enhanced.

Photoconductors and photodiodes are operated by connecting an external power source while phototransistors work by exploiting the built-in electric field across the junction which acquires high values close to the junction, thereby sweeping the positively and negatively carriers to opposite sides.

4.5 GRAPHENE PHOTODETECTORS

4.5.1 ADVANTAGES AND DISADVANTAGES OF GRAPHENE AS A PHOTODETECTION MATERIAL

Advantages of using graphene are the following:

(i) Graphene photodetectors show super-fast response. Intrinsic bandwidth of field-effect transistor (FET)-based graphene photodetectors may be >500 GHz (Xia et al. 2009).
(ii) Graphene photodetectors have a broad spectral response.
(iii) 2.3% of light falling on 0.33-nm thick graphene layer is absorbed by it. This is appreciably high for a material whose thickness is a monoatomic layer. Equivalent absorbance is displayed by 20 nm thick Si or 5 nm thick GaAs (Bernardi et al. 2013).
(iv) Mechanical flexibility of graphene makes possible the realization of flexible electronics.

Drawbacks of graphene are:

(i) Poor photoresponsivity of graphene ~10 mAW^{-1}.
(ii) The lifetime of carriers in graphene is small~ 1×10^{-12}s.
(iii) Graphene being a gapless semiconductor, the dark current of a graphene photodetector is very high.

Some examples of graphene photodetectors will be described in the ensuing subsections.

4.5.2 ULTRAFAST GRAPHENE FET PHOTODETECTOR

Graphene field-effect transistors are made in the familiar back-gated configuration (Xia et al. 2009) by depositing mechanically exfoliated graphene over silicon dioxide on a high-resistivity silicon substrate (1–10 kΩ-cm) to reduce the parasitic capacitance (Figure 4.1). The contacts are made using Ti/Pd/Au. When excited with 1.55 μm laser, the devices show ultrafast response at frequencies of 40 GHz for modulation of light. The photocurrent is efficiently produced without application of source–drain

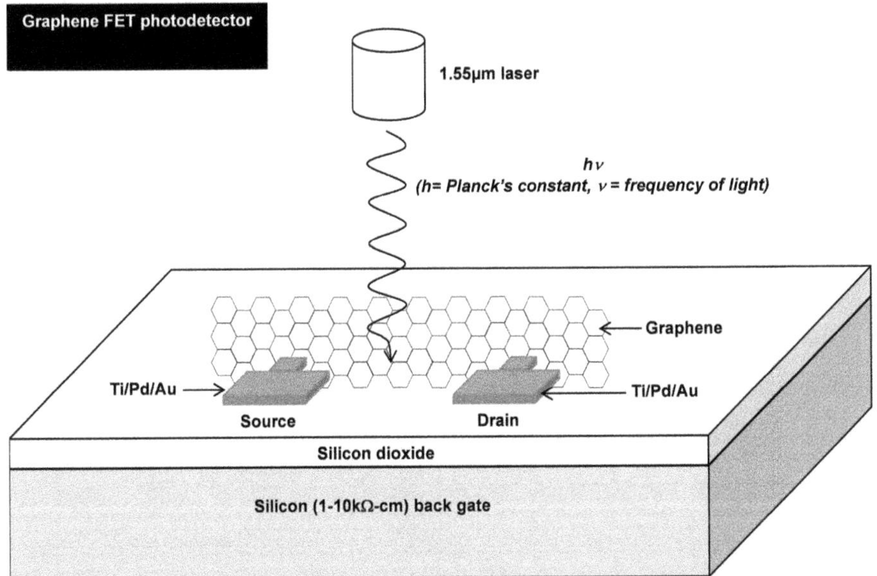

FIGURE 4.1 Graphene FET photodetector. The device has a silicon back gate over which there is a thin layer of silicon dioxide dielectric. Graphene is placed on the surface of silicon dioxide. On the two opposite corners of graphene layer, Ti/Pd/Au contact pads are made for source and drain connections. Light of energy $h\nu$ is falling on graphene from 1.55 µm laser; h is Planck's constant, and ν is the frequency of light.

bias because the photogenerated carriers are transported under the influence of the moderate internal electric fields at the metal electrode–graphene interfaces. In the photodetection region, the internal quantum efficiency is 6–16%. Graphene FET photodetectors promise >500 GHz intrinsic bandwidth (Xia et al. 2009).

4.5.3 Metal–Graphene–Metal Photodetector Deployment in a 10-Gbits⁻¹ Optical Data Link

In the graphene FET photodetector, the built-in electric field providing separation of optically produced charge carriers is confined to a small region surrounding the metal electrode–graphene interface. But carriers are also generated at places far away from these interfaces. Such carriers are lost by recombination and fail to contribute to the photocurrent. To surmount this limitation, a metal–graphene–metal (MGM) photodetector structure having several interdigitated fingers with an asymmetric metallization scheme using two different metals Ti and Pd is used (Mueller et al. 2010). Figure 4.2 shows the metal–graphene–metal photodetector. The asymmetry of metallization disrupts the mirror symmetry of built-in electric field so that the separate photocurrents are added together instead of cancelling out.

1-, 2-, or 3-Layer graphene flakes checked using Raman spectroscopy are transferred over SiO₂/Si substrates. The interdigitated electrodes comprise two groups of fingers. One group is made with Ti/Au and the other group with Pd/Au. Photocurrent

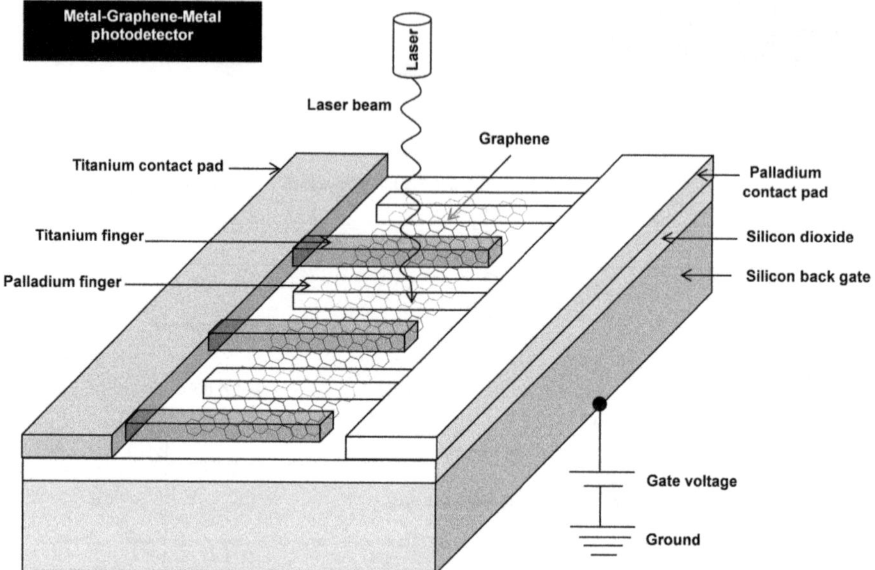

FIGURE 4.2 Metal–graphene–metal photodetector. Graphene is laid over the SiO_2 surface of an SiO_2/Si substrate. The graphene is contacted with two groups of fingers and contact pads, one gray and the other white with alternating gray and white fingers. Gray fingers and contact pads represent titanium while the white fingers and contact pads are made of palladium. A laser beam strikes the graphene film from top. The gate bias is applied to the silicon back gate from the positive terminal of the battery. The negative end of the battery supplying gate bias is grounded.

imaging of the MGM photodetector is done with 632.8 nm He–Ne laser excitation at gate biases ranging from −60 V to +60 V and zero source–drain bias. High-frequency characterization is carried out at a data rate of 10 Gbits^{-1} under 1.55 μm laser excitation. This photodetector is used in a 10-Gbits^{-1} optical data link at 1.55 μm wavelength giving error-free detection of signal (Mueller et al. 2010).

4.5.4 ALL-GRAPHENE PHOTODETECTOR

A heterostructure is made between pristine few-layered graphene (FLG) and FeCl$_3$-intercalated few-layered graphene (FeCl$_3$-FLG), which is named as graphexeter (Withers et al. 2013). Figure 4.3 illustrates the all-graphene photodetector. A photovoltage is generated at the interface of this FLG /FeCl$_3$-FLG heterostructure similar to that at the interface of graphene with a metal such as graphene/Au. This happens because graphene is doped to a high carrier concentration ~ 9×10^{14}cm^{-2} by FeCl$_3$ intercalation reducing its sheet resistance to a few ohms per square and making it behave as a good conductor and interconnect material. So, the costly and opaque gold film can be substituted by inexpensive and transparent FeCl$_3$-FLG. Additionally, FeCl$_3$-FLG offers the advantage of mechanical flexibility. Replacement of metal film with FeCl$_3$-FLG is a step toward all-graphene transparent photodetectors. Note that no photovoltage is produced at the interface of FeCl$_3$-FLG with Au.

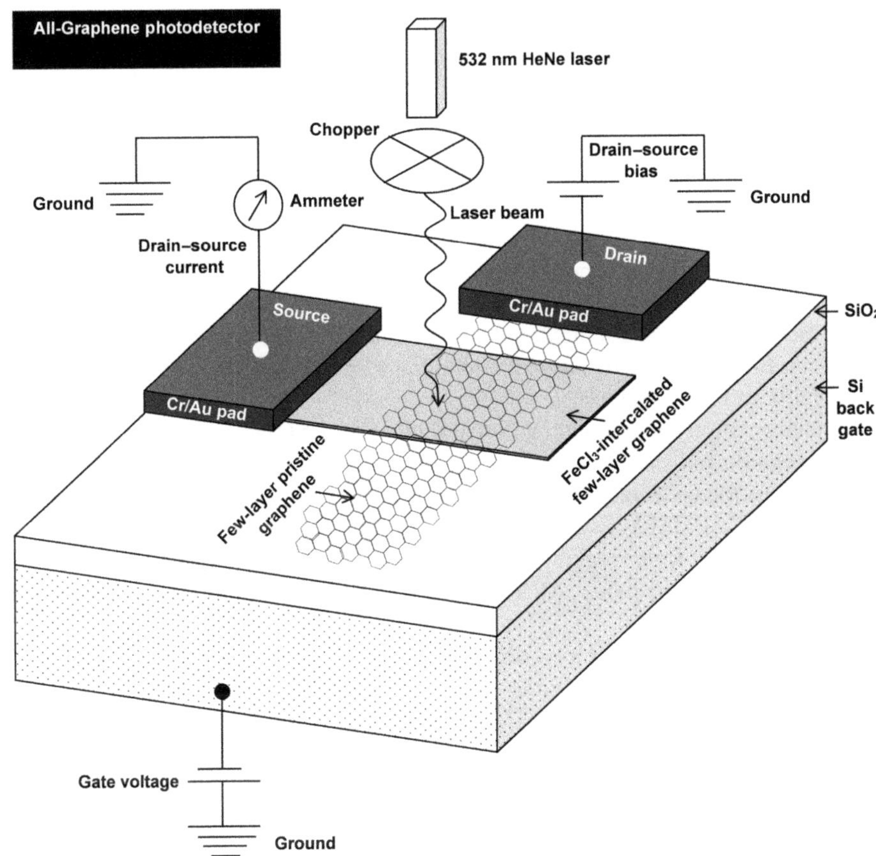

FIGURE 4.3 All-graphene photodetector. A heterostructure is formed between few-layered pristine graphene and FeCl$_3$-intercalated few-layered graphene. Since both layers are made from graphene, hence, it is an all-graphene device. The heterostructure is fabricated on an SiO$_2$/Si substrate. The heavily doped Si substrate acts as a back gate. Source and drain terminals and their Cr/Au contact pads are shown. Gate and drain-source voltage sources are indicated. Their opposite terminals are grounded. The drain–source current is measured by an ammeter connected to the source contact pad. The other end of the ammeter is grounded. The photodetector is illuminated with a 532-nm He–Ne laser through a chopper working at a given chopping frequency and forming a part of the photovoltage measurement circuit.

To make this sensor, an SiO$_2$/Si substrate is overlaid with few-layered pristine graphene. The number of graphene layers and their assembling order are discerned by Raman spectroscopy and optical contrast. Few-layered pristine graphene/SiO$_2$/Si substrate is placed with anhydrous FeCl$_3$ powder at 360°C temperature and 2×10^{-4} Torr pressure when FeCl$_3$ molecules enter the layers of few-layered graphene to dope it heavily, thus rendering it highly conducting. This is FeCl$_3$-FLG. Over the FeCl$_3$-FLG film, pristine few-layered graphene is placed. Cr/Au contact pads are made for connection to the upper pristine few-layered graphene and lower FeCl$_3$-FLG film.

When the chemical potential of pristine graphene is swept through charge neutrality point, the photovoltage shows a change of sign. The photovoltage produced is ascribed to photothermoelectric effect. Appearance of a maximum photovoltage ~0.1 VW^{-1} is shown at the pristine $FLG/FeCl_3$-FLG interface. The maximum Seebeck coefficient is 20 μVK^{-1}. This value is for trilayer graphene with ABA or bernal stacking order (Withers et al. 2013).

4.5.5   50 Gbits⁻¹ Graphene Photodetector

A silicon photonic waveguide is fabricated on a silicon-on-insulator wafer with 220 nm thick Si on 3 μm thick buried oxide (Schall et al. 2014). It consists of a silicon strip of width 400 nm with two grating couplers for 1550 nm wavelength and transverse electric mode. A grating coupler is an optical component with a periodically varying refractive index. It is used for coupling a light signal from free space into the waveguide and conversely.

A planarization layer of hydrogen silsesquioxane (HSQ) is spin coated on the silicon waveguide and cured for hardening. Separately, graphene is grown by chemical vapor deposition on a copper substrate. The graphene film on the copper substrate is transferred by a PMMA-assisted technique over the HSQ layer. The soft HSQ layer prevents graphene from cracking at the sharp step regions of silicon waveguide with the underlying buried oxide. Al/Ni contacts are made by evaporation and lift-off lithography.

The photodetector shows a −3 dB bandwidth of 41 GHz in bias-free operation in the C-band at 1550 nm. It works satisfactorily at 50 Gbits⁻¹ data rates (Schall et al. 2014).

4.5.6   Graphene Photodetector Integrated on a Silicon Microring Resonator

As a graphene photodetector has a low responsivity, the poor absorption of light by graphene, which is ~ 2.3% for normal incidence, needs to be maximized (Schuler et al. 2021). To this end, a graphene photothermoelectric detector is integrated with a silicon microring resonator fabricated on a silicon-on-insulator wafer (Figure 4.4). The graphene photodetector is aligned and connected on both sides with the waveguide of the resonator. A ring resonator consists of a loop-shaped optical waveguide along with a mechanism to access the loop. When the phase shift of the waves in the loop becomes an integral multiple of 2π, constructive interference takes place between the waves. Then, the cavity is said to be in resonance. So, the resonator acts as a resonant cavity.

The high-energy density within the cavity augments the interaction of light with matter by an order of magnitude. By optimizing the coverage of the single-layer graphite channel on the resonator, the absorption of light by it is drastically improved. Furthermore, the resonator has a wavelength selectivity. This selectivity helps in using graphene photodetector for wavelength division multiplexing (WDM). Signals of different wavelengths are combined at the transmitter and separated at the receiver. The benefit derived is that the data rate of an optical channel is considerably increased.

The graphene photodetector is a layered materials heterostructure (LMH). The single-layer graphene channel is encapsulated in hBN to increase the

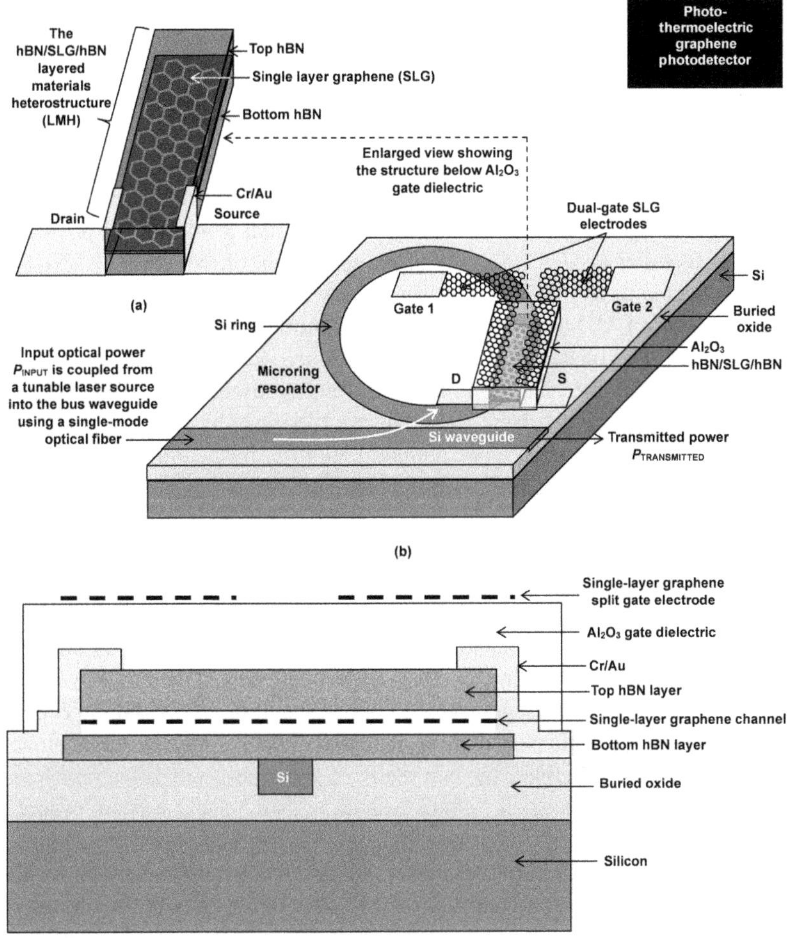

FIGURE 4.4 Photo-thermoelectric graphene photodetector integrated with a silicon microring resonator: (a) magnified view of the FET part underneath the Al_2O_3 gate insulator, (b) the complete photodetector combined with a microring resonator, and (c) schematic cross-section of the graphene photodetector. Part (a) shows three layers of the layered materials heterostructure (LMH): a bottom hBN layer, a single-layer graphene film, and a top hBN layer. Cr/Au contacts are made for source and drain connections. Part (b) shows a microring resonator made in a single-on-insulator wafer. There is a silicon wafer with a buried oxide layer. A silicon waveguide and a silicon ring are seen. Over the silicon ring is placed the structure described in part (a) comprising hBN-SLG-hBN and Cr/Au contacts. The top hBN layer is covered with Al_2O_3 extending to a small region over Cr/Au contact pads resulting in partial overlapping. On the top of the Al_2O_3, two graphene films are seen with a small gap between them. This is the single-layer graphene split gate. Cr/Au contacts are made to the two parts of the gate shown as gates 1 and 2. Part (c) shows the different layers constituting the graphene photodetector: silicon/buried oxide containing the silicon waveguide/lower hBN layer/single-layer graphene/ upper hBN layer/Cr–Au contacts/Al_2O_3/graphene gates 1 and 2. The Cr/Au contacts for graphene gates 1 and 2 are not shown.

photothermoelectric voltage. This is achieved by provision of a high carrier mobility $\sim10^4$ cm^2 V^{-1} s^{-1} resulting in enhanced peak Seebeck coefficient $\sim$ 200 μVK^{-1}. Additional tuning of the Seebeck coefficient in the adjoining regions of the device is done using dual-gate single-layer graphene electrodes. An Al$_2$O$_3$ film is deposited to separate these electrodes from the LMH.

Single-layer graphite and hBN flakes are obtained by micromechanical cleavage of bulk graphite and hBN single crystals, respectively. The flakes are picked up by polycarbonate/polydimethylsiloxane (PC/PDMS) stamp and a micro-manipulator. They are stacked together at 50°C. The target photonic chip on which the LMH is to be mounted is heated to 180°C. The interfaces of LMH are thereby cleaned. At the same time, alignment of LMH with silicon waveguide is done. Then PC is laminated on the target substrate. The blisters formed at the interface between single-layer graphene and hBN are pushed out. The PC film is removed by dissolution in chloroform. The channel geometry of the graphene photodetector is defined by electron beam lithography using a PDMS etch mask. The defined pattern is formed on the LMH via two steps of reactive ion etching. Then electron beam lithography is performed for contact metallization. Cr/Au is deposited by electron beam evaporation. By lift-off in acetone, the exposed edges of the single-layer graphene channel are contacted. A 1-nm thick aluminum seed layer is deposited on the top hBN layer by thermal evaporation. Atomic layer deposition is used to form 20 nm thick Al$_2$O$_3$ layer as a gate dielectric. Separately, a single-layer graphene (SLG) film is formed on a copper substrate by chemical vapor deposition giving SLG/Cu. This is transferred on a PDMS membrane to get: PDMS/SLG/Cu. Etching away the copper, we have: PDMS/SLG. The SLG is transferred on the photonic chip and the backside PDMS is removed in acetone. The split-gate geometry is defined by two steps of electron-beam lithography and O$_2$ plasma etching. Metallization for contacts is done using electron beam evaporation. The source and drain contacts are made by optical lithography and HF etching.

Adjustment of the SLG coverage over the resonator helped in optimization of round-trip propagation losses leading to > 90% absorption of light. In addition, hBN encapsulation ensured a high carrier mobility. Both these factors combined together to yield a responsivity$\sim$ 90 VW^{-1} (Schuler et al. 2021).

4.6 MOS$_2$ PHOTODETECTORS

4.6.1 Advantages and Disadvantages of MoS$_2$ as a Photodetection Material

Advantages of MOS$_2$ are:

(i) In the monolayer limit, the bandgap of MOS$_2$ changes from indirect to direct bandgap resulting in intense light–matter interaction.

(ii) Its optical absorption is an order of magnitude larger than that of silicon or gallium arsenide. A 1 nm thick MOS$_2$ layer absorbs 5–10% of incident light (Bernardi et al. 2013).

Disadvantages of MOS_2 are:

(i) Its modest electron mobility ~ 200 $cm^2V^{-1}s^{-1}$ results in slow response.
(ii) Its quantum efficiency is low, lying in the range 10^{-4}–10^{-2} for TMDs. A large fraction of photogenerated electrons and holes are lost by nonradiative recombination processes.

4.6.2 N-MoS$_2$/P-MoS$_2$ Bilayer Solution-Processed Photodetector

This photodetector is fabricated by a facile solution-based method (Ye et al. 2015). It is shown in Figure 4.5. Liquid ultrasound exfoliation technique is applied to synthesize nanosheets of MoS_2. For this synthesis, MoS_2 powder is dispersed in ethyl cellulose $(C_{20}H_{38}O_{11})$-isopropanol $[(CH_3)_2\ CHOH]$ solution, sonicated, and centrifuged. Deionized water is mixed with the collected supernatant liquid, and the mixture is centrifuged. The collected lower precipitation material is dried and dispersed in ethanol. The dispersion is centrifuged after adding NaCl aqueous solution to it. The precipitated MoS_2 is washed with deionized water. It is collected using vacuum filtration. The MoS_2 nanosheet formed has the appearance of a fine black powder.

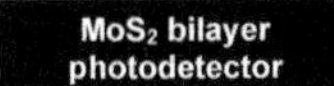

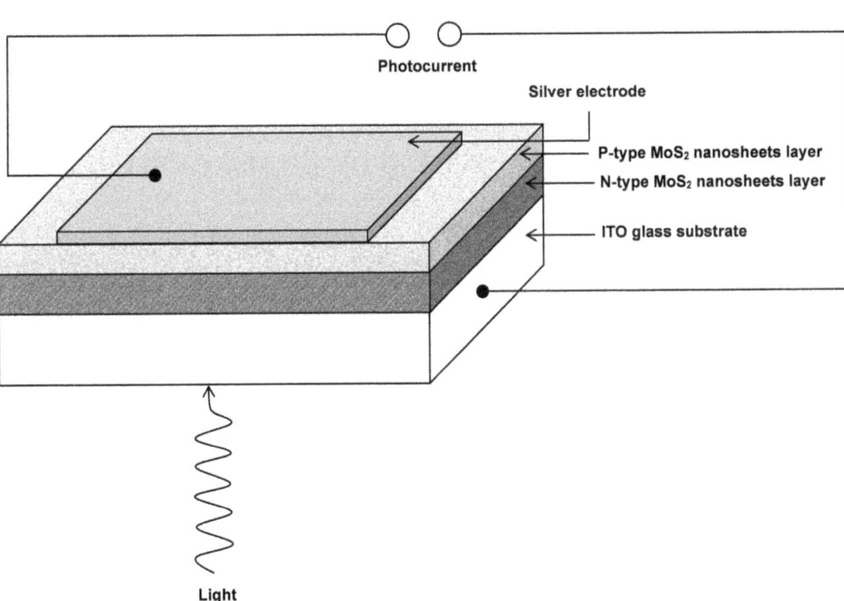

FIGURE 4.5 N-type MoS_2/P-type MoS_2 bilayer photodetector. On an ITO glass substrate, a stack of layers is shown in the sequence: ITO glass, N-type nanosheets layer, P-type nanosheets layer, silver electrode. Light is falling on ITO glass substrate and photocurrent is measured between the ITO glass substrate and silver electrode.

The pristine nanosheet has N-type behavior. The P-type nanosheet is formed by converting the N-type nanosheet into P-type. This is achieved by subjecting N-type nanosheet to UV-ozone plasma exposure.

The N-MoS$_2$/P-MoS$_2$ bilayer photodetector is fabricated on an ITO glass substrate. First N-type MoS$_2$ nanosheets in isopropanol are spin coated on the substrate and annealed in air at 150°C. Next, P-type MoS$_2$ nanosheets in isopropanol are deposited over N-type MoS$_2$ nanosheets layer by spin coating. After annealing at 150°C, a silver film is thermally evaporated on the P-type nanosheets layer. The overall structure of the resulting photodetector is ITO glass substrate/ N-type nanosheets layer/P-type nanosheets layer/silver electrode, forming a P–N junction device with a built-in potential.

For testing, the photodetector is illuminated with 150W Xe lamp. At 0 V and 1 V DC bias, photocurrent values in nA range are measured. The spectral photoresponse extends over a broad range of visible illumination from 350 to 600 nm (Ye et al. 2015).

4.6.3 Back-Gated, Metal–MoS$_2$ Photodetector

In the SiO$_2$/Si substrate, the silicon is heavily doped to serve as a back gate electrode (Wang et al. 2015a). Monolayer MoS$_2$ exfoliated from a bulk crystal is transferred on the SiO$_2$ surface (Figure 4.6). Cr/Au contacts are deposited and patterned by electron beam lithography forming two metal–semiconductor Schottky junctions. An optical pulse from a Ti: sapphire laser is mechanically chopped, wavelength halved (frequency doubled), and split into two pulses by a 50/50 beam splitter. A time delay Δt is introduced between the two pulses using a linear translation stage. The two time-delayed pulses are focused on a metal–semiconductor junction of the device. The DC photovoltage produced across the photodetector is measured as a function of Δt using a lock-in amplifier. The experiment shows that the photodetector has an exceedingly short intrinsic response time ~3 ps suggesting that the photodetection bandwidth is 300 GHz broad. The quick response of the photodetector is attributed to the short electron–hole and exciton lifetimes in MoS$_2$. The devices realized by an easy fabrication process are useful for affordable superfast optical communication links (Wang et al. 2015a).

4.6.4 Ferroelectric-Driven MoS$_2$ FET Photodetector

The commonly used gate dielectric materials in traditional MoS$_2$ FETs are SiO$_2$, Al$_2$O$_3$, HfO$_2$, etc. As a photodetector, these MoS$_2$ FETs show low photoresponsivity, e.g., 7.5 mAW^{-1}, which is below the value desired for practical use. To increase the sensitivity, a large drain–source voltage V_{SD} together with a high gate voltage V_G are applied. The large V_{SD} raises the dark current while the high V_G produces gate–source leakage current. These effects lead to self-heating of the channel. The power dissipation increases and the performance of the photodetector deteriorates. These problems are solved by making an FET using a ferroelectric material P(VDF-TrFE) for the gate (Wang et al. 2015b).

Figure 4.7 shows the MoS$_2$ photodetector driven by P(VDF-TrFE). The remnant polarization of P(VDF-TrFE) suppresses the dark current in the channel. In addition, a high electric field ~ 10^9Vm^{-1} exists in the channel within a range of several

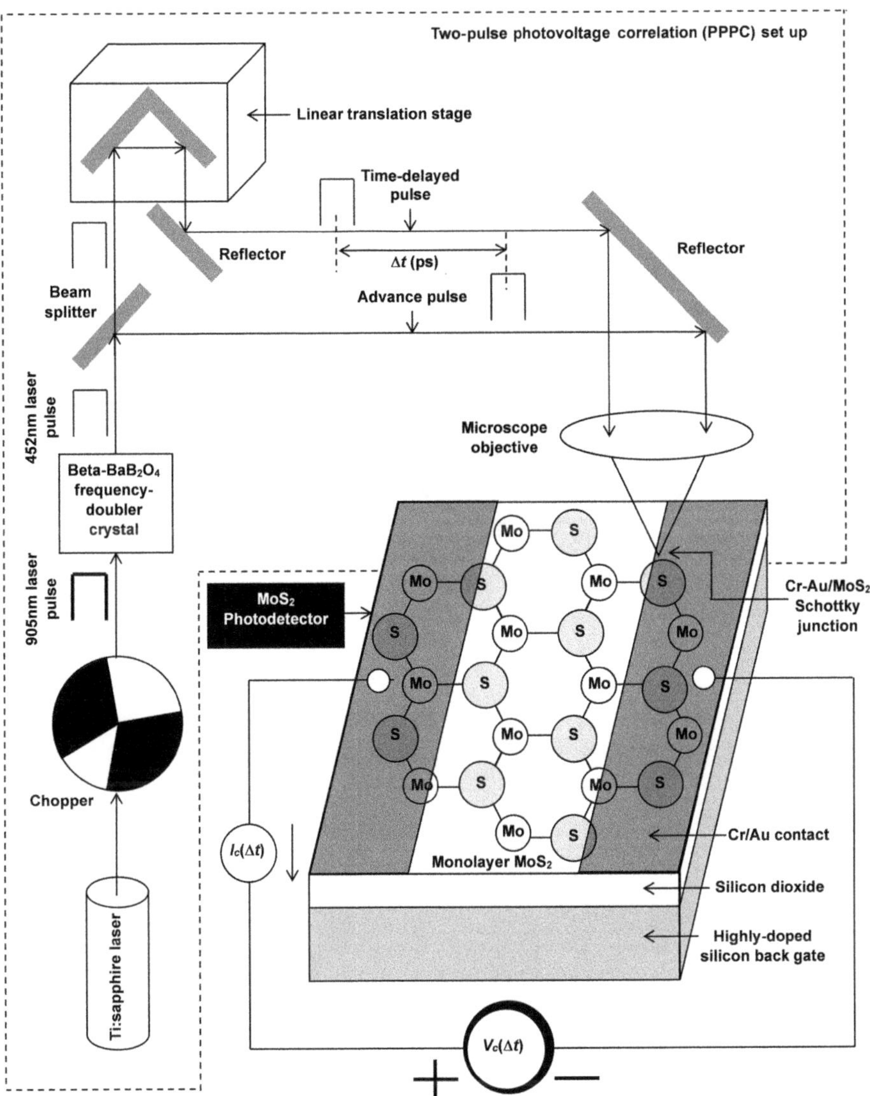

FIGURE 4.6 Back-gated metal–monolayer MOS$_2$ photodetector and the two-pulse photovoltage correlation experiment. Over an SiO$_2$/Si substrate, there is a single-layer MoS$_2$ film. Cr/Au contact electrodes are made on both sides of the MoS$_2$ film. The silicon substrate being highly doped acts as a back gate. Time-delayed optical pulses (advance pulse and time-delayed pulse) are obtained using a 905-nm Ti: sapphire laser source, a chopper, a beta-BaB$_2$O$_4$ frequency doubler crystal giving 452 nm laser pulse, a beam splitter, a linear translation stage, and reflectors, all together constituting a two-pulse photovoltage correlation (TPPC) set up, enclosed within dotted lines. These pulses are focused on a Cr/Au-MoS$_2$ Schottky junction of the photodetector through a microscope objective. Two-pulse correlation photovoltage $V_c(\Delta t)$ across the MoS$_2$ photodetector and the corresponding current $I_c(\Delta t)$ flowing through it are measured.

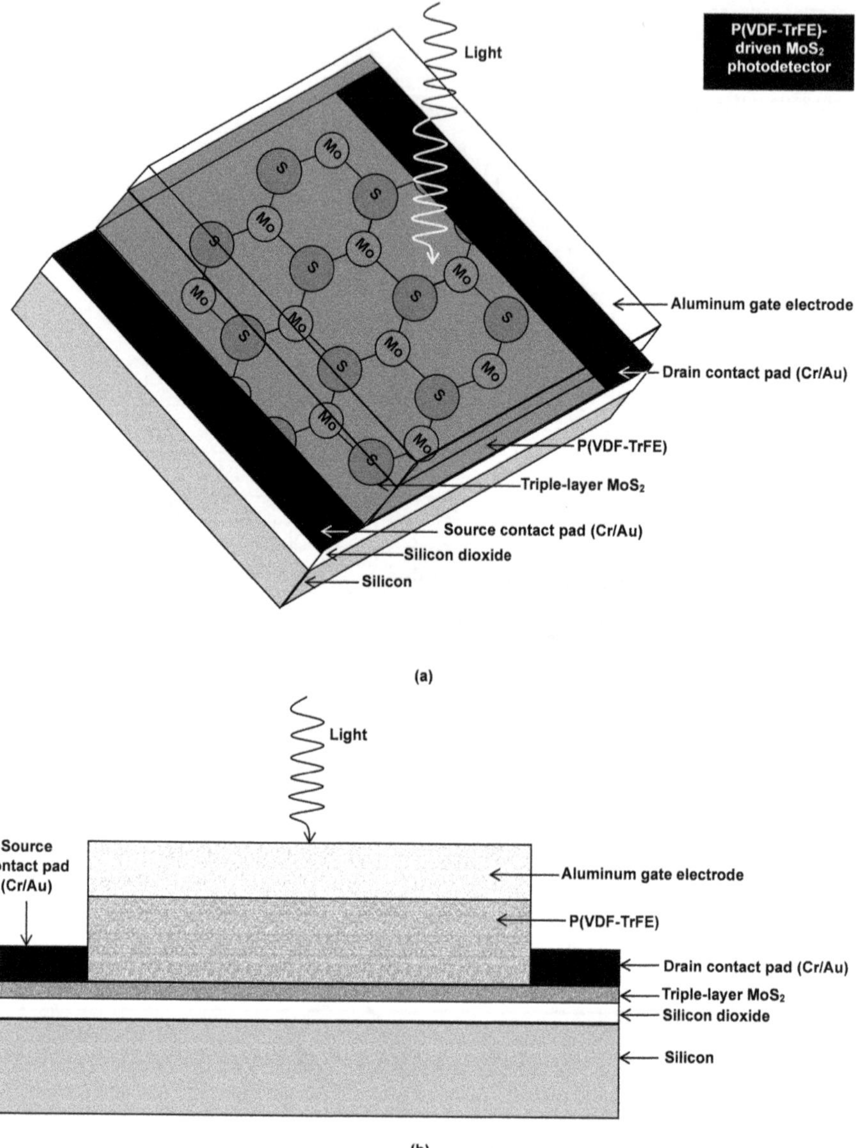

FIGURE 4.7   Triple-layer MoS_2 photodetector driven by P(VDF-TrFE) ferroelectric: (a) 3D view and (b) cross-sectional diagram. Part (a) shows an SiO_2/Si substrate on the surface of which triple-layer MOS_2 film is laid out and Cr/Au source/drain contact pads are formed on its two sides. Over the MoS_2 film, there is a P(VDF-TrFE) layer. An aluminum gate electrode is deposited on the surface of the P(VDF-TrFE) ferroelectric gate dielectric. The light beam is incident on the Al film. Part (b) shows the same sequence of layers, stated from bottom upward as: silicon, silicon dioxide, triple-layer MoS_2 with Cr/Au source/drain contact pads on its two sides, P(VDF-TrFE) ferroelectric layer over MoS_2, and Al film on P(VDF-TrFE). Light is falling on the aluminum film surface.

nanometers because of this remnant polarization. This field being significantly larger than the electric field of a conventional MOS_2 FET completely depletes the FET channel. Consequently, the FET shows a much higher sensitivity without applying any gate bias.

Exfoliation using Scotch tape is applied to transfer few-layered MoS_2 film on an SiO_2/Si substrate in which the silicon is degeneratively doped. Cr/Au electrodes are made by metal deposition and lift-off lithography. Then the P(VDF-TrFE) film is coated, and over it a semitransparent film of very thin aluminum is deposited. The photoresponsivity of this detector is 2570 AW^{-1}. Its detectivity is 2.2×10^{12} Jones, which is $cm \cdot Hz^{1/2}/W$. The wavelength range of its photoresponse is very wide ~0.85–1.55 μm from visible to near-infrared. The photodetector is appropriate for optical communication (Wang et al. 2015b).

4.7 WSE$_2$ PHOTODETECTORS

4.7.1 ADVANTAGES AND DISADVANTAGES OF WSE$_2$ AS A PHOTODETECTION MATERIAL

Advantages of WSe_2 are:

(i) It has a high optical absorption with absorption coefficient of 10^5 cm^{-1} at 532 nm.
(ii) It provides spectral response over a broad range of wavelengths from deep UV to infrared.
(iii) It shows good stability in air ambience unlike black phosphorous.

The main disadvantage of WSe_2 is that mobility of holes in WSe_2 is ~118 $cm^2V^{-1}s^{-1}$. Consequently, the photodetector devices are slow.

4.7.2 WSE$_2$/H-BN-BASED PHOTOTRANSISTOR

A WSe_2 phototransistor is improvised by the integration of a standard WS_2 phototransistor with an in-plane lateral photodiode (Ghosh et al. 2021).

Starting with an Si/P^+ Si wafer substrate, two side gates and one middle gate are patterned by e-beam lithography, followed by Cr/Au deposition and lift-off (Figure 4.8). The two side gates are shorted together by a metal line (not shown).

Scotch tape is used to micro-mechanically exfoliate hBN flake. The hBN flake is transferred to a PDMS stamp. The PDMS stamp is fixed to a glass slide, and the glass slide is fastened to a micromanipulator. The SiO_2/Si substrate with middle gate/side gate (MG/SG) pattern is kept on a heater. Then hBN side of the PDMS stamp is placed over the MG/SG pattern, and the assembly is heated to 60°C to weaken the binding of hBN to PDMS stamp. After cooling to 50°C, the glass slide and PDMS stamp are removed, leaving hBN over the MG/SG pattern. A similar process is adopted for transferring the WSe_2 flake on hBN flake taking care that the WSe_2 flake covers the full MG/SG area without touching the extended MF leads moving to the

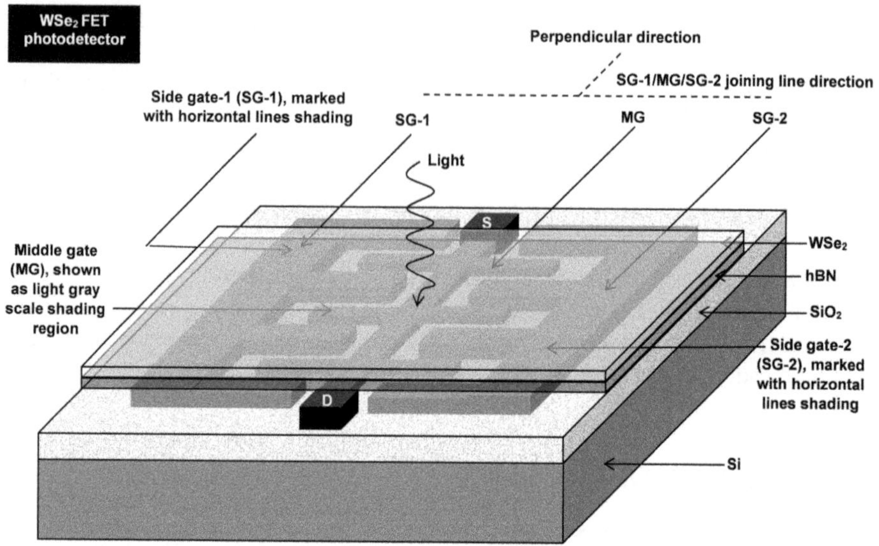

FIGURE 4.8 WSe$_2$ field-effect phototransistor with side and middle gates. The diagram shows an SiO$_2$/Si substrate upon which three interdigitated metal gates, two side gates and one middle gate, are fabricated. Each finger of the middle gate lies between two fingers of the side gate and vice versa. The gate pattern is covered first with the gate dielectric hBN and then the semiconducting channel material WSe$_2$ to form the FET structure. At both ends of the middle gate, the source/drain contacts are seen along the line perpendicular to the direction: side gate-middle gate-side gate. Both directions are indicated by dotted lines. Light is falling on the WSe$_2$ layer.

source/drain contact pads. The source/drain contacts are defined by e-beam lithography on WSe$_2$ overlapping the MG. The next step is Cr/Pt/Au sputtering and photoresist lift-off.

Thus, a 2D material (WSe$_2$/hBN)-based phototransistor is realized in which each gate influences the carrier density in the WSe$_2$ layer above it. Side gates control the carrier density on the sides of the WSe$_2$ layer. They also affect the potential barriers between the middle channel and sides of the WSe$_2$ layer. In this way, the side gates affect the operation of middle FET. By electrostatic doping, lateral P–N junctions are created at the two interfaces between middle gate and side gate along the sides of the middle channel between source and drain. Integration of the electrostatically tunable lateral, in-plane P-N homojunctions with the phototransistor enables the modulation of the photocarrier population as well as channel width independently of the phototransistor, thereby imparting a high responsivity to the device. Responsivity of the photodetector is 170 AW^{-1} at 300 W laser power. Typical operating speed is ~14–16 μs. Flicker noise-limited specific detectivity is 1.1×10^{12} Jones (Ghosh et al. 2021).

4.7.3 HIGH-TEMPERATURE WSe$_2$ FET PHOTODETECTOR WITH H-BN/GRAPHITE FLAKE HETEROSTRUCTURE PROTECTION

MoS$_2$ suffers degradation by oxidation in air. The oxidation begins at 300°C. Oxidation of the edge of WSe$_2$ starts around the same temperature. For high-temperature operation, a field-effect transistor is made using WSe$_2$ (Zou et al. 2022). It is encapsulated with h-BN. Contacts of source and drain terminals are made with graphite. Figure 4.9 shows the photodetector designed for high-temperature applications.

The device is fabricated on a cleaved mica substrate cleaned in acetone, alcohol, and distilled water. Then the constituent layers, namely, bottom h-BN layer, WSe$_2$ film, graphite source/drain electrodes, and the top h-BN layer, are mechanically exfoliated and transferred on the mica substrate. The transfer is done using a PDMS film. The top Pt gate electrode is formed by photolithographically defining the electrode pattern, depositing platinum and lifting off the photoresist. Annealing is done in a nitrogen ambience at 250°C for ½ h.

This photodetector with h-BN/graphite heterostructure protection works up to 700°C in air. In vacuum, it works up to 1000°C. It shows negative photoconductivity at high temperature. When the device is operated in the negative photoconductivity mode,

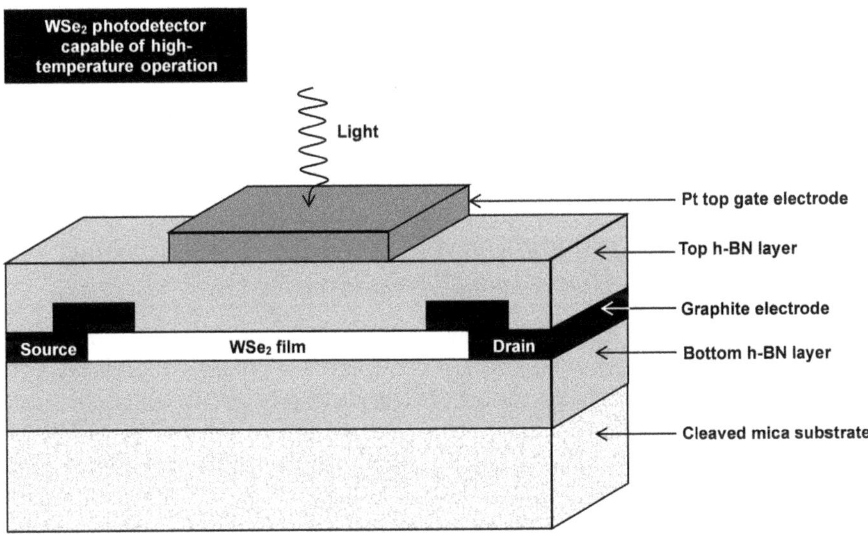

FIGURE 4.9 h-BN encapsulated WSe$_2$ FET photodetector with high-temperature operation capability. The diagram shows a cleaved mica substrate. On this substrate, there is a bottom h-BN layer (thickness 50 nm). Above this h-BN layer is the 10–20-nm thick Wse$_2$ film with graphite source/drain electrodes at its two ends. The WSe$_2$ film and the graphite electrodes are covered with a 50-nm thick top h-BN layer. On the upper surface of the top-BN layer, a platinum electrode is formed. This is the platinum top gate. Light is falling on the platinum top gate electrode.

the photoresponsivity is 2.2 $\times 10^6$ AW^{-1}. This value is five orders of magnitude higher than that of the contemporary high-temperature photodetectors (Zou et al. 2022).

4.8 BLACK PHOSPHOROUS PHOTODETECTORS

4.8.1 Advantages and Disadvantages of Black Phosphorous as a Photodetection Material

Advantages of black phosphorous are:

(i)   Its carrier mobility is high ~10^4 cm^{2}V^{-1}s^{-1} in bulk and 10^3 cm^{2}V^{-1}s^{-1} in flakes.
(ii) Its bandgap depends on the number layers. The bandgap is 0.3 eV in bulk state and increases to 1.8–2.0 eV for monolayer thickness.
(iii) Its spectral response is wider than that of TMDs. While TMDs with bandgaps of 1–2 eV are restricted between UV and visible portions of the spectrum of light, black phosphorous is useful for near- and mid-infrared detection.

Shortcomings of black phosphorous include:

(i)   It exhibits poor stability in air. Therefore, protection with an encapsulating film is necessary.
(ii) High contact resistances are obtained with metal films owing to the formation of Schottky barriers. These oppose the transport of photocarriers, thereby decreasing the photoresponsivity.

4.8.2 Widely Tunable h-BN/BP/h-BN-Sandwiched Heterostructure Mid-Infrared Dual-Gate FET Photodetector

This photodetector is formed by interposing black phosphorous between hexagonal boron nitride layers (Chen et al. 2017). It is displayed in Figure 4.10. It is intended for applications farther from the cut-off wavelength of pristine black phosphorous.

BP and hBN flakes exfoliated from bulk crystals are assembled to form hBN/BP/hBN heterostructures by polymer-free dry transfer process in an argon-filled glove box and transferred to SiO$_2$/Si substrates inside the glove box. The hBN dielectric plays the role of providing an ultra-clean interface for collection of photogenerated carriers. Moreover, it thwarts the oxidation of black phosphorous. The hBN/BP/hBN/SiO$_2$/Si structures are annealed at 633 K in H$_2$/Ar atmosphere for 6 h. The crystal direction of BP is characterized. PMMA is spin coated on hBN and an interdigitated electrode pattern is defined by e-beam lithography. After reactive ion etching of the exposed hBN layer in CHF$_3$/O$_2$ plasma, Cr/Au evaporation is done, and another hBN flake is transferred over the entire surface. Electron-beam lithography and Pt evaporation are done for top Pt gate metallization.

It may be noted that the top hBN comprises two layers. The first hBN layer is part of the hBN/BP/hBN sandwich. The second hBN layer transferred after the

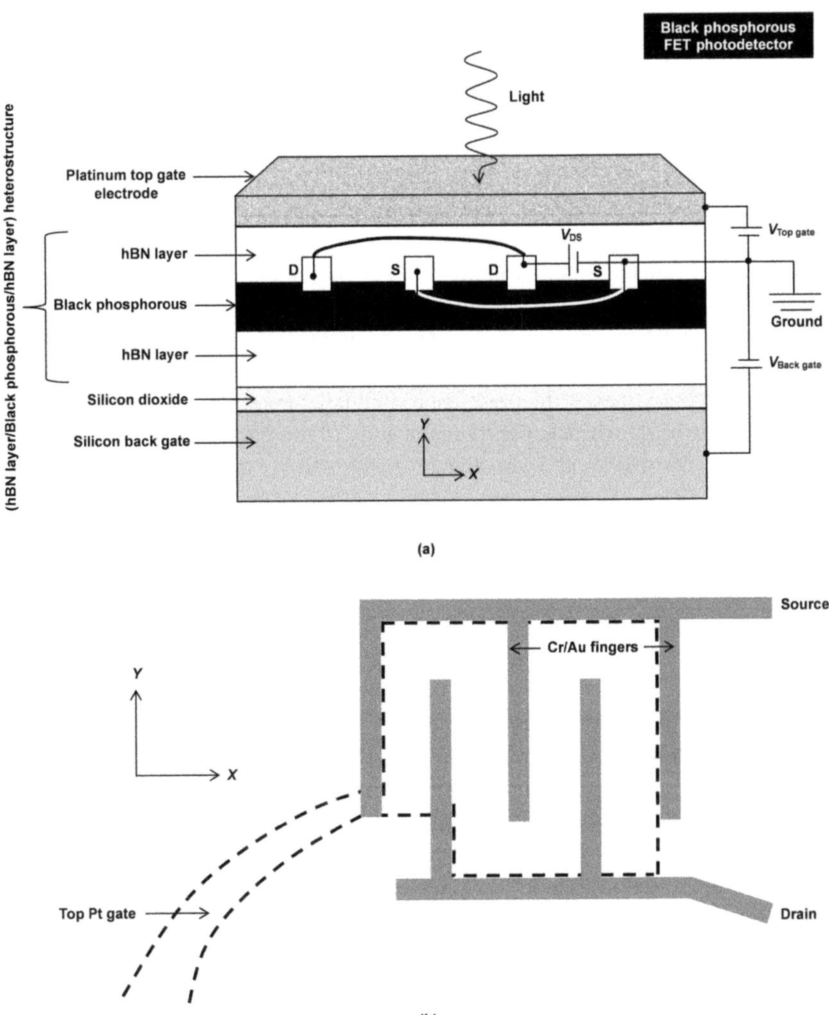

FIGURE 4.10 Tunable black phosphorous mid-infrared FET photodetector: (a) cross-sectional diagram with electrical connections and (b) interdigitated source–drain geometry. Part (a) shows a silicon dioxide/silicon substrate in which silicon acts as the back gate electrode. X- and Y-directions are indicated. Above silicon dioxide is an hBN/black phosphorous/hBN-sandwiched heterostructure. Alternating source and drain electrodes are seen. It is also shown how they are electrically connected. The source electrodes are shorted together. The drain electrodes are also connected together. The drain–source supply V_{DS} is connected between source and drain electrodes. The source electrode is grounded. Above the top hBN flake is a partially transparent platinum film acting as the top gate electrode. The batteries $V_{Top\ gate}$ connected between top Pt gate and ground terminal and $V_{Back\ gate}$ connected between silicon back gate and ground terminal are used for biasing the respective gates. Light falls on the Pt electrode. Part (b) shows the geometrical layout of the device. There are two sets of fingers arranged such that fingers of one set lie between those of the other set in an interlaced or intertwined form. One set of Cr/Au fingers is connected to source terminal and the other set of fingers to drain terminal. The top Pt gate area is enclosed by dashed lines.

chromium/gold fingers have been deposited envelopes these fingers as also the channel to prevent electrical shorting between the Cr/Au fingers and the Pt gate electrode. The task of Cr/Au electrodes is collection of the photocurrent. They are aligned along the Y-axis of black phosphorous flakes. Hence, current is collected along the X-axis. High carrier mobility is thereby obtained, appreciably enhancing the photoresponsivity.

On application of a vertical electric field, the photoresponse of the detector is extended from 3.7 to > 7.7 µm by taking advantage of Stark effect. So, the photodetector is widely tunable. The Stark effect is a phenomenon taking place in an electric field in analogy to the Zeeman effect in a magnetic field. It involves the splitting of spectral lines of atoms and molecules into multiple components when placed in an electric field. This happens in the same way as in the Zeeman effect, which entails the splitting of spectral lines in the presence of magnetic field.

At 77 K, the peak extrinsic photoresponsivity of the photodetector is 518 mAW^{-1} at 3.4 µm; it is 30 mAW^{-1} at 5 µm, and 2.2 mAW^{-1} at 7.7 µm (Chen et al. 2017).

4.8.3 BP In-Plane PN Homojunction Photodetector

A photodetector is developed using a BP in-plane PN homojunction delineated by ferroelectric domains (Wu et al. 2022). An SiO$_2$/Si substrate is used (Figure 4.11). Few-layered BP made by mechanical exfoliation is transferred over the substrate. The Cr/Au source/drain electrodes for few-layered BP are deposited and defined by electron-beam lithography. A 100-nm thick P(VDF-TrFE) film is spin coated over the prepared structure.

P(VDF-TrFE) is a ferroelectric copolymer exhibiting intense spontaneous polarization. Piezoresponse force microscopy (PFM) is a mode of atomic force microscopy by which the domains of ferroelectric materials can be manipulated. So, PFM is utilized with P(VDF-TrFE) for P–N junction definition. To make a junction, different regions of the ferroelectric copolymer surface are polarized upward and downward by contacting the sharp conductive tip of PFM. The resulting ferroelectric domains define an in-plane P-N homojunction in few-layered BP.

The electron–hole pairs produced in the few-layered BP by optical excitation are separated by the built-in electric field of the P–N homojunction. This photovoltaic effect is further supported by the photothermal effect to yield a high photoresponsivity of 1.06 AW^{-1}, and a detectivity of 1.27×10^{11} cm Hz$^{1/2}$W^{-1} (Wu et al. 2022).

4.9 DISCUSSION AND CONCLUSIONS

Progress in the fabrication of photodetectors with different 2D materials in various device configurations was appraised in relation to the achieved response parameters and characteristics in each case (Figure 4.12). In particular, the 2D photodetectors being free of dangling bonds are safeguarded from the surface recombination problem which significantly elevates the dark current. Encouragingly, 2D materials offer seamless integration with several material platforms, such as flexible materials, silicon, and III–V compound semiconductors. Among these, 2D material photodetectors

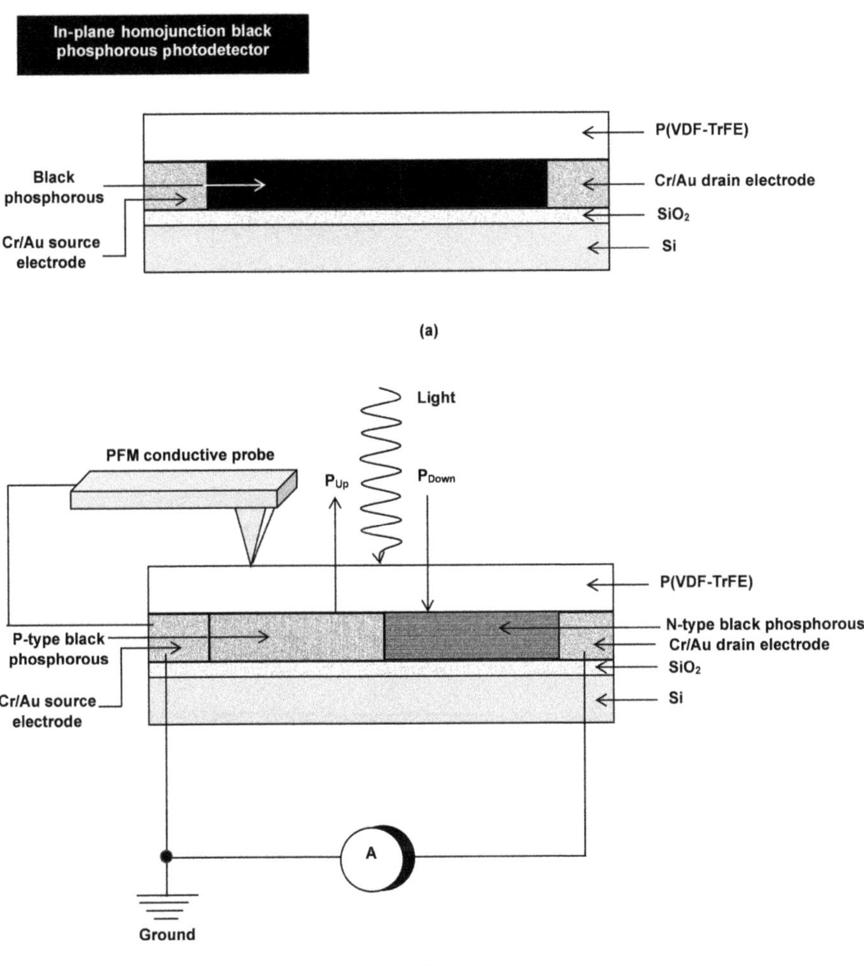

FIGURE 4.11 In-plane homojunction made in few-layered BP for photodetection. Part (a) shows an SiO_2/Si substrate. Over the SiO_2 film lies few-layered BP flanked on its two sides with Cr/Au contact electrodes. The top surface of the device is covered with a P(VDF-TrFE) ferroelectric film. Part(b) shows the same sequence of layers except that the few-layered BP is subdivided into two parts. The left part is doped P-type and the right part is doped N-type so that a P–N junction is created at the interface between the two parts. P_{up} and P_{Down} represent upward and downward polarizations of the ferroelectric copolymer, respectively, as done with piezoresponse force microscopy (PFM) conductive probe and indicated by the arrows pointing in those directions. Light is falling on the surface of the transparent P(VDF-TrFE) layer. The photocurrent produced is measured by an ammeter connected between the source and drain electrodes.

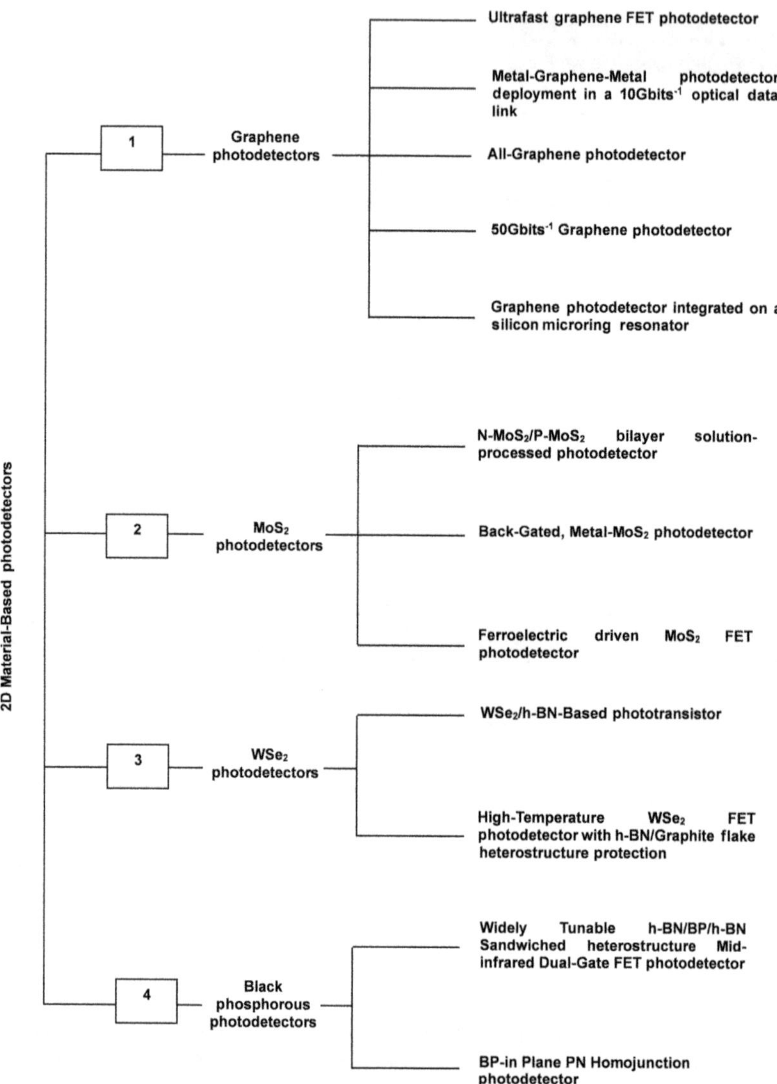

FIGURE 4.12   Graphical summary of 2D material photodetectors described in this chapter. In the diagram, the photodetectors are subdivided into four subgroups concerned with photodetectors made from graphene, MoS_2, WSe_2, and black phosphorous. The graphene subgroups of photodetectors considered are graphene FET photodetector (Xia et al. 2009), metal–graphene–metal photodetector (Mueller et al. 2010), all-graphene photodetector (Withers et al. 2013), 50 Gbits^{-1} graphene photodetector (Schall et al. 2014), and graphene photodetector integrated on silicon microring resonator (Schuler et al. 2021). The MoS_2 subgroups include N-MoS_2/P-MoS_2 photodetector (Ye et al. 2015), back-gated metal–MoS_2 photodetector (Wang et al. 2015a), and ferroelectric driven MoS_2 FET photodetector (Wang et al. 2015b). The devices placed in the WSe_2 photodetector subgroups are WSe_2/h-BN-based phototransistor (Ghosh et al. 2021) and WSe_2 FET photodetector protected with h-BN/graphite flake heterostructure (Zou et al. 2022). Photodetectors mentioned in the black phosphorous subgroup are: h-BN/BP/h-BN heterostructure dual-gate FET photodetector (Chen et al. 2017) and BP-in plane PN homojunction photodetector (Wu et al. 2022).

made on Si platform are of immense interest because of the opportunities for optoelectronics and microelectronics integration. Apart from the provision of microelectronic readout circuitry by silicon, cost-effective photonic-integrated circuits developed in silicon photonics technology are available owing to its compatibility with complementary metal–oxide semiconductor (CMOS) process (Liu et al. 2021).

After understanding the principles of optical detection by semiconductor materials and their application to development of 2D materials devices, including the trends in photodetector device configurations, and noting the key performance parameters achieved by characterization of these photosensors, we direct our attention to another category of sensors in which 2D materials have made a profound impact. These are the 'magnetic-field sensors', which will be the topic of next chapter.

REFERENCES

Bernardi M., M. Palummo, and J. C. Grossman 2013 Extraordinary sunlight absorption and one nanometer thick photovoltaics using two-dimensional monolayer materials, *Nano Letters*, 13, 3664–3670.

Chen X., X. Lu, B. Deng, O. Sinai, Y. Shao, C. Li, S. Yuan, V. Tran, et al. 2017 Widely tunable black phosphorus mid-infrared photodetector, *Nature Communications*, 8(1672), pp. 1–7.

Ghosh S., A. Varghese, K. Thakar, S. Dhara and S. Lodha 2021 Enhanced responsivity and detectivity of fast WSe$_2$ phototransistor using electrostatically tunable in-plane lateral p-n homojunction, *Nature Communications*, 12(3336), pp. 1–9.

Gupta S., S. N. Shirodkar, A. Kutana, and B. I. Yakobson 2018 In pursuit of 2D materials for maximum optical response, *ACS Nano*, 12(11), 10880–10889.

Long M., P. Wang, H. Fang, and W. Hu 2019 Progress, challenges, and opportunities for 2D material-based photodetectors, *Advanced Functional Materials*, 29(1803807), pp. 1–28.

Liu C., J. Guo, L. Yu, J. Li, M. Zhang, H. Lii, Y. Shi et al. 2021 Silicon/2D-material photodetectors: From near-infrared to mid-infrared, *Light: Science & Applications*, 10(123), pp. 1–21.

Mueller T., F. Xia, and P. Avouris 2010 Graphene photodetectors for high-speed optical communications, *Nature Photonics*, 4, pp. 297–301.

Schall D., D. Neumaier, M. Mohsin, B. Chmielak, J. Bolten, C. Porschatis, A. Prinzen, C. Matheisen, et al. 2014 50 GBit/s photodetectors based on wafer-scale graphene for integrated silicon photonic communication systems, *ACS Photonics*, 1, pp. 781–784.

Schuler S., J. E. Muench, A. Ruocco, O. Balci, D. V. Thourhout, V. Sorianello, M. Romagnoli, K. Watanabe, et al. 2021 High-responsivity graphene photodetectors integrated on silicon microring resonators, *Nature Communications*, 12(1), pp. 1–9.

Taffelli A., S. Dirè, A. Quaranta and L. Pancheri 2021 MoS$_2$ based photodetectors: A review, *Sensors*, 21(2758), pp. 1–22.

Wang H., C. Zhang, W. Chan, S. Tiwari and F. Rana 2015a Ultrafast response of monolayer molybdenum disulfide photodetectors, *Nature Communications*, 6(8831), pp. 1–6.

Wang X., P. Wang, J. Wang, W. Hu, X. Zhou, N. Guo, H. Huang, et al. 2015b Ultrasensitive and broadband MoS$_2$ photodetector driven by ferroelectrics, *Advanced Materials*, 27(42), 6575–6581.

Withers F., T. H. Bointon, M. F. Craciun, and S. Russo 2013 All-graphene photodetectors, *ACS Nano*, 7(6), 5052–5057.

Wu S., Y. Chen, X. Wang, H. Jiao, Q. Zhao, X. Huang, X. Tai, et al. 2022 Ultra-sensitive polarization-resolved black phosphorus homojunction photodetector defined by ferroelectric domains, *Nature Communications*, 13(3198), pp. 1–9.

Xia F., T. Mueller, Y.-m. Lin, A. Valdes-Garcia, and P. Avouris 2009 Ultrafast graphene photo-detector, *Nature Nanotechnology*, 4, 839–843.

Ye J., X. Li, J. Zhao, X. Mei and Q. Li 2015 A facile way to fabricate high-performance solution-processed n-MoS$_2$/p-MoS$_2$ bilayer photodetectors, *Nanoscale Research Letters*, 10(454), pp. 1–7.

Zhao M., W. Xia, Y. Wang, M. Luo, Z. Tian, Y. Guo, W. Hu, and J. Xue 2019 Nb$_2$SiTe$_4$: A stable narrow-gap two-dimensional material with ambipolar transport and mid-infrared response, *ACS Nano*, 13(9), pp. 10705–10710.

Zheng L., L. Zhongzhu, and S. Guozhen 2016 Photodetectors based on two dimensional materials, *Journal of Semiconductors*, 37(9), 091001-1–091001-11.

Zou Y., Z. Zhang, J. Yan, L. Lin, G. Huang, Y. Tan, Z. You and P. Li 2022 High-temperature flexible WSe$_2$ photodetectors with ultrahigh photoresponsivity, *Nature Communications*, 13(4372), pp. 1–9.

5 2D Materials-Based Magnetic Sensors

A magnetic sensor is a device which can:

(i)   detect the presence of a magnetic field in its surrounding regions,
(ii) measure the strength and direction of the perceived magnetic field, and
(iii) produce an electrical signal representing the strength of magnetic field and its variations.

Magnetic sensors are used in industrial process control instrumentation, robotics and factory automation, data recording and memory storage; automotive, consumer and biomedical electronics; and security applications, e.g., detection and localization of metallic objects, navigation and position tracking (Ripka 2013). This chapter describes magnetic nanosensors using 2D materials.

5.1 TYPES OF MAGNETIC SENSORS

5.1.1 INDUCTION COIL SENSOR

An induction coil is a simple magnetic sensor. When a magnet approaches the coil, the magnetic flux density in the coil increases inducing an electromotive force in the coil in a direction whose magnetic flux density opposes the increase in magnetic flux density in the coil. The reverse happens when the magnet is withdrawn. This is the principle employed in conventional inductive coil sensors. In another method, the magnetic field is obtained from the change in resonant frequency of a coil, as we shall see in this chapter.

5.1.2 HALL-EFFECT SENSOR

A Hall-effect magnetic field sensor works on the well-known Hall-effect principle. The Hall effect is an electromagnetic phenomenon that takes place in a current-carrying thin-film conductor or semiconductor when it is placed in a magnetic field acting perpendicular to the current flow direction. The effect is observed as the production of a transverse potential difference known as the Hall voltage across the conductor or semiconductor. The direction of Hall voltage depends on the directions of the current and magnetic field. It acts perpendicularly to both the current and the magnetic field directions.

DOI: 10.1201/9781003330585-5

5.1.3 Magnetoresistive Sensor

A magnetoresistive sensor measures a magnetic field from the change in electrical resistance of a thin film of a ferromagnetic material exposed to the magnetic field. Following effects are applied in these sensors:

(i) Anisotropic magnetoresistance (AMR) effect, the galvanomagnetic effect of dependence of electrical resistance of the material on the relative angle between the directions of electric current and magnetization.

(ii) Giant magnetoresistance (GMR) effect, a quantum-mechanical effect concerning large resistance change in a multilayer stack consisting of alternate ferromagnetic and non-magnetic layers under the influence of a magnetic field.

(iii) Tunnel magnetoresistance (TMR) effect, a quantum-mechanical effect about magnetic field-induced change in resistance of a tunnel junction comprising a few nm thick insulating layer between two ferromagnetic layers.

(iv) Extraordinary magnetoresistance (EMR), a geometric magnetoresistance effect exhibited by hybrid structures which can produce 10^6 % resistance change in large magnetic fields at room temperature.

5.1.4 Quantum Magnetic Sensor

A quantum magnetic sensor harnesses the quantum-mechanical phenomena such as the Zeeman effect in which a spectral line is split up into two or more components of different frequencies when a source of light is placed in an external magnetic field. The magnetic field is extracted by observing the splitting of the spectral line.

5.2 MAGNETIC MICROWIRE INDUCTION COIL SENSOR WITH A FERROMAGNETIC VSE$_2$ MONOLAYER CORE

5.2.1 Induction Coil and the VSe$_2$ Core

A coil is made with $Co_{69.25}Fe_{4.25}Si_{13}B_{12.5}Nb_1$ wire of diameter ~60 µm (Jimenez et al. 2020). The coil consists of 15 turns. Its length is 10 mm and the internal diameter is 5 mm. Inside the coil, a core is inserted. This core is made of a single atomic layer thick VSe_2 film grown epitaxially on an MoS_2 substrate. The VSe_2 core has a length of 14 mm and a width of 4 mm. The soft ferromagnetic microwire coil with the ferromagnetic VSe_2 film core inserted inside it constitutes the induction coil sensor as shown in Figure 5.1(a).

5.2.2 Modeling the Induction Coil Sensor

The induction coil sensor is modeled by representing it as the lumped form of a non-perfect inductor (Figure 5.1(b)), keeping in view that winding of the cylindrical conductors in close vicinity introduces parasitic elements. The elements introduced are parasitic resistance $R_{\text{Parasitic}}$ and parasitic capacitance $C_{\text{Parasitic}}$. Hence, the model of

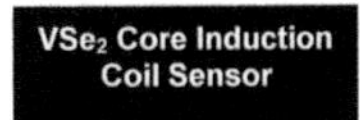

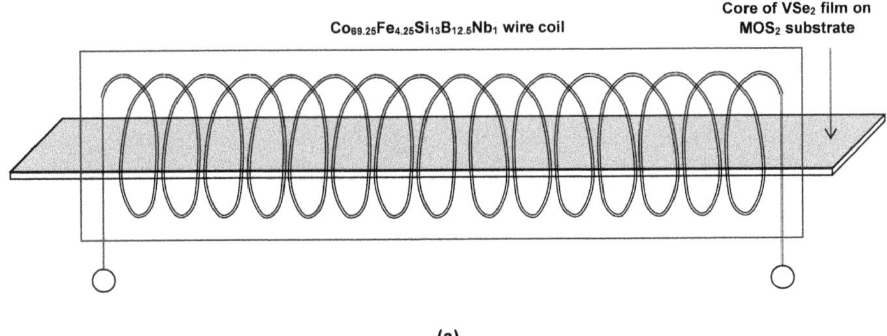

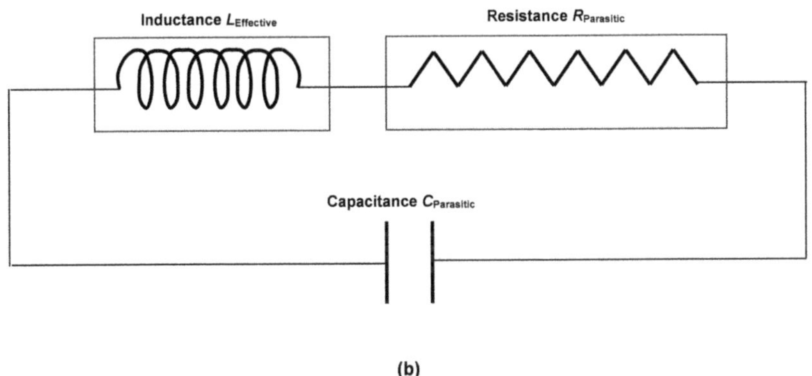

FIGURE 5.1 Induction coil sensor: (a) construction of the sensor and (b) its equivalent circuit model. Part (a) shows a magnetic microwire coil made of 15 turns of $Co_{69.25}Fe_{4.25}Si_{13}B_{12.5}Nb_1$ wire into which a VSe_2 film deposited on a MoS_2 substrate is inserted as a core. Part (b) shows a circuit consisting of an inductance $L_{Effective}$ connected in series with a resistance $R_{Parasitic}$, and this serial $L_{Effective}$–$R_{Parasitic}$ combination is connected in parallel with a capacitor $C_{Parasitic}$.

the induction coil sensor comprises a series combination of an ideal inductor $L_{Effective}$ with resistance $R_{Parasitic}$ in parallel connection with capacitance $C_{Parasitic}$. The effective inductance $L_{Effective}$ is the inductance of the induction coil sensor taking into consideration that both the wire of the coil and the core inserted in the coil are ferromagnetic materials leading to an effective permeability $\mu_{Effective}$ of the wire and the core, and the inductance of the sensor is determined by this effective permeability, i.e., $L_{Effective}$ is the inductance of a coil having a permeability $\mu_{Effective}$.

The coil has an impedance Z_{Coil} made of reactive components (inductive and capacitive) X_L and X_C. At resonance, inductive reactance X_L is equal to capacitive reactance X_C, and the phase difference between them is equal to π radians. Then Z_{Coil}

increases to a large value and negligible current flows through the coil. The resonance frequency is given by

$$v_0 = \frac{\sqrt{1 - \left(\dfrac{R_{\text{Parasitic}}^2 C_{\text{Parasitic}}}{L_{\text{Effective}}}\right)}}{2\pi\sqrt{L_{\text{Effective}}C_{\text{Parasitic}}}} \tag{5.1}$$

When the induction coil sensor is placed in a DC magnetic field H, its resonance frequency v_0 changes, and the change in resonance frequency is related to magnetic field intensity. The sensitivity of the sensor is defined as

$$S = \frac{dv_0}{dH} \tag{5.2}$$

5.2.3 MAGNETIC RESPONSE OF THE SENSOR

The sensitivity is measured by exposing the sensor to a DC magnetic field produced by a Helmholtz coil in the plane of the VSe_2 core and perpendicular to the axis of the coil. For every value of magnetic field, it is swept between ±30 Oe in 1 Oe steps. Measurements are done with an impedance analyzer between 1MHz and 200 MHz. The sensitivity is found to be 1.6×10^7 HzOe^{-1} (Jimenez et al. 2020).

5.3 HALL MAGNETIC FIELD SENSORS

5.3.1 THE HALL EFFECT AND THE HALL PROBE

As mentioned in Section 5.1.2, the basis of Hall sensors is the Hall effect, which states that: when a current-carrying semiconductor slab is subjected to a magnetic field acting in a direction perpendicular to and out of plane to the current, positive charge carriers are deflected to one side of the semiconductor slab and negative charge carriers to the opposite side, creating an electric field in the transverse direction (Collomb et al. 2021).

The Hall effect in a specimen in the form of a finite slab of a material is shown in Figure 5.2(a), where the Hall voltage is observed in the perpendicular direction to the Hall current. The Hall sensor is fabricated as a Hall probe shaped like a Greek cross (Figure 5.2(b)). When a magnetic field acts perpendicular to the plane of the cross, and the Hall current I_H flows along one arm of the cross, the Hall voltage V_H is measured along the perpendicular arm of the cross.

5.3.2 GRAPHENE HALL-EFFECT MAGNETOMETER

5.3.2.1 Construction of the Hall Sensor

Single-layer graphene film is formed on an Si/SiO$_2$ substrate by chemical vapor deposition (Izci et al. 2018). The samples are annealed at 300°C for 3 h to make

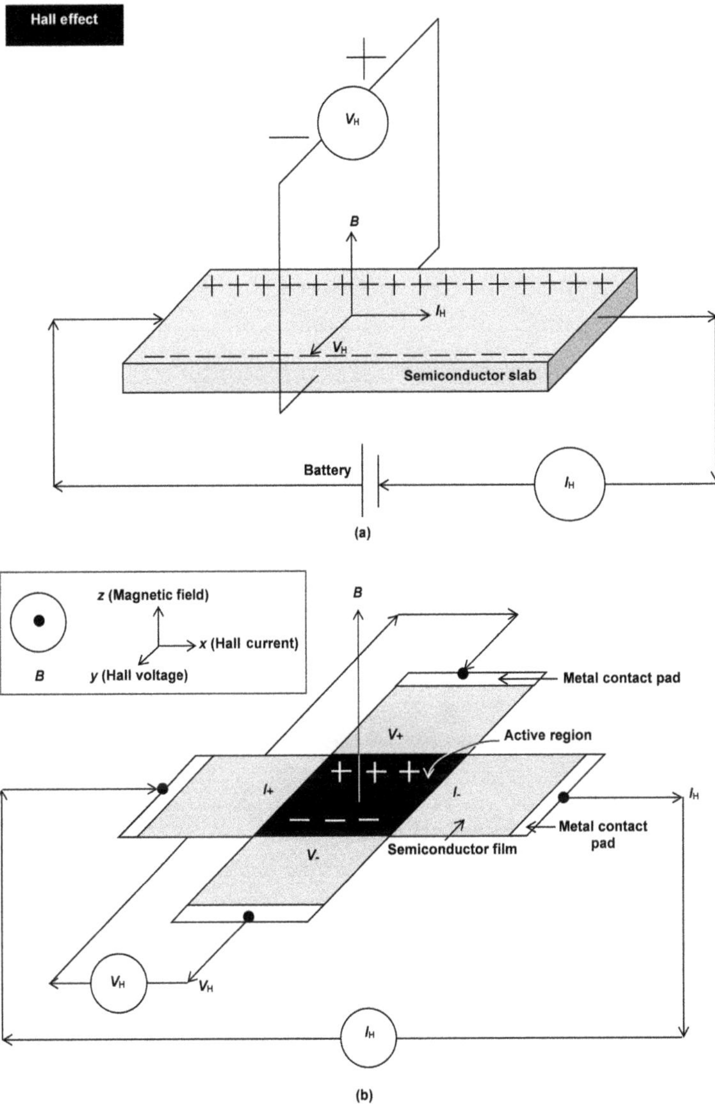

FIGURE 5.2 (a) Principle of Hall effect. The diagram shows a current flowing from the positive pole of the battery into a slab of semiconductor material. An ammeter is connected to the circuit to measure this current. It is designated as the Hall current I_H. A magnetic field B is seen acting in a direction perpendicular to the plane of the semiconductor slab. Interaction between the current and the magnetic field causes the positive and negative charge carriers to build up on opposite sides of the semiconductor slab leading to the generation of an electric potential V_H known as the Hall voltage, as measured by a voltmeter. (b) Representative shape of a Hall sensor as a Greek cross. The active region of the device is shown as a black-colored area. The Hall current I_H, Hall voltage V_H and magnetic field B are represented by arrows. The semiconductor film and the metal contacts are marked. The currents I_+, I_-, and the voltages V_+, V_- are indicated. Inset shows the relative directions of Hall current, Hall voltage, and magnetic field along the x, y, and z axes, respectively. The encircled dot indicates that the magnetic field lines are pointing outward from the plane of the diagram.

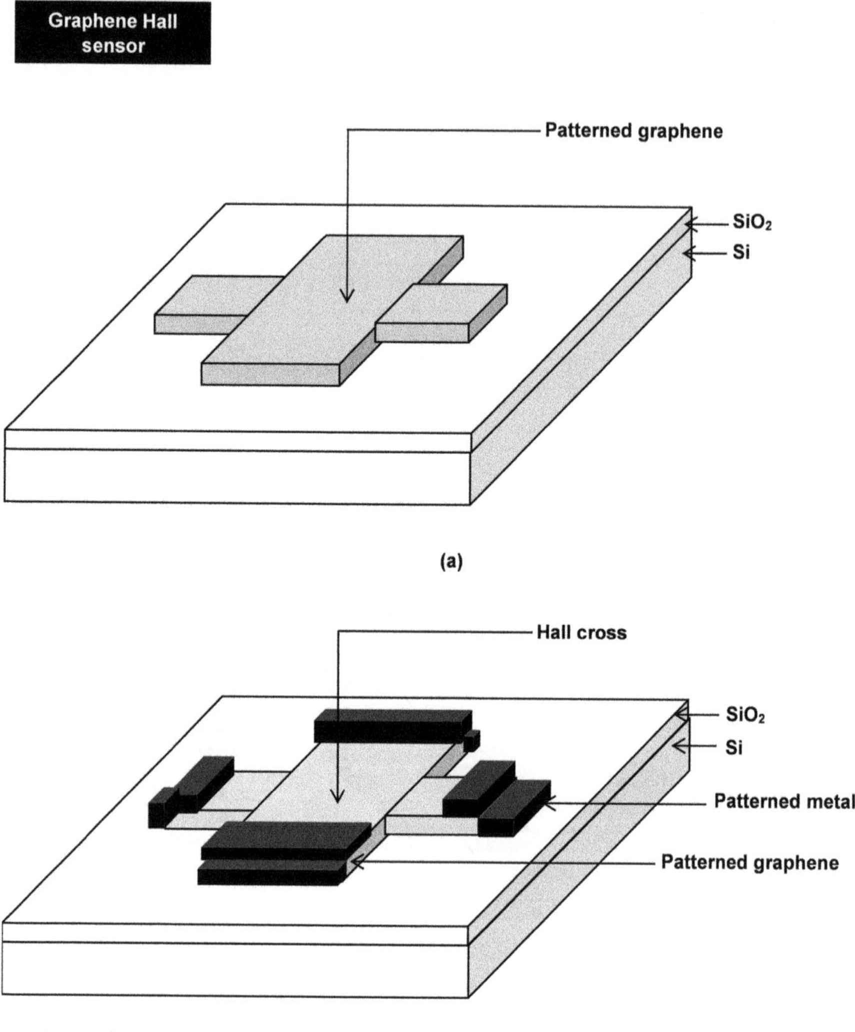

FIGURE 5.3   Sketch of the graphene Hall sensor: (a) before metallization and (b) after metallization. Part (a): on an SiO₂/Si substrate, there is a patterned graphene film. Part (b): over the graphene film, Au/Cr metal contact pattern is defined. The Hall cross is clearly seen.

graphene resistant to delamination. An array of 7 devices with Au/Cr contacts is made. The active region of each device has an area of 10 μm ×10 μm. Figure 5.3 shows the Hall sensor.

5.3.2.2   Microfabrication of the Hall Sensor

The Hall sensor is made by microfabrication techniques. The process is illustrated in Figure 5.4 for a single device with Hall cross geometry. The process steps are subdivided into two parts:

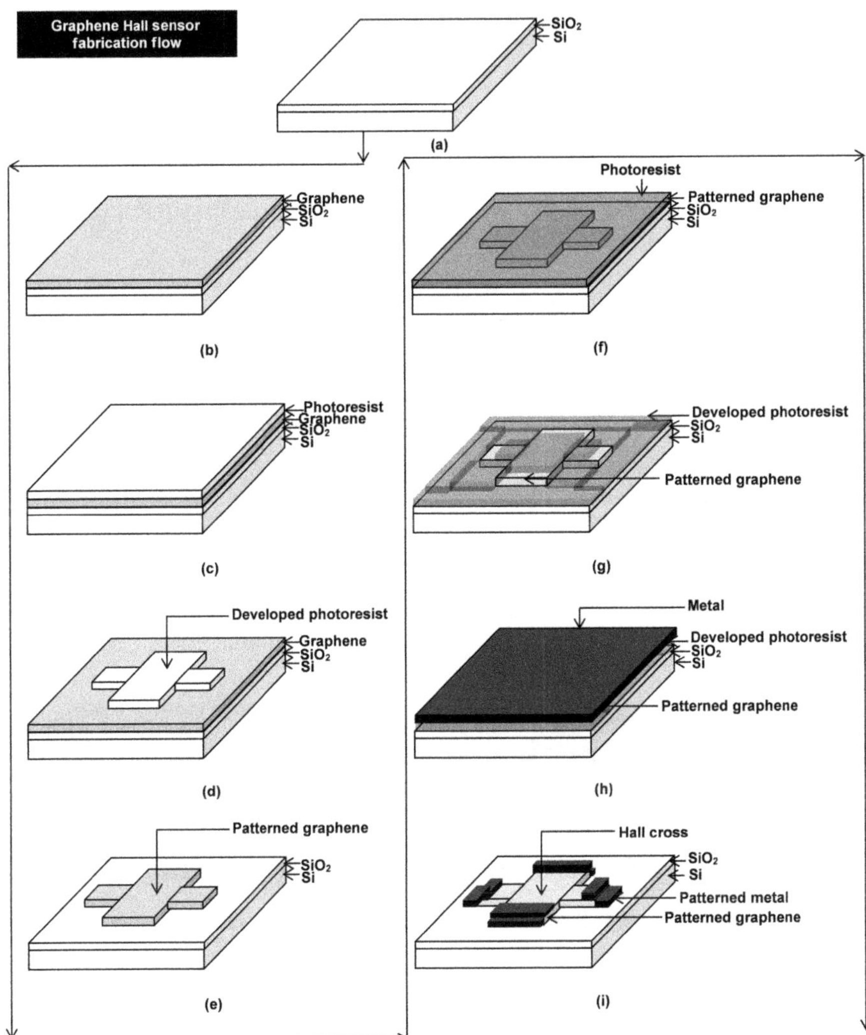

FIGURE 5.4 Fabrication process of graphene Hall sensor. Various process stages are shown: (a) the SiO$_2$/Si substrate, (b) graphene film formation by CVD, (c) photoresist coating over graphene, (d) developed photoresist after lithography, (e) graphene patterning by etching, (f) photoresist coating on graphene pattern, (g) developed photoresist in which the photoresist is removed from portions where the metal film is to be formed and photoresist is retained in portions where there will be no metal, (h) deposition of Cr and Au films over the developed photoresist, and (i) stripping off the photoresist to get the metal pattern; the Hall cross is visible.

(i) Deposition and patterning graphene (Figure 5.4(b)–(e)): graphene deposition in step (b) is followed by photoresist coating in step (c), photoresist exposure through a mask and developing in step (d), and graphene etching in oxygen plasma in step (e).

(ii) Au/Cr contact formation (Figure 5.4(f)–(i)): the graphene pattern is coated with photoresist in step (f), the photoresist is exposed through a mask and developed in step (g), Au/Cr metallization is carried out in step (h), and lifting off the photoresist is done in step (i).

5.3.2.3 Sensitivity Parameters of the Hall Sensor

The absolute sensitivity S_A of the Hall sensor is given by

$$S_A = \frac{\text{Change in output voltage}\left(V_H\right)}{\text{Applied magnetic field}\left(B_Y\right)\text{in a given applied biasing current}\left(I_X\right)} = \left|\frac{V_H}{B_Y}\right| \quad (5.3)$$

Its current-related sensitivity S_I is

$$S_I = \frac{\text{Absolute sensitivity}\left(S_A\right)}{\text{Applied biasing current}\left(I_X\right)} \quad (5.4)$$

Its unit is $VA^{-1}T^{-1}$. The maximum current-related sensitivity achieved with this Hall sensor is $2540\ VA^{-1}T^{-1}$. The magnetic field resolution is $162nT\left(\sqrt{Hz}\right)^{-1}$ (Izci et al. 2018).

5.3.3 HBN-ENCAPSULATED ULTRACLEAN GRAPHENE HALL SENSOR

5.3.3.1 The Hall Cross: Two-Gate and Three-Gate Structures

The active region of a Hall sensor is a cross-shaped area with low carrier density and high mobility. Contrarily to the decrease in mobility at low carrier concentrations observed in 2D semiconductors, it increases in graphene because long-range impurity scattering is reduced. Consequently, the detection limit of magnetic field is lowered. The required low-density, high-mobility regime for graphene Hall sensor is achieved by encapsulating graphene with hexagonal boron nitride (hBN). Additionally, the use of a few-layer graphite as gate electrode enables fabrication of ultraclean graphene sensors.

A Hall sensor is made on an SiO_2/Si substrate with three/two gate electrodes (Figure 5.5) for tuning the carrier density to optimize the limit of magnetic field detection (Schaefer et al. 2020). The layers in the Hall cross for three-gate and two-gate structures are:

(i) Three gate structure: from top downward, the layers are: hexagonal boron nitride (hBN)/few-layer graphite (FLG)/hexagonal boron nitride (hBN)/ monolayer graphene (MLG)/hexagonal boron nitride (hBN)/few-layer graphite (FLG)/SiO_2/Si. The three gates are: FLG top gate, FLG bottom

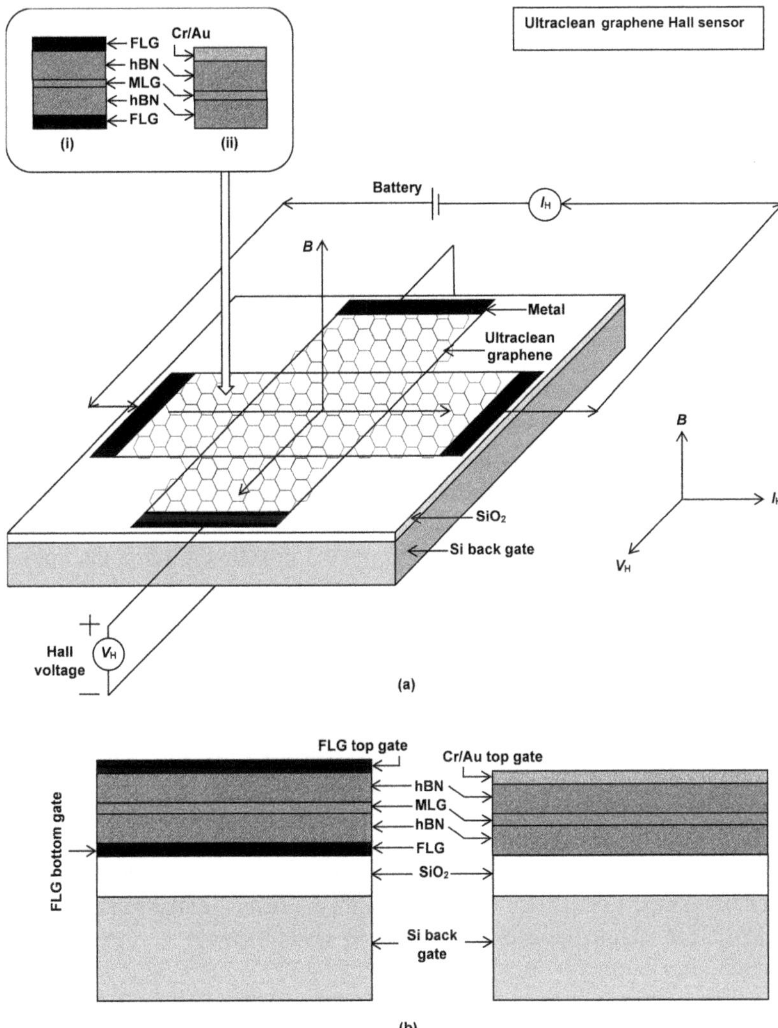

FIGURE 5.5 Ultraclean graphene Hall sensor: (a) The Hall sensor is made with ultraclean graphene. The sensor consists of an SiO$_2$/Si substrate with a Si back gate. Over the substrate is a graphene film patterned in the shape of two perpendicular rectangular strips intersecting with each other to form a Hall cross. Metal contact pads are made at the four edges of graphene strips. A magnetic field is acting in a direction perpendicular to the plane of the substrate. The Hall current I_H flows from a battery into the graphene strip. The Hall voltage V_H is measured across the perpendicular graphene strip. Two types of ultraclean graphene structures are shown in the inset sequenced from top downwards as: (i) few-layer graphite (conductor), hexagonal boron nitride (dielectric), monolayer graphene (semiconductor)/ hexagonal boron nitride (dielectric), few-layered graphite (conductor). (ii) Cr/Au film, hexagonal boron nitride, monolayer graphene, and hexagonal boron nitride. (b) The gate stacks of structures (i) and (ii) showing in (i) the few-layer graphite top gate, the few-layer graphite bottom gate, and the Si back gate, and in (ii), the Cr/Au top gate and the Si back gate; there is no few-layer graphite gate in this configuration.

gate, and Si back gate. The FLG top gate is used to adjust the carrier density in the active area of Hall sensor. The FLG bottom gate is grounded to screen the electric field from Si back gate. The Si back gate is given a bias of 40 V to produce a high carrier density in the graphene portion of the leads by electrostatic induction. As a result, the resistances of the leads and edge contacts are decreased leading to reduction of voltage noise.

(ii) Two-gate structure: from top downward, the layers are: Ti–Au–Pt/hexagonal boron nitride (hBN)/monolayer graphene (MLG)/hexagonal boron nitride (hBN)/SiO_2/Si. The two gates are Ti–Au–Pt top gate and Si back gate. The functions of the gates are the same as in (i) except that the bottom gate is not available.

5.3.3.2 Sensor Fabrication

The sensors are made from exfoliated graphene by assembling the layers in the heterostructures by a dry-transfer technique. A thin sheet of poly (bisphenol A carbonate) (PC) on a PDMS stamp serves as the transfer slide. The flakes are picked up one-by-one at 80°C. The substrate is heated to 180°C before the stack is released. In this way, any bubbles confined between the flakes are impelled towards the boundaries upon fixing. The PC is removed by dissolution in chloroform for 4 h. After rinsing with isopropyl alcohol and drying in nitrogen, the samples are annealed in a vacuum at 300°C for 3 h to get rid of traces of polymer. For contact formation, Cr/Pd/Au or Cr/Au metallization is done by electron beam evaporation.

5.3.3.3 Detection Limit

For a 1-μm Hall sensor, the detection limit is 700 nT $(Hz)^{-1/2}$ at 1 kHz at room temperature. At a temperature of 4.2 K, it is 80 nT $(Hz)^{-1/2}$. While this sensor performs comparably to the top Hall sensors, improved readout may bring its detection limit at par with planar niobium-based SQUID magnetometers at frequencies in kHz range. The SQUIDs show 403 nT $(Hz)^{-1/2}$ for a 0.25-μm diameter circular sensitive area, and 3 nT $(Hz)^{-1/2}$ for a 1 μm diameter area (Schaefer et al. 2020).

5.4 EXTRAORDINARY MAGNETORESISTANCE SENSORS

Extraordinary magnetoresistance is a geometrical effect observed in a metal-hybrid system as an extremely high change in its resistance on applying a transverse magnetic field (Sun and Kosel 2013). An EMR device is a hybrid semiconductor–metal structure. It consists of a high-mobility semiconductor and an adjacent metallic shunt. The resistance of this structure changes by an exceedingly large amount when it is subjected to a magnetic field applied in a direction orthogonal to the plane of current flow. The magnetic field dependence of electrical resistance of the structure makes it useful as a magnetic field sensor for measuring a magnetic field acting in a direction at right angles to the plane of the device.

5.4.1 REASON FOR THE LARGE CHANGE IN RESISTANCE OF EMR DEVICE

The large resistance change in an EMR device takes place by deflection of current, resulting in redistribution of the current flow paths between the semiconductor and the metal portions of the device. Because the resistance of the semiconductor is much larger than that of the metal, the redistribution of current flow paths within the device is able to cause a large resistance change. Without magnetic field, the current flows mainly through the metallic portion. This is the low-resistance condition of the device. Since the current is shunted by the metal, the metallic portion is referred to as a metal shunt. In the presence of magnetic field, the current flow is restricted within the semiconductor region. This is the high-resistance state of the device. So, the ratio of resistance of an EMR device in magnetic field to that without magnetic field becomes colossal ~ several orders of magnitude, helping in the detection of magnetic field with increased accuracy.

5.4.2 Origin of EMR Effect

The EMR effect is based on the Lorentz force. As we know, Lorentz force is the combined electric and magnetic force acting on a charged particle moving through an electric field $\mathbf{E}$ and magnetic field $\mathbf{B}$. The EMR effect originates in the same way as the Hall effect. Fundamentally, it arises from a change in the path of current flow rather than a change in the conductivity of the metal or the semiconductor.

5.4.3 Advantages of EMR Devices

(i) EMR devices are intrinsically low-noise devices because they are made of nonmagnetic materials. Hence, magnetic noise is absent. The tunnel magnetoresistance or giant magnetoresistance devices are comparatively noisier.

(ii) Thermal noise in EMR devices is smaller than in Hall sensors because of the low resistance of metallic shunt.

(iii) The saturation magnetic field of EMR devices is 1T. So, these devices can work in a broad range of magnetic fields.

5.4.4 Structures of EMR Sensors

(i) van der Pauw structure: It consists of a circular semiconductor disc with a concentric metal disc embedded inside (Figure 5.6(a)). The structure is provided with two metallic contacts I_+ and I_- for current injection and two metallic contacts V_+ and V_- for measuring the voltage.

(ii) Bar-type structure: It consists of a semiconductor bar shunted with a rectangular metallic strip (Figures 5.6(b), (c)). Four metal contacts I_+, I_-, V_+, and V_- are made, two for injecting current and two for detecting voltage.

5.4.5 Four-Wire Measurements

Conventionally, the EMR devices are characterized by placing the current electrodes at the two edges and the voltage electrodes between them in an I–V–V–I arrangement. It has been shown by finite-element modeling that I–V–I–V resistance measurement will provide higher sensitivity (Boone et al. 2007). So, in some diagrams, the I–V–I–V arrangement is displayed.

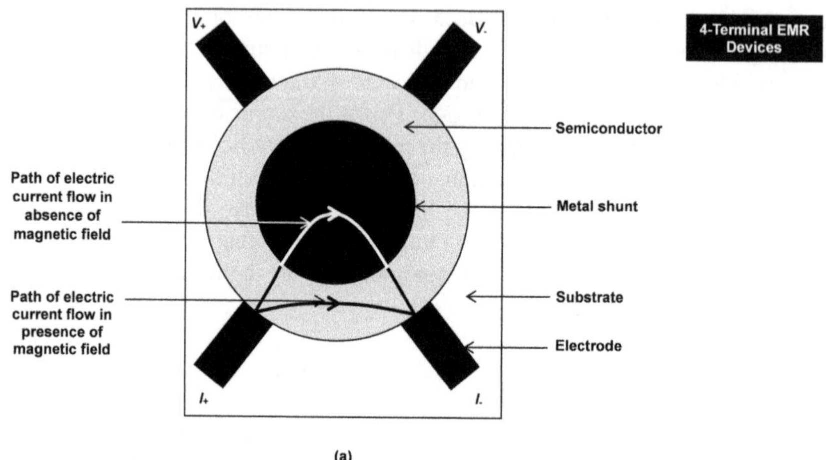

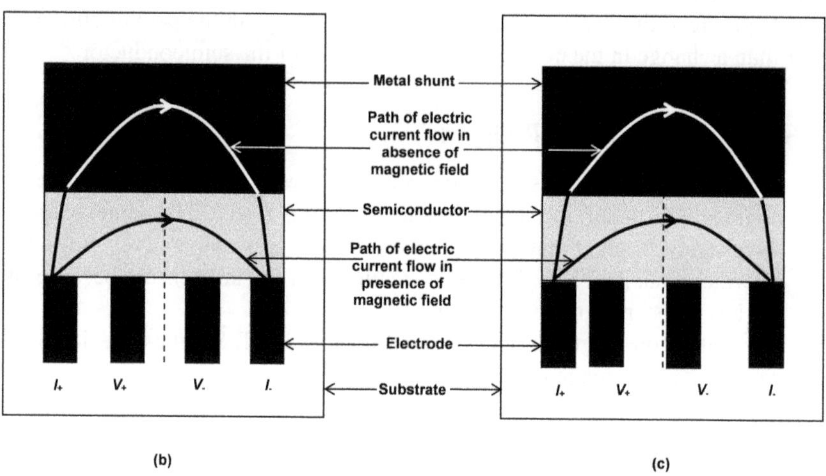

FIGURE 5.6 Traditional four-terminal EMR devices: (a) van der Pauw geometry and (b) and (c) bar geometries: (b) symmetric geometry and (c) asymmetric geometry. Part (a) shows a circular disc of a semiconductor material formed on an insulating substrate. Inside the semiconductor disc is a concentrically placed metallic disc of a smaller diameter. It is called the metallic shunt. There are four electrodes at the four corners of the outer semiconductor disc. The electrodes I_+, I_- are used for injecting current while the electrodes V_+, V_- are voltage probes. Part (b) shows a rectangular strip of semiconductor material on a substrate. Above the semiconductor strip is a metallic strip labelled as the metallic shunt. Also, there are four electrodes I_+, I_-, V_+, and V_- performing identical tasks to the electrodes in part (a). The dashed line dissects the rectangular strip into two equal halves. Distances of electrodes I_+, V_+ from the dashed line are equal to the distances of electrodes I_-, V_- from this line, thus constituting a symmetric placement. Part (c) is similar to part (b) with the only difference that the distance of electrode V_+ from the dashed line is much larger than the distance of electrode V_- from this line rendering this electrode placement asymmetric with respect to the central dashed line. In all parts (a), (b), and (c), the paths of electric current flow are shown without and with applied magnetic field.

5.5 GRAPHENE EMR DEVICES

5.5.1 van der Pauw Geometry Graphene Magnetoresistance Device

5.5.1.1 van der Pauw Geometry

Figure 5.7 illustrates the van der Pauw magnetoresistance sensor made on an SiO_2/ Si substrate. The silicon back gate is biased to control the characteristics of the sensor. An I–V–V–I measurement circuit is shown. In this circuit, the current is injected from two neighboring electrodes I_+, I_- and the voltage is measured across two adjoining electrodes V_+, V_-. The radius of the metal shunt is denoted by r_a, and that of the semiconductor disc by r_b.

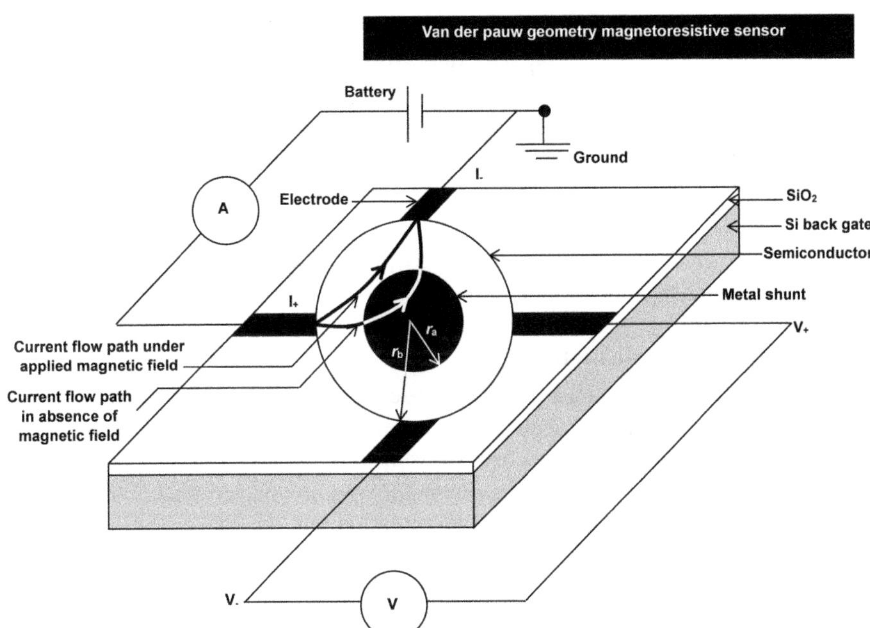

FIGURE 5.7 Magnetoresistance sensor made in van der Pauw geometry. The diagram shows an SiO_2/Si substrate on which is made a circular semiconductor disc of radius r_b. Inside this disc, there is another circular disc of smaller radius r_a. It is the metallic shunt. Four electrodes I_+, I_-, V_+, and V_- are made at angles of 0°, 90°, 180°, and 270° around the outer semiconductor disc. Electrode I_+ is connected to an ammeter. The opposite terminal of the ammeter is joined to the positive terminal of the battery. The negative terminal of the battery is joined to I_- electrode and is also grounded. A voltmeter is connected between electrodes V_+, V_-. Trajectories of current flow under different conditions are shown. When there is no magnetic field, electric current flows from the electrode I_+ through the semiconductor, then metal, again semiconductor and returns to electrode I_-. When a magnetic field is applied in a direction out of the plane of the EMR device, the path of current changes. It flows from I_+, through the semiconductor, and back to I_- bypassing the metal shunt. Also shown is a silicon back gate for controlling the carrier density in the semiconductor disc.

If I is the injected current, a resistance R is defined as

$$R = \frac{V_+ - V_-}{I} \tag{5.5}$$

If B is the magnetic field, the magnetoresistance MR of the sensor is

$$MR = \frac{R(B) - R(B=0)}{R(B=0)} \tag{5.6}$$

The sensitivity S of the device is given by

$$S = \frac{\text{Response of output voltage } V \text{(in volts)}}{\text{normalized by the injected current } I \text{(in amperes)}} \Big/ \text{Magnetic field } B \text{(in Tesla)} \tag{5.7}$$

and is stated in units of $VA^{-1}T^{-1}$ or $\Omega\ T^{-1}$. Four-probe measurements offer two advantages:

(i) The errors arising from contact resistances are removed.
(ii) Errors from resistances of arms connecting the contact pads with graphene film are eliminated.

5.5.1.2 Device Fabrication

A graphene magnetoresistance device has been realized in van der Pauw geometry (Lu et al. 2011). For the fabrication of the device, the graphene film is exfoliated from natural graphite and placed over the substrate (Figure 5.8). Raman spectroscopy is applied to confirm that the graphene film is a monolayer thick. The graphene film is shaped by dry etching in an oxygen plasma into the shape of van der Pauw geometry. The pattern for the concentric metal shunt as well as the metal contacts is defined in the photoresist by electron beam lithography. Then 3 nm Ti/60 nm Au metal layers or Pd layers are formed by electron beam evaporation. The photoresist is lifted off leaving behind the defined metallic structures.

5.5.1.3 Effect of Magnetic Field on Resistance

For a back gate voltage of -8 V on a device with a r_a/r_b ratio of 3/4, the magnetoresistance is very small for magnetic fields of ±0.8T and increases up to 550,000% at 9T (Lu et al. 2011).

5.5.2 Bar-Shaped Back-Gate Tunable Graphene Magnetometer

5.5.2.1 Construction and Operation of the Sensor

The generic construction of the bar-shaped sensor is shown in Figure 5.9 with the electrical connections and I–V–I–V measurement circuit. The sensor is made on an SiO_2/Si substrate. It has a Si back gate to which a voltage $V_{\text{Back gate}}$ is applied to control the carrier concentration and hence conductivity in the semiconductor.

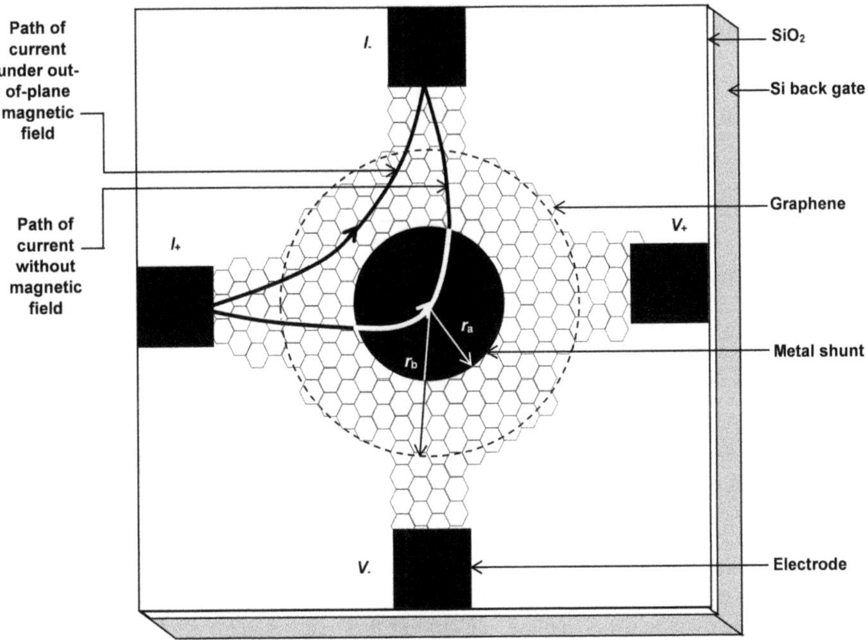

FIGURE 5.8 Graphene magnetoresistance device in van der Pauw geometry. On a SiO$_2$/Si substrate, there is a circular-shaped graphene film of radius r_b with graphene projections in four directions mutually orthogonal to each other. At the edges of the graphene, projections are metal pads named as I_+, I_-, V_+, and V_-. A concentric metal disc of radius $r_a < r_b$ is seen within the graphene film. It is the metal shunt. Without any magnetic field, the current flow path is from I_+, through semiconductor and metal shunt, then again semiconductor and back to I_-. In an applied magnetic field, the current flows along the path I_+/semiconductor/I_-. It does not flow through the metal shunt. The silicon back gate allows modulation of carrier density in the semiconductor disc.

The device has a silicon back gate. The back gate voltage $V_{\text{Back gate}}$ is varied to tune the device characteristics. By such tuning, the effects of device variability are overcome. This variability is concerned with the variations from device to device owing to differences in material properties as well as process limitations. The gate voltage is adjusted to set the device at the required operating condition.

5.5.2.2 Magnetoresistance Formula

The biasing voltage V_b and the biasing current I_b are applied at the terminal I_+ while the terminal I_- is kept at the ground potential. By measuring the potentials at the V_+ and V_- leads, the non-local voltage $V_{\text{Difference}}$ is determined in the absence of magnetic field as:

$$V_{\text{Difference}}\left(B = 0\right) = V_+ - V_- \tag{5.8}$$

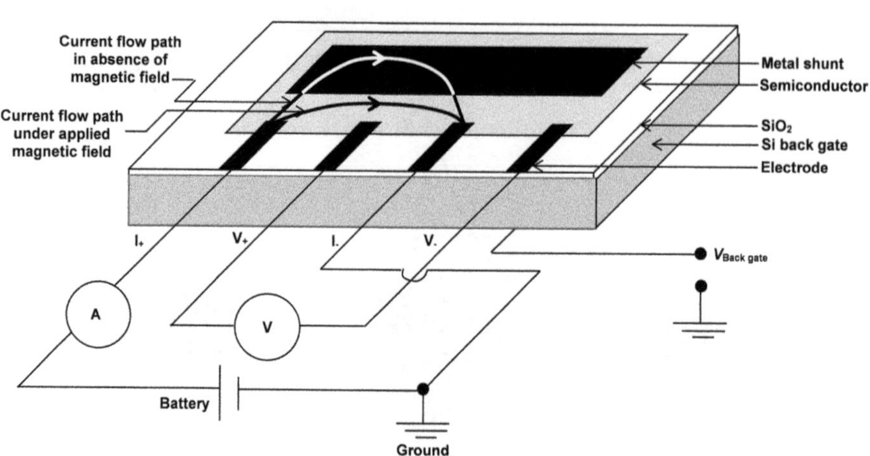

FIGURE 5.9 Magnetoresistance device in bar configuration. A rectangular strip of semiconductor film can be seen on an SiO$_2$/Si substrate. The Si substrate is used as a back gate. Inside the semiconductor strip, there is a rectangular metallic strip acting as the metal shunt. The electrode I_+ is connected to an ammeter. The opposite terminal of ammeter is joined to the positive pole of the battery whose negative pole is joined to I_- electrode. The negative pole of the battery is grounded. At $B = 0$, the current follows the path: I_+, semiconductor, metal, semiconductor, I_-. When B is applied, the current flow path changes to: I_+, semiconductor, I_- avoiding the metal shunt. The silicon back gate is used for obtaining the required carrier concentration in the semiconductor strip.

Then the non-local resistance without magnetic field is

$$R_{\text{Non-local}}\left(B=0\right) = \frac{V_{\text{Difference}}}{I_b} \tag{5.9}$$

and the two-terminal resistance without magnetic field is found through the current leads I_+, I_- is

$$R_{\text{Two-terminal}}\left(B=0\right) = \frac{V_b}{I_b} \tag{5.10}$$

The device is placed in the external magnetic field B. As done in the experiment without magnetic field, the non-local resistance in the presence of magnetic field is found as

$$R_{\text{Non-local}}\left(B\right) = \left| \frac{V_{\text{Difference}}}{I_b} \right|_B \tag{5.11}$$

The non-local magnetoresistance *MR* is extracted from these measurements using the formula:

$$MR = \frac{R_{\text{Non-local}}\left(B\right) - R_{\text{Non-local}}\left(B = 0\right)}{R_{\text{Two-terminal}}\left(B = 0\right)}$$
(5.12)

5.5.2.3 Fabrication of Bar-Geometry Sensor

A nanoscale graphene magnetic field sensor is fabricated in the bar geometry (Pisana et al. 2010). The substrate used for the fabrication of EMR sensor comprises a 300-nm thick silicon dioxide film grown over a silicon substrate (Figure 5.10). The silicon substrate is N-doped. Single-layer and bilayer graphene flakes are obtained by cleaving highly oriented pyrolytic graphite. They are identified by Raman spectroscopy. For making the leads, PMMA is coated over the substrate with graphene film. Patterns for leads are defined in PMMA by electron beam lithography. Then 2.5 nm Ta/20 nm Au film is deposited over the pattern. Upon lifting off the PMMA, the metal is retained in the regions not covered with PMMA.

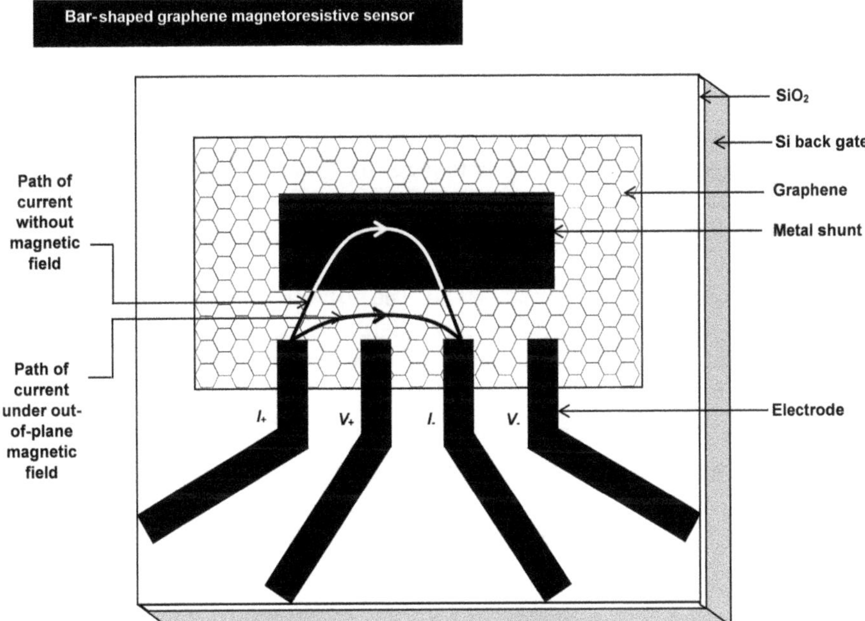

FIGURE 5.10 Graphene magnetoresistance device in the shape of a bar. On an SiO₂/Si substrate, there is a rectangular strip of graphene film with a rectangular metal strip inside to act as a metal shunt. The graphene film has four electrodes. In the absence of magnetic field, the current flows from I_+ to graphene, then returns via the metal shunt through graphene to *I-*. On applying a magnetic field, the current flows from I_+ through graphene to I_- without passing through the metal shunt. The Si back gate helps in achieving the desired carrier density in graphene film.

5.5.2.4 Magnetic Field Sensitivity

The sensitivity of EMR device is the slope of non-local voltage difference $V_{\text{Difference}}$ over a magnetic field span of ± 0.5 T. It is normalized by current and expressed in the units of $VA^{-1}T^{-1}$. The sensitivity depends on the back gate voltage. At a $V_{\text{Back gate}}$ of -30 V, the sensitivity is $110\ VA^{-1}T^{-1}$. Its value is $-250\ VA^{-1}T^{-1}$ at $V_{\text{Back gate}} = +5$ V. It changes to $-120\ VA^{-1}T^{-1}$ at $+30$ V back gate voltage. The transforming values show that the device is tunable over a broad range of back gate voltages. This graphene extraordinary magnetoresistance device consolidates the Hall effect with enhanced geometric magnetoresistance (Pisana et al. 2010).

5.6 MAGNETIC QUANTUM SENSORS

Quantum sensors are small devices using extremely small amounts of energy and matter to measure infinitesimally small changes in environmental parameters by applying quantum mechanical phenomena. Magnetic fields can be measured by quantum sensors through the Zeeman effect.

5.6.1 ZEEMAN EFFECT

Atomic spectra consist of lines called spectral lines. These lines represent transitions of electrons between different energy levels. A spectral line is a bright or dark line of a single frequency or wavelength. It is emitted (or absorbed) when an electron changes its energy level from a higher to lower (or lower to higher) energy level.

Zeeman effect or Zeeman spectral splitting is the magneto-optical phenomenon concerned with the splitting of spectral lines of atomic spectra into several components of different frequencies when the source of light is placed in an external magnetic field (van den Bosc 1957, Birtwistle 2009). The distance between these components is a signature of the presence of magnetic field and its intensity. So, the Zeeman effect is a powerful tool for investigation and measurement of a magnetic field.

5.6.1.1 Normal Zeeman Effect

The electron has an angular momentum due to its orbital motion. This motion imparts a magnetic dipole moment to the electron. Since the electron has a magnetic dipole moment, it will interact with an external magnetic field. Visualizing the dipole as a current loop of area A carrying a current I, the interaction energy of a magnetic dipole of moment $\mu = IA$ with an external magnetic field B depends on its orientation θ with the field and is given by the scalar product

$$U(\theta) = -\mu.\mathbf{B} = \mu B \cos\theta = -IAB\cos\theta \qquad (5.13)$$

where θ is the angle between the magnetic field B and magnetic dipole moment μ. The interaction energy due to the electron's orbital angular momentum is given by

$$U = m_l \mu_B B \qquad (5.14)$$

where m_l is the magnetic quantum number determining the spatial orientation of an orbital of given energy specified by the principal quantum number n and shape specified by azimuthal quantum number l, and μ_B is the Bohr magneton:

$$\mu_B = \frac{e\hbar}{2m_e} \tag{5.15}$$

e is the electronic charge, m_e is the electron mass, and $\hbar$ is reduced Planck's constant. For any value of l, m_l varies from $-l$ to $+l$ so that it has $2l+1$ values.

Consider the emission spectrum of hydrogen atom for the transition from energy level E_2 with quantum numbers $n = 2$, $l = 1$, $m_l = -1$, 0, +1 to energy level E_1 with quantum numbers: $n = 1$, $l = 0$, $m_l = 0$ (Ling 2022). When the magnetic field is off, $B = 0$, energy of the electron in state $n = 2$ has the energy E_2 irrespective of the value of l or m_l. But when the magnetic field is switched on, $B \neq 0$, the energy level $n = 2$ is split into three sublevels:

(i) $n = 2,\ l = 1,\ m_l = 0, E = E_2$
It is the original energy level because an electron in the $l = 0$ state has zero magnetic moment and therefore is not affected by a magnetic field.

(ii) $n = 2,\ l = 1,\ m_l = -1, E = E_2 - \mu_B B$
It is an energy level below E_2 by $\mu_B B$

(iii) $n = 2,\ l = 1,\ m_l = +1, E = E_2 + \mu_B B$
It is an energy level above E_2 by $\mu_B B$

Thus, three spectral lines are seen when the magnetic field is switched on against a single line when it was switched off. The separation ΔE of the new lines with respect to the original line E_2 is proportional to the magnetic field B for fields < several tenths of 1 Tesla. It results from electron transitions between singlet states, the states in which all electrons are paired and the spin quantum number $m_s = 0$. A singlet spectral line viewed perpendicular to the field is decomposed into three components. The central un-shifted component with the electric vector vibrating parallel to the field is called the π-component. The other two components equally spaced on opposite sides of the central component and with the electric vector vibrating perpendicular to the field are known as the σ components.

This splitting of spectral lines in a magnetic field into three sublevels shown in Figure 5.11 is traditionally called the normal Zeeman effect. Here only orbital angular momentum is taken into account.

5.6.1.2 Anomalous Zeeman Effect

It is the Zeeman effect taking into consideration that electrons have an intrinsic spin. It occurs when the total spin of the initial state, the final state, or both states is nonzero. In this case, the interaction energy of the electron magnetic dipole moment with the external magnetic field is modified to

$$U = \left(m_l + g m_s\right)\mu_B B \tag{5.16}$$

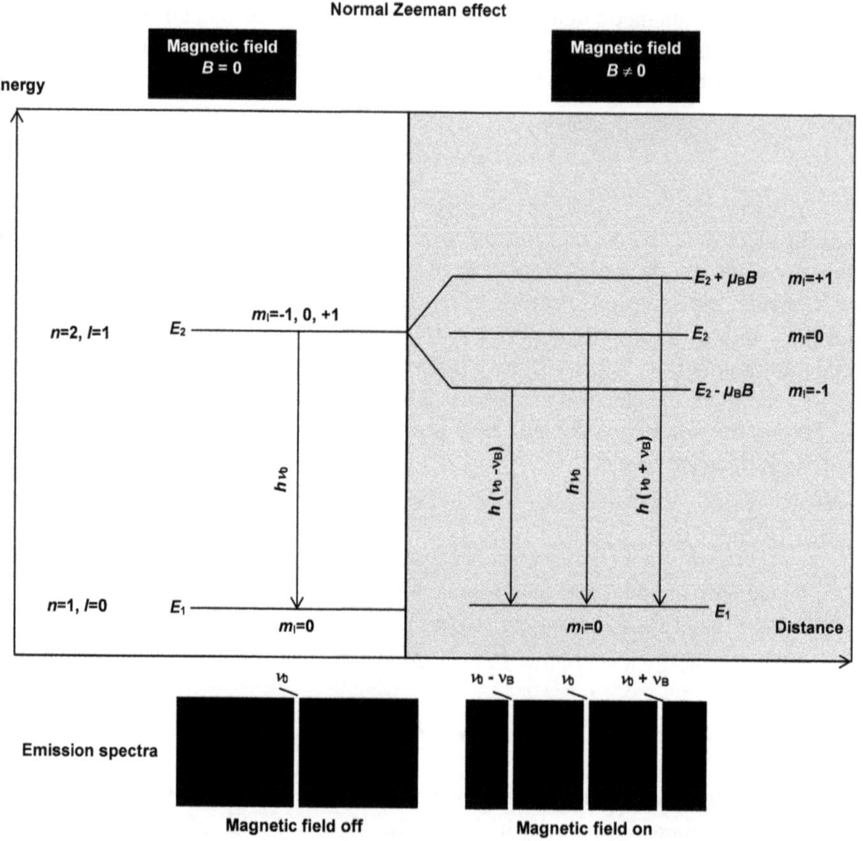

FIGURE 5.11 Normal Zeeman effect. Top left: energy level diagram when $B = 0$ showing energy level E_2 with quantum numbers $n = 2$, $l = 1$, $m_l = -1$, 0, +1; energy level E_1 with quantum numbers: $n = 1$, $l = 0$, $m_l = 0$; and electron falling from energy level E_2 to energy level E_1 losing energy $h\nu_0$. Bottom left: spectrum containing one bright line at frequency ν_0. Top right: energy level diagram when $B \neq 0$ showing energy level $E_2 + \mu_B B$ with quantum numbers $n = 2$, $l = 1$, $m_l = +1$; energy level E_2 with quantum numbers $n = 2$, $l = 1$, $m_l = 0$; energy level $E_2 - \mu_B B$ with quantum numbers $n = 2$, $l = 1$, $m_l = -1$; electron falling from energy level $E_2 + \mu_B B$ to energy level E_1 by losing energy $h\nu_0 + \mu_B B$; electron falling from energy level E_2 to energy level E_1 by losing energy $h\nu_0$; and electron falling from energy level $E_2 - \mu_B B$ to energy level E_1 by losing energy $h\nu_0 - \mu_B B$. Bottom right: spectrum containing three bright lines at frequencies $\nu_0 - \nu_B$, ν_0, $\nu_0 + \nu_B$.

where m_s is the spin quantum number; $m_s = +\frac{1}{2}$ or $-\frac{1}{2}$, and the electron g-factor $= 2$. The energy levels of hydrogen atom for $n = 2$, $l = 1$ state are:

(i) $n = 2, l = 1, m_l = -1, m_s = -\frac{1}{2}$,

$E_{2,1,-1,-1/2} = E_2 - \left(-1 + 2 \times -1/2\right)\mu_B B = E_2 + 2\mu_B B$

(ii) $n = 2,\ l = 1,\ m_l = -1,\ m_s = +\tfrac{1}{2},$

$$E_{2,1,-1,+1/2} = E_2 - \left(-1 + 2 \times 1/2\right)\mu_B B = E_2 - 0 \times \mu_B B = E_2$$

(iii) $n = 2,\ l = 1,\ m_l = 0,\ m_s = -\tfrac{1}{2},\ E_{2,1,0,-1/2} = E_2 - \left(0 + 2 \times -1/2\right)\mu_B B = E_2 + \mu_B B$

(iv) $n = 2,\ l = 1,\ m_l = 0,\ m_s = +\tfrac{1}{2},\ E_{2,1,0,+1/2} = E_2 - \left(0 + 2 \times 1/2\right)\mu_B B = E_2 - \mu_B B$

(v) $n = 2,\ l = 1,\ m_l = +1,\ m_s = -\tfrac{1}{2},$

$$E_{2,1,+1,-1/2} = E_2 - \left(+1 + 2 \times -1/2\right)\mu_B B = E_2 - 0 \times \mu_B B = E_2$$

(vi) $n = 2,\ l = 1,\ m_l = +1,\ m_s = +\tfrac{1}{2},$

$$E_{2,1,+1,+1/2} = E_2 - \left(+1 + 2 \times 1/2\right)\mu_B B = E_2 - 2\mu_B B$$

States (ii) and (v) have the same energy E_2. So, six possible combinations of three m_l values and two m_s values lead to five distinct energy levels shown in Figure 5.12 under the influence of a magnetic field. Actually, the energy levels of $n = 2, l = 1$ state resolve into three sublevels. Each of these three sublevels is further separated into two sublevels leading to six sublevels. But two sublevels coincide in energy. Hence, there are altogether five distinct sublevels only.

5.7   ZEEMAN EFFECT-BASED DIAMOND QUANTUM SENSOR

5.7.1   PRINCIPLE OF THE DIAMOND QUANTUM SENSOR

A quantum sensor utilizing the Zeeman effect will operate by observing the splitting of energy levels of electrons in an unknown magnetic field and extract the field value from the extent of this splitting. But making a sensor with a large vacuum system and big lasers is not practicable. So, attempts are made to miniaturize the device. Nitrogen vacancy (NV) centers in CVD diamond are utilized for precision measurements of magnetic fields with sub-picotesla resolution by illuminating it with a laser and monitoring its fluorescence.

A laser beam strikes the NV centers in diamond (Wojciechowski et al. 2019, Hruby et al. 2022). The incident beam uplifts electrons from the ground state to an excited state in the NV center. When these excited electrons return to the ground state, red light is emitted. The optical transitions between <0> and <±1> spin sublevels of the ground state of NV center differ in brightness. By applying a microwave field with resonant frequency, the electrons are driven from <0> to <±1> spin sublevels leading to a decrease in luminous intensity. In the presence of a magnetic field, splitting occurs between <±1> spin sublevels of NV center due to the Zeeman effect. Accordingly, there is a splitting between microwave resonant frequencies. Thus, the magnetic field is measured by optical detection of magnetic resonance (ODMR) described in the section below.

The NV center in diamond can be used at room temperature under ambient pressure showing high sensitivity up to $60\ \text{pT}(\sqrt{s})^{-1}$. A diamond crystal with an NV center embedded close to its surface can be mounted on the scanning tip of an atomic force microscope (AFM) and used as a rugged and stable device for magnetic field mapping.

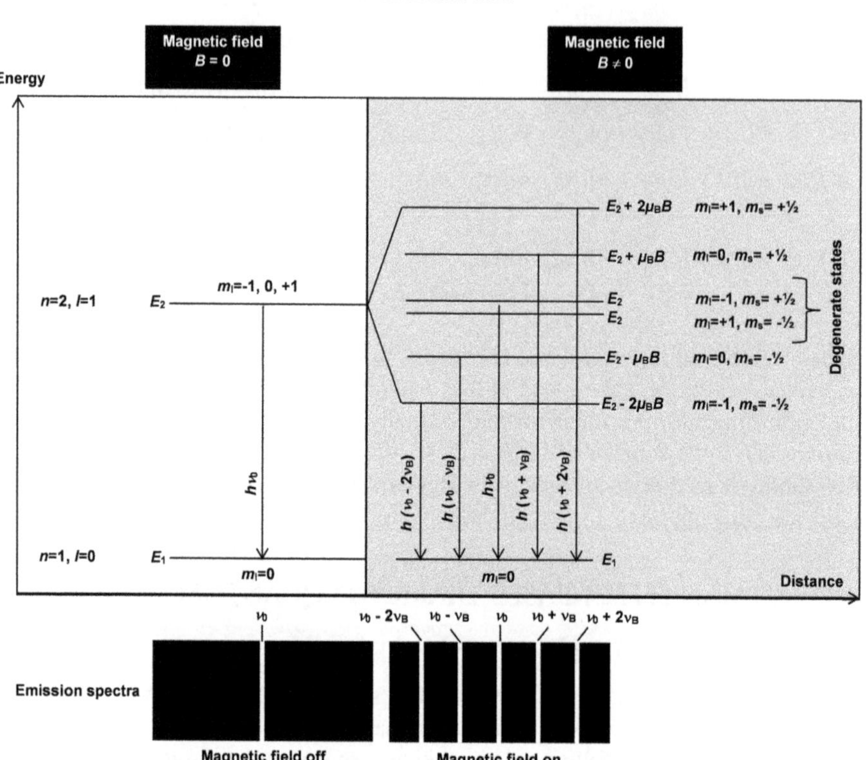

FIGURE 5.12 Anomalous Zeeman effect. Top left: energy-level diagram when $B = 0$ showing energy level E_2 with quantum numbers $n = 2$, $l = 1$, $m_l = -1$, 0, +1; energy level E_1 with quantum numbers: $n = 1$, $l = 0$, $m_l = 0$; and an electron falling from energy level E_2 to energy level E_1 by losing energy $h\nu_0$. Bottom left: spectrum containing one bright line at frequency ν_0. Top right: energy-level diagram when $B \neq 0$ showing the energy level $E_2 + 2\mu_B B$ with quantum numbers $n = 2$, $l = 1$, $m_l = +1$, $m_s = +\frac{1}{2}$; the energy level $E_2 + \mu_B B$ with quantum numbers $n = 2$, $l = 1$, $m_l = 0$, $m_s = +\frac{1}{2}$; the energy level E_2 with quantum numbers $n = 2$, $l = 1$, $m_l = -1$, $m_s = +\frac{1}{2}$; energy level E_2 with quantum numbers $n = 2$, $l = 1$, $m_l = +1$, $m_s = -\frac{1}{2}$; the energy level $E_2 - \mu_B B$ with quantum numbers $n = 2$, $l = 1$, $m_l = 0$, $m_s = -\frac{1}{2}$; the energy level $E_2 - 2\mu_B B$ with quantum numbers $n = 2$, $l = 1$, $m_l = -1$, $m_s = -\frac{1}{2}$; an electron falling from energy level $E_2 + 2\mu_B B$ to energy level E_1 by losing energy $h\nu_0 + 2\mu_B B$; an electron falling from energy level $E_2 + \mu_B B$ to energy level E_1 by losing energy $h\nu_0 + \mu_B B$; an electron falling from energy level E_2 to energy level E_1 by losing energy $h\nu_0$; an electron falling from energy level $E_2 - \mu_B B$ to energy level E_1 by losing energy $h\nu_0 - \mu_B B$; and an electron falling from energy level $E_2 - 2\mu_B B$ to energy level E_1 by losing energy $h\nu_0 - 2\mu_B B$. The two E_2 states having the same energy are degenerate states. Bottom right: spectrum containing five bright lines at frequencies $\nu_0 - 2\nu_B$, $\nu_0 - \nu_B$, ν_0, $\nu_0 + \nu_B$, $\nu_0 + 2\nu_B$.

5.7.2 Limitations of the NV Center Diamond Sensor, and the Need for 2D Material Sensor

The diamond sensor faces shortcomings arising from the 3D nature of the host diamond crystal in two ways:

(i) Hindrance to sensitivity from inability to place the sensor in close vicinity of the target sample.

(ii) Inability to produce ultrathin and flexible diamond layers to establish intimate contact of sensor with probed specimens and integration with them.

An obvious strategy is to take the help of spin defects in a 2D material obtained by exfoliation to the monolayer stage because these spin defects remain photostable in nanosheets of 2D materials of monolayer thickness. The spin defects in 2D materials serve as a means of overcoming the deficiencies of 3D materials.

5.8 QUANTUM SENSOR USING NEGATIVELY CHARGED BORON VACANCIES IN hBN

5.8.1 V_B^- Defects

Negatively charged boron vacancies in hBN consist of a missing boron atom (Figure 5.13(b)) which is replaced by an extra electron in the hBN crystal lattice (Figure 5.13(a)). There are three equivalent nitrogen atoms surrounding the missing boron atom in the hBN lattice.

The negatively charged boron vacancies are denoted by the symbol V_B^-. These vacancies show similar magneto-optical properties as the NV centers in diamond. Over and above, they endow better flexibility and versatility. Besides they are easy to use by permitting initialization, manipulation, and readout at room temperature. They also offer the advantage of being placed in contiguity with the specimen to be studied. The nanoscale proximity ensures high-resolution sensing (Gottscholl et al. 2021).

A V_B^- center shows a triplet ground state (Gottscholl et al. 2020). Triplet is the quantum state of a system having two unpaired electrons (Turro 1969). Its spin quantum number is 1. It allows three values of angular momentum. These values are −1, 0, and +1. So, the spectral lines for this state are split into three lines showing threefold splitting.

5.8.2 V_B^- Defect Generation Methods

These defects have been generated by various techniques. The thoroughgoing experiments done and the extensive characterization of defects by various workers are summarized below.

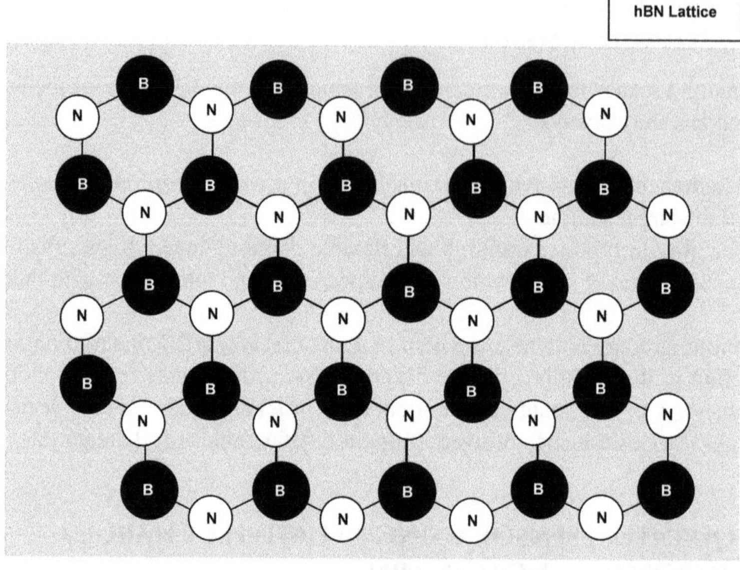

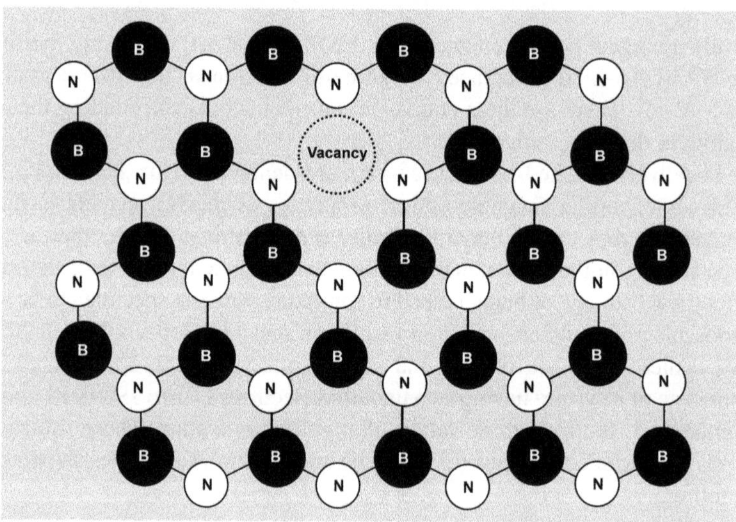

FIGURE 5.13 hBN lattice: (a) without negatively charged boron vacancy and (b) with negatively charged boron vacancy. The diagram (a) shows one layer of hexagons in which boron and nitrogen atoms are placed alternately along the six corners of each hexagon. The boron and nitrogen atoms are bound together by covalent bonds. The diagram (b) shows a similar structure but one boron atom is missing from its regular position. Also missing are the bonds of the absent boron atom with adjoining nitrogen atoms. Note that the different layers of 2D hBN containing covalently bonded nitrogen and boron atoms are attached to each other through van der Waals bonds.

5.8.2.1 Neutron Irradiation

hBN in the form of an ultrafine powder with particle size ~70 nm is irradiated with neutrons in a nuclear reactor in cadmium capsules, which block thermal neutrons and allow entry of energetic neutrons only (Toledo et al. 2018). The irradiation is done for different times from 2 h to 16 h obtaining integrated dose from 3×10^{17} to 2.3×10^{18} n-cm^{-2}. The color of the samples changes to pink accompanied by a near-infrared photoluminescence band centered at 820 nm. Further, the color center induced by irradiation is associated with a spin of 1. The 820 nm-centered band is characteristic of V_B^- centers.

Although neutron irradiation is a primary technique for boron vacancy generation, it needs to be carried out in a nuclear reactor. So, it is inconvenient and expensive.

5.8.2.2 High-Energy Electron Irradiation

V_B^- defects are produced in hBN by irradiation with high-energy electrons (Murzakhanov et al. 2021). $1 \times 1 \times 0.15$ mm^3 hBN single crystals are irradiated with electrons of energy 2 MeV to a net dose of 6×10^{18}cm^{-2} at room temperature. The samples are not annealed. The near-IR photoluminescence band at a maximum wavelength of 810 nm along with spin-Hamiltonian parameters from ESR spectroscopy studies indicate the creation of V_B^- centers in the hBN lattice.

5.8.2.3 Focused Ion Beam

hbN flakes of 100–200 nm thickness, tape-exfoliated from hBN crystals and transferred over SiO$_2$/Si substrates, are exposed to Xe$^+$ ion beam of 30 keV energy in a focused ion beam (FIB) system at fluences of 1×10^{14}-1×10^{17} ions-cm^{-2} (Kianinia et al. 2020).

A confocal scan of the implanted regions under 532 nm laser excitation shows photoluminescence from the regions centered around 820 nm, which is a property of V_B^- centers.

Optically detected magnetic resonance (ODMR) measurements are performed by using a copper wire suspended near the sample as an antenna to deliver microwave energy to the sample under 532 nm laser excitation without any external magnetic field. Photoluminescence monitoring in the 720–900 nm wavelength range reveals that the fluorescence decreases at microwave frequencies ~3.41 and 3.51 GHz, in consistency with spin-optical dynamics of V_B^- defects.

Easy control and positioning capability of FIB allow patterning of V_B^- defect arrays. However, the choice of ion sources is limited since most commercial FIB equipment use Ga ion source. Other ions are infrequently used.

5.8.2.4 Ion Implantation

From an hBN sample of lateral size 1 mm, monocrystalline hBN is exfoliated using tape (Guo et al. 2022). The flakes of thickness 10–100 nm obtained by exfoliation are transferred to a silicon substrate. They are loaded in an ion implanter. The ion implantation is done with large-area–parallelized beams of ions having 30 keV energy at a fluence of 1×10^{14} ions cm^{-2}. Different ion sources are used, e.g., Ar$^+$, He$^+$, C$^+$, N$^+$. The Ar$^+$ ions produce shallow defects while He$^+$ ions yield deeper defects. By the impact of high-energy ions, the B–N bonds are broken, dislodging boron atoms from lattice sites. Intense photoluminescence is observed in ion implanted samples in the range 700–1000 nm centered at 820 nm.

Commercial availability of ion implanter makes the method convenient although expensive. Moreover, much lower energy (30 keV) is required for ion implantation than used in electron irradiation method (2 MeV). Furthermore, various ion species are available while the same choice is not offered by electron irradiation or focused ion beam technique.

5.8.2.5 Femtosecond Laser Writing

Tape-exfoliated hBN flakes are mounted on a glass coverslip. The laser source used is Ti: Sapphire femtosecond laser of wavelength 800 nm, pulse duration of 50 fs, pulse repetition rate of 1 kHz, and maximum pulse energy of 13 mJ (Gao et al. 2021b). The number of pulses is varied by an optical shutter. First, the hBN specimen is placed at the laser focus and its surface is oriented perpendicular to the direction of laser beam. For defect generation, the specimen is shifted 4 μm away from the focal spot. Here the beam waist at the surface of hBN is 3 μm.

In confocal imaging study, emission spectra are observed to be centered at 820 nm. Microwave energy is supplied to the laser-irradiated area through a waveguide. On excitation with a 532 nm laser without applying any external magnetic field, and counting the photons while sweeping the microwave frequency from 3.15 to 3.75 GHz, the photon counts fall at two frequencies, 3.42 and 3.52 GHz, as usually happens in the case of V_B^- defects.

5.8.3 Energy States of V_B^- Defects and Their Features

The energy-level structure of V_B^- contains four states (Liu et al. 2022):

  (i)   a triplet ground state
 (ii)   a triplet excited state, and
(iii)   two singlet intermediate states

Figure 5.14 shows the simplified diagram of energy levels of V_B^-. States (i) and (ii) contain three spin states, two of which are $|m_s = \pm 1\rangle$ having parallel electron spins: up for +1 and down for −1. The third is the $|m_s = 0\rangle$ state with antiparallel electron spins. The $|m_s = \pm 1\rangle$ states are at a higher energy than the $|m_s = 0\rangle$ state.

On applying a magnetic field in a direction perpendicular to the 2D plane of hBN and parallel to its hexagonal c-axis, the $|m_s = \pm 1\rangle$ states are split by the Zeeman effect but $|m_s = 0\rangle$ state remains unaffected. In the energy-level diagram, ISC is the intersystem crossing between $S = 1$ and $S = 0$ states. The energy necessary to cause transitions between $|m_s = 0\rangle$ and $|m_s = \pm 1\rangle$ states is supplied by the applied microwave signal which modulates the populations of $|m_s = 0\rangle$ and $|m_s = \pm 1\rangle$ states. This modulation

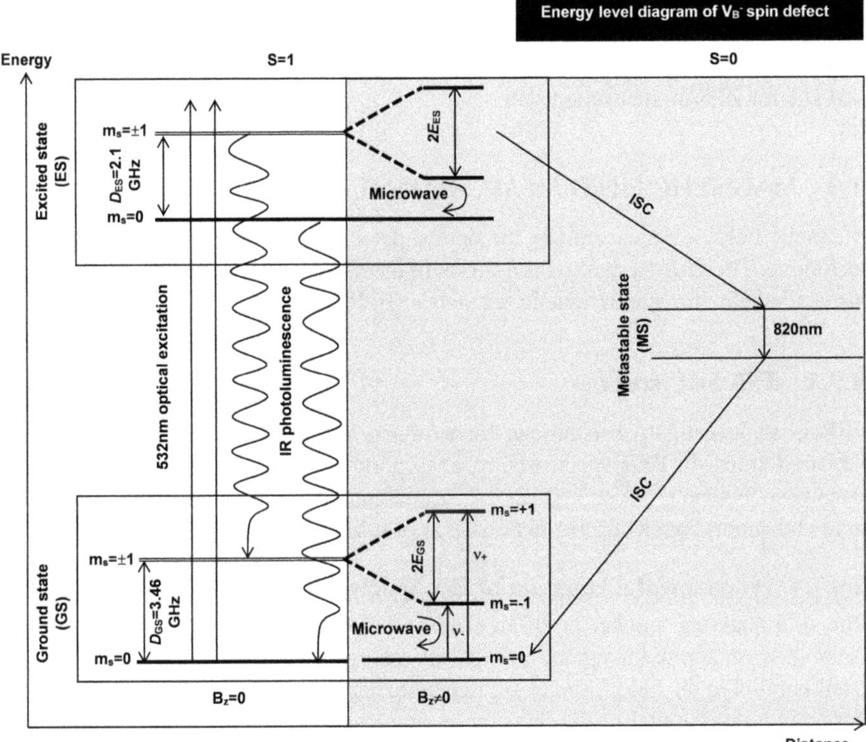

FIGURE 5.14 Energy-level diagram of V_B^- spin defect at $B = 0$ and $B{\neq}0$. Energy of electrons is plotted against distance. Both the ground state (GS) and the excited state (ES) are triplet states having lower energy level with spin quantum number $m_s = 0$ and upper degenerate states at the same energy with $m_s = \pm1$. Besides the ground and excited states, there is a metastable state (MS) through which excited electrons decay by intersystem crossing (ISC) emitting an infrared photon at 820 nm as shown by wavy lines. Electrons are pumped from the ground state to the excited state by illumination of hBN crystal with a 532 nm laser. This transition is shown by upward arrows. Zero-field splitting parameter D_{GS} for the ground state is 3.46 GHz while the zero-field splitting parameter D_{ES} for the excited state is 2.1 GHz (Mathur et al. 2022). When $B{\neq}0$, the degeneracy on $m_s = \pm1$ spin states is lifted, and they are subdivided into separate $m_s = +1$ and $m_s = -1$ sublevels by the Zeeman effect. The magnitude of separation between $m_s = +1$ and $m_s = -1$ sublevels depends on the intensity of magnetic field B. Greater the field B, more the splitting is. The energy difference between the spin sublevels $m_s = +1$ and -1 is $2E_{GS}$ for the ground state and $2E_{ES}$ for the excited state. Transitions take place between $m_s = +1$ and $m_s = -1$ sublevels, as shown by arrows. Microwave field is applied for these transitions. As these microwave-induced transitions occur in the ground state as well as in the excited state, they are shown by writing microwave with circular arrows. On the top of the triplet ground and excited state $S = 1$ is written to show that they have spin 1. On the top of the metastable $S = 0$ is written showing that it is a zero spin state.

changes the fluorescence intensity. The intensity changes are observed when recoding the fluorescence spectra, as elaborated in the section below. D is the zero-field splitting (ZFS) between $m_s = 0$ and $m_s = \pm 1$ states. D_{GS} stands for ZFS in the ground state and D_{ES} for ZFS in the excited state.

5.9 MAGNETIC FIELD MEASUREMENT

Magnetic fields are measured by the optical detection magnetic resonance (ODMR) technique. The ODMR method is a subset of electron resonance (ESR) spectroscopy, also called electron paramagnetic resonance (EPR) spectroscopy.

5.9.1 ESR SPECTROSCOPY

ESR is an investigative technique for studying paramagnetic materials exhibiting unpaired spins. In ESR spectroscopy, absorption spectra are obtained by inducing transitions between the spin states of paramagnetic solids by applying a magnetic field and supplying electromagnetic energy at microwave frequencies.

5.9.1.1 Fundamental Equation of ESR Spectroscopy

The spin quantum number m_s of an electron has both magnitude (½) and direction (+ or -). In an applied magnetic field B, the magnetic moment of the electron aligns itself parallel to the field for $m_s = +½$ magnetic component of the spin quantum number s, and anti-parallel to the field for $m_s = -½$ magnetic component. Each aligned state is associated with a specific energy in accordance with the Zeeman effect. The energy for an aligned state is given by

$$E_{ms} = m_s g_e \mu_B B \tag{5.17}$$

where g_e is the Landé g-factor $= 2.0023$ for a free electron, and μ_B is the Bohr magneton.

So, the energy of the parallel state is

$$E_{+1/2} = +\frac{1}{2} g_e \mu_B B \tag{5.18}$$

Likewise, the energy for antiparallel aligned state is

$$E_{-1/2} = -\frac{1}{2} g_e \mu_B B \tag{5.19}$$

Therefore, the energy difference between the two states is

$$\Delta E = E_{+1/2} - E_{-1/2} == +\frac{1}{2} g_e \mu_B B - \left(-\frac{1}{2} g_e \mu_B B_0 \right) = g_e \mu_B B \tag{5.20}$$

implying that the energy separation between the states is proportional to the intensity of magnetic field B. Hence, the electron spin is altered by absorption or emission of a photon of frequency ν satisfying the equation

$$h\nu = \Delta E = g_e \mu_B B \tag{5.21}$$

where h is Planck's constant. This equation relating to the variation in frequency ν of the photon with magnetic field B forms the basis of magnetic field measurements using ESR.

The energies of the two states $E_{+\frac{1}{2}}$ and $E_{-\frac{1}{2}}$ diverge apart as the magnetic field intensity increases (Figure 5.15). The frequency ν for the energy separation $\Delta E = h\nu$

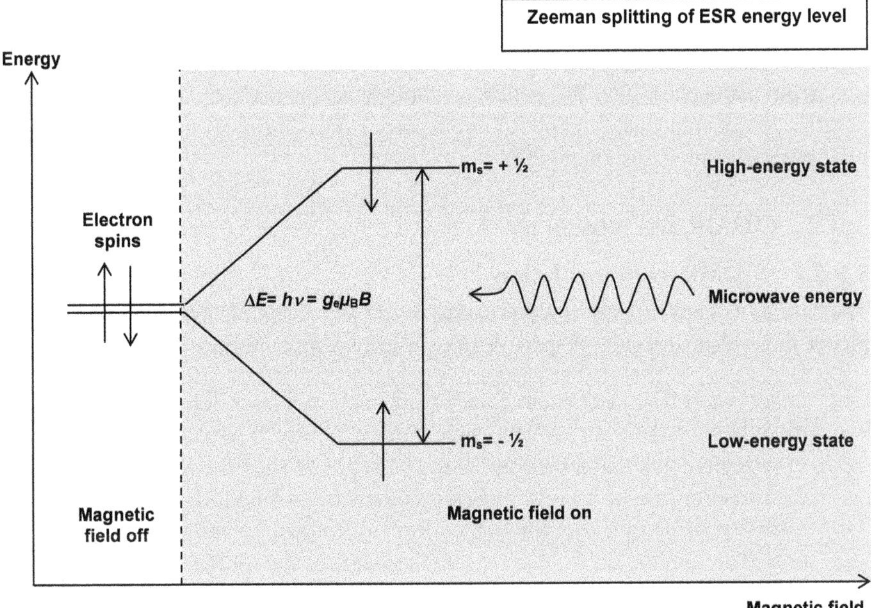

FIGURE 5.15   Zeeman splitting of the ESR energy level under the influence of a magnetic field. Energy of electrons is plotted against distance through the crystal. On the left side of the dashed line, the magnetic field is in the 'off condition'. There are two energy levels of electrons at the same energy. These energy states are said to be in degeneracy. They are degenerate states. The electron spins are indicated by upward- and downward-directed arrows. On the right side of the dashed line, the magnetic field is in the 'on condition'. The energy levels of the two states spread out toward opposite sides as the degeneracy is lifted and the states at the same energy split up into spin sublevels $m_s = +\frac{1}{2}$ and $m_s = -\frac{1}{2}$. The spin $m_s = +\frac{1}{2}$ is the high-energy state and is shown by an upward arrow while the spin $m_s = -\frac{1}{2}$ is the low-energy state and is shown by a downward arrow. The difference in energy between the two states is $\Delta E = h\nu = g_e\mu_B B$. The amount of splitting increases as the magnetic field rises. Transitions are induced from the spin sublevel $m_s = -\frac{1}{2}$ to the sublevel $m_s = +\frac{1}{2}$ by supplying microwave energy.

generally falls in the microwave portion of the electromagnetic spectrum for magnetic field magnitudes of a few tesla.

5.9.1.2 Difficulties Faced in ESR Measurements, and the Need for ODMR Spectroscopy

The ESR using microwaves only is confronted with two major bottlenecks preventing its utilization for practical magnetometry:

(i) Here spin polarization is dependent upon the Boltzmann distribution which means that the sample must be cooled down to very low temperatures otherwise very high magnetic fields are required.

(ii) Microwave detection is more difficult than detection of optical radiation, as easily done with an ordinary silicon photodiode.

These obstacles are removed in a modified variant of ESR, the optically detected magnetic resonance (ODMR), which is a double resonance technique using optical polarization of spin states and photoluminescence measurements in conjunction with microwave driving (Carbonera 2009).

5.9.2 ODMR SPECTROSCOPY

5.9.2.1 ODMR Instrumentation

The sample is kept in a microwave cavity inside a cryostat placed between the pole pieces of an electromagnet (Figure 5.16). It receives three inputs:

(i) *Laser beam*: The laser beam comes from a 532-nm laser. It passes through a modulator, usually an acousto-optic modulator which works on the change in refractive index of a medium in the presence of sound waves. As a result, the medium acts as a grating causing diffraction of light. By changing the intensity of sound, the amount of light diffracted is varied for modulation. The modulated light traverses the path: beam splitter $\rightarrow$ microscope objective $\rightarrow$ sample.

(ii) *Magnetic field*: This comes from an electromagnet mounted on an X–Y stage for changing its position with respect to the sample.

(iii) *Microwave field*: This comes from a microwave generator such as a klystron which produces microwave radiations of desired frequencies. The microwave radiations pass through an amplifier before falling on the sample.

Light emitted by the sample moves along the path: microscope objective$\rightarrow$mirror $\rightarrow$beam splitter. Then it enters a long-pass filter which passes light of wavelength $>$ a cut-off wavelength, and blocks short wavelengths. The next filter is a short-pass filter which allows light of wavelength $<$ cut-off wavelength and stops long wavelengths. The filtered light is incident on the avalanche photodiode and the generated photocurrent signal is fed to the computer.

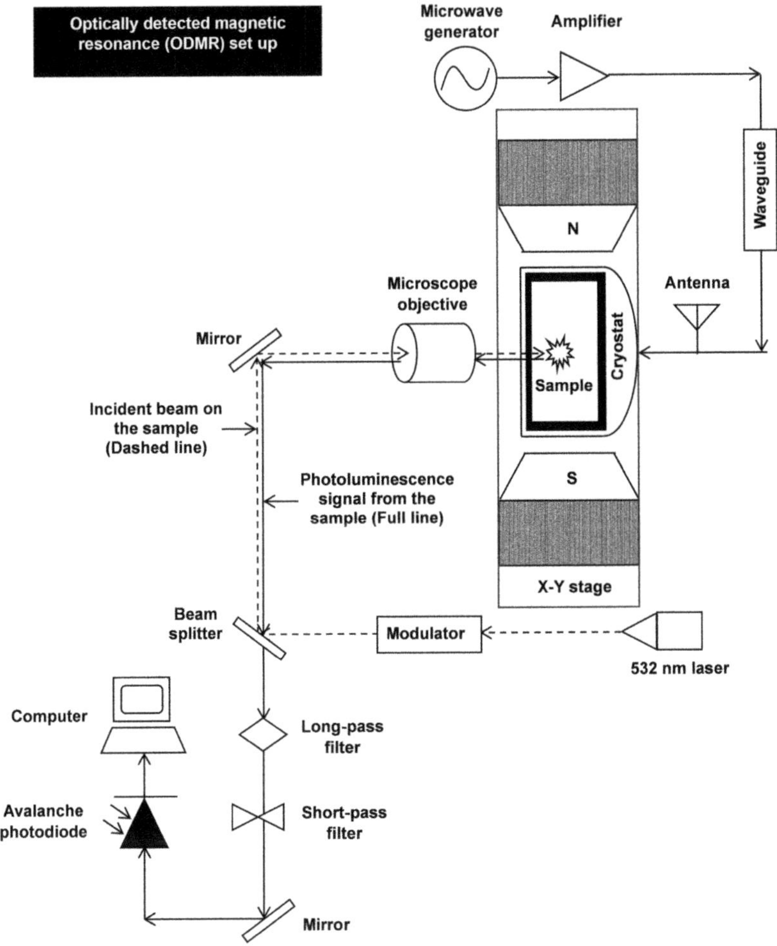

FIGURE 5.16 Block diagram of the setup for optical detection resonance measurements. Microwave signal from a generator is fed to a high-power amplifier. The amplified microwave signal is delivered through a waveguide to an antenna for transmission to the sample kept in a microwave cavity enclosed within a cryostat for cooling the sample to low temperatures. A permanent magnet or electromagnet with north pole N and south pole S is used to apply a magnetic field on the sample. The magnet can be moved on an X–Y stage to bring it close to the sample or moved away from it to control the field strength. A 532-nm laser beam is used for optical pumping. An acousto-optical modulator is used for varying the output power of the laser. The modulated laser beam strikes a beam splitter decomposing the incident beam into a reflected beam and a transmitted beam. The reflected beam falls on a mirror and passes through a microscope objective to irradiate the sample. The photoluminescence signal produced in the sample moves through the microscope objective and is guided by the mirror toward the beam splitter. The light emerging from the beam splitter passes through long-pass and short-pass filters to strike a mirror which reflects it towards the avalanche photodiode. The long-pass and short-pass filters are used to ensure that the signal of correct wavelength, i.e., the photoluminescence signal from the sample, falls on the photodiode so that the response measured is due to this signal only. The personal computer processes the received current signals. It handles, manages, and stores measured data.

5.9.2.2 Recording ODMR Spectra

ODMR spectra acquisition involves three principal operations explained below:

(i) Optical pumping with 532 nm green light from triplet ground state to triplet excited state: when the energy of light is equal to the energy difference between ground and excited states, resonance takes place. Light is absorbed promoting the electrons from the ground state to the excited state.
There are two possibilities (Mathur et al. 2022, Mu et al. 2022):

(a) The spin is rotated from $|m_s = 0\rangle$ to $|m_s = \pm1\rangle$ state in the ground state and then optically lifted to the excited state. In the excited state, the spin is in $|m_s = \pm1\rangle$ state.

(b) The spin is not rotated from $|m_s = 0\rangle$ to $|m_s = \pm1\rangle$ state in the ground state, i.e., it is in the $|m_s = 0\rangle$ spin sublevel, and is optically lifted in this form to the excited state. In the excited state, the spin is rotated to $|m_s = \pm1\rangle$ state. So, in the excited state, the spin is in $|m_s = \pm1\rangle$ state.

In both cases, spin transitions occur in the ground and excited states, as indicated by writing microwave with circular arrow in the diagram in the ground as well as excited states.

By vibrational relaxation, the promoted electron returns to the $|m_s = 0\rangle$ ground state through $|m_s = 0\rangle$ singlet states via intersystem crossing. As optical pumping is continuously carried out, the V_B^- system is polarized into the $|m_s = 0\rangle$ ground state. This is the initialization step.

(iia) Zero external field: This measurement is necessary to examine the zero-field splitting, which is the splitting caused by spin–spin interaction even in the absence of the magnetic field.

(iib) Microwave driving from $|m_s = 0\rangle$ state to $|m_s = +1\rangle$ state and to $|m_s = -1\rangle$ state: Microwave frequency is swept across a range of values. When the microwaves have frequency ν_- corresponding to the energy difference between the sublevels $|m_s = 0\rangle$ to $|m_s = -1\rangle$, electrons are driven from $|m_s = 0\rangle$ spin sublevel to $|m_s = -1\rangle$ spin sublevel. Similarly, when microwaves have frequency ν_+ corresponding to the energy difference between the sublevels $|m_s = 0\rangle$ to $|m_s = +1\rangle$, electrons are driven from $|m_s = 0\rangle$ spin sublevel to $|m_s = +1\rangle$ spin sublevel. As the electron drops from the $|m_s = +1\rangle$ spin sublevel or $|m_s = -1\rangle$ spin sublevel to $|m_s = 0\rangle$ spin sublevel, the fluorescence intensity decreases at two frequencies ν_- and ν_+ because the emitted photon is not in the visible range but in the infrared region. Thus, in both cases, depletion of the ground $|m_s = 0\rangle$ state leads to a decrease in the intensity of photoluminescence.

Minimum photoluminescence condition is noticed when the applied microwave field is in resonance with spin transitions. The electron spins rotate from $|m_s = 0\rangle$ to $|m_s = \pm1\rangle$ state from where they decay by ISC to ground state. As these are non-radiative transitions, the intensity of light is reduced.

Maximum photoluminescence condition is observed when the applied microwave field is not in resonance with spin transitions. Electron spins remain in the $|m_s = 0\rangle$ state. The zero magnetic field spectrum is shown in Figure 5.17(a).

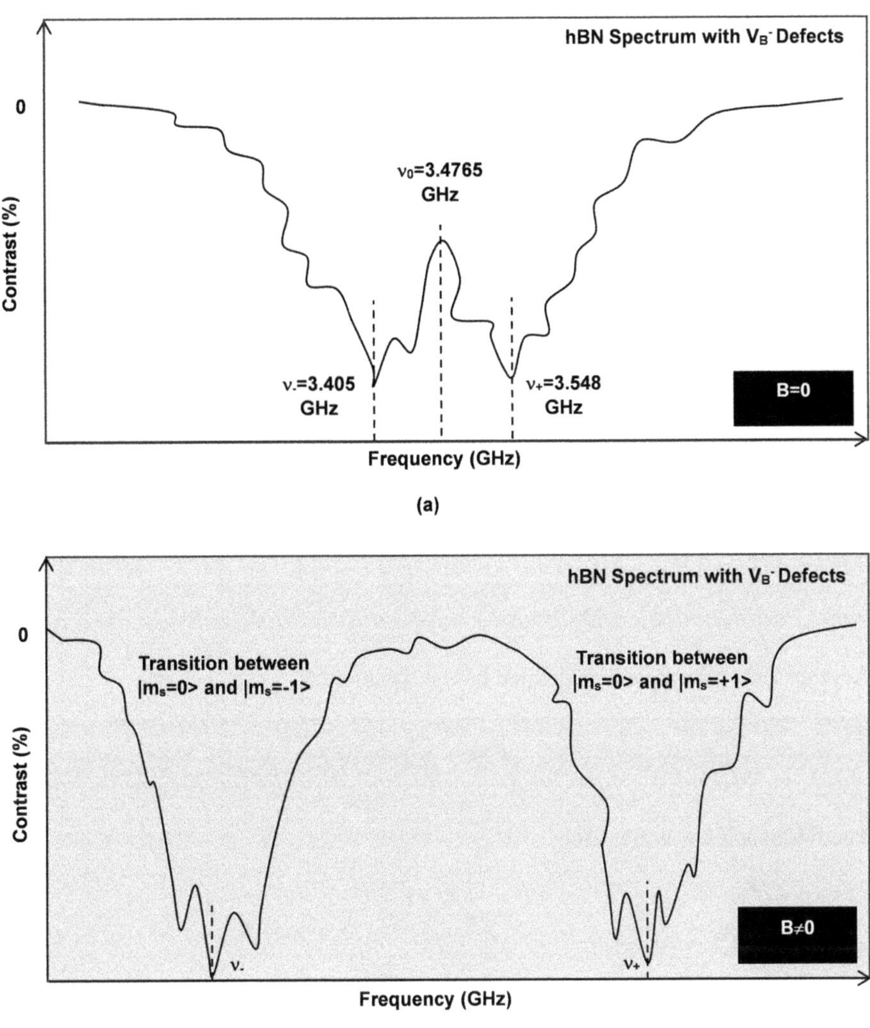

FIGURE 5.17 Spectra of hBN with V_B^- defects under: (a) zero magnetic field and (b) non-zero magnetic field. Part (a) shows a graph of contrast = ΔPL/PL expressed in percentage and plotted against frequency in gigahertz. The graph shows two valleys or dips at a lower frequency $v_- = 3.405$ GHz and at a higher frequency $v_+ = 3.548$ GHz centered around a mean frequency $v = (v_- + v_+)/2 = 3.4765$ GHz. Part (b) shows the splitting of $m_s = \pm1>$ states into spin sublevels $|m_s = +1>$ and $|m_s = -1>$ under magnetic field by the Zeeman effect.

(iiia) Applying a magnetic field for $|m_s = \pm1>$ states splitting by the Zeeman effect: When the magnetic field acts on the V_B^- center, it lifts the degeneracy of $|m_s = \pm1>$ state causing splitting into spin sublevels $|m_s = +1>$ and $|m_s = -1>$. Figure 5.17(b) shows the ODMR spectrum when a magnetic field is applied.

(iiib) Microwave driving from $|m_s = 0\rangle$ state to $|m_s = +1\rangle$ state and to $|m_s = -1\rangle$ state: This is the same step as step (iib) except that in step (iib), there was no external magnetic field, whereas in step (iiib), there is an external magnetic field.

5.9.2.3 Resonance Frequencies for ODMR Spectra without and with Externally Applied Magnetic Field

(i) Zero-field splitting: Without a magnetic field, i.e., for $B = 0$, the ODMR spectrum shows resonances at two frequencies ν_- and ν_+ for the same reasons as explained for the ODMR spectrum with applied magnetic field, point (iib) in Section 5.9.2.2. The frequencies ν_- and ν_+ are centered around a frequency ν_0 given by (Guo et al. 2022)

$$\nu_0 = \frac{D}{h}$$

(5.22)

where D is a ZFS parameter defined as

$$D = \frac{\nu_+ + \nu_-}{2}$$

(5.23)

Another ZFS parameter E is defined by the equation

$$E = \frac{\nu_+ - \nu_-}{2}$$

(5.24)

The frequencies ν_+ and ν_- are

$$\nu_+ = \frac{D+E}{h}$$

(5.25)

and

$$\nu_- = \frac{D-E}{h}$$

(5.26)

By combining eqs. (5.25) and (5.26), we get

$$\nu_{+,-} = \frac{D \pm E}{h}$$

(5.27)

(ii) Splitting under magnetic field: In the case $B \neq 0$, the resonance frequencies are expressed as

$$\nu_+ = \frac{D}{h} + \sqrt{E^2 + g_e^2 \mu_B^2 B^2}$$

(5.28)

and

$$v_- = \frac{D}{h} - \sqrt{E^2 + g_e^2 \mu_B^2 B^2}$$

(5.29)

Eqs. (5.28) and (5.29) are combined together in a single equation

$$v_{+,-} = \frac{D}{h} \pm \sqrt{E^2 + g_e^2 \mu_B^2 B^2}$$

(5.30)

In terms of v_0 (eq. 5.22), we can write

$$v_{+,-} = v_0 \pm \frac{1}{h} \sqrt{E^2 + g_e^2 \mu_B^2 B^2}$$

(5.31)

These equations are applied to determine unknown magnetic fields by finding v_+ and v_- from the ODMR spectra in the given magnetic fields, as shown in Figure 5.18. Figure 5.18(a) shows the ODMR spectra at two magnetic fields B_1 and $B_2 > B_1$. For both the magnetic fields, the lower and upper resonance frequencies are marked on the diagram. It is seen that $v_-|_{B2} < v_-|_{B1}$ and $v_+|_{B2} > v_+|_{B1}$. The v_- frequency shifts toward the left (decreases) while the v_+ frequency moves rightward (increases) as the magnetic field increases. So, the graph of magnetic field with microwave frequency separates into two lines, one directed to the left for v_- and the other directed toward the right for v_+, as shown in Figure 5.18(b). Either v_- line or v_+ line can be used to find an unknown magnetic field together with zero-field spectrum of Figure 5.17(a). The mathematical equations used for magnetic field calculation are also given in Figure 5.18(b).

5.10 V_B^- DEFECT MAGNETIC IMAGING SENSOR

This sensor has bright prospects for research on 2D materials because it provides an easy way to investigate the properties of heterostructures formed between it and 2D magnetic materials. As an example, a heterostructure is formed between ^{10}hBN with spin defects and $CrTe_2$ (Kumar et al. 2022).

5.10.1 $CrTe_2$-^{10}hBN HETEROSTRUCTURE

In this heterostructure, $CrTe_2$ is a 2D magnetic material and ^{10}hBN is used for magnetic imaging of $CrTe_2$.

$CrTe_2$ displays ferromagnetic properties which are preserved even when it is thinned down up to the monolayer stage. Hence, it is called a van der Waals ferromagnetic material. It has a high Curie temperature ~310 K. Bulk $1T$-$CrTe_2$ is mechanically exfoliated to obtain flakes having thickness around a few tens of nanometers. The flakes are transferred to SiO_2/Si substrates for final placement on hBN flakes.

A monoisotopic ^{10}hBN crystal is irradiated with thermal neutrons. The dose is 2.6×10^{16} neutrons cm^{-2}. The isotope crystal is selected because of its high neutron

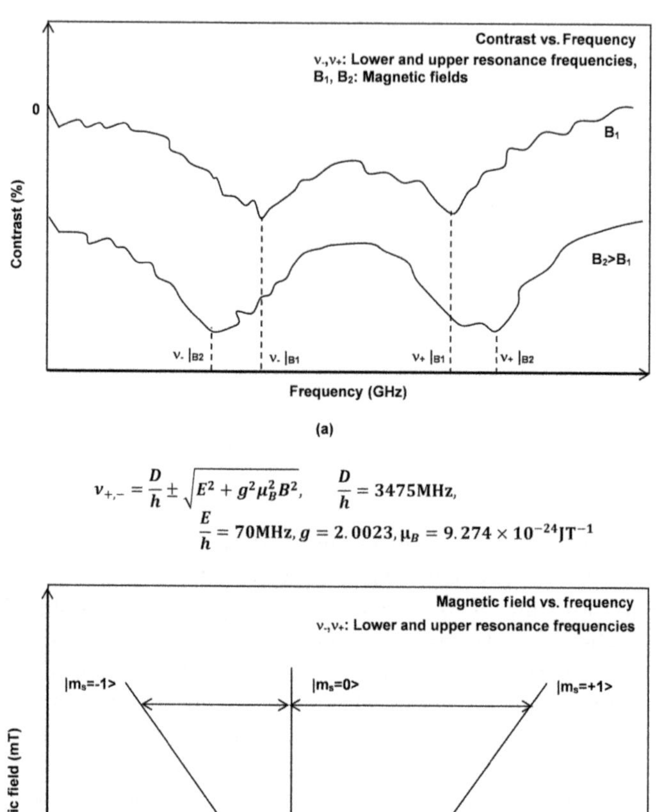

$$\nu_{+,-} = \frac{D}{h} \pm \sqrt{E^2 + g^2 \mu_B^2 B^2}, \qquad \frac{D}{h} = 3475\,\text{MHz},$$

$$\frac{E}{h} = 70\,\text{MHz}, g = 2.0023, \mu_B = 9.274 \times 10^{-24}\,\text{JT}^{-1}$$

FIGURE 5.18 Effect of increasing magnetic field on the lower and upper resonance frequencies of ODMR spectra: (a) spectra at two magnetic fields B_1 and $B_2 > B_1$, and (b) magnetic field variation with microwave frequency from ODMR spectra. (a) Contrast (%) is plotted against frequency (GHz) for magnetic fields B_1 and $B_2 > B_1$. For the smaller magnetic field B_1, the lower resonance frequency is $\nu_-|_{B1}$ and the upper resonance frequency is $\nu_+|_{B1}$. The corresponding frequencies for the larger magnetic field B_2 are $\nu_-|_{B2}$ and $\nu_+|_{B2}$. (b) Magnetic field in mT is plotted against microwave frequency in GHz for both frequencies ν_+, ν_-. The lines for frequencies ν_+, ν_- move away in opposite directions: line for ν_+ toward the right and line for ν_- toward the left. The line for frequency ν_+ is produced by spin transitions from $|m_s = +1>$ to $|m_s = 0>$ sublevel, as indicated by double-sided arrows. Similarly, the line for frequency ν_- is generated by spin transitions from $|m_s = -1>$ to $|m_s = 0>$ sublevel, as shown by double-sided arrows. Equations for $\nu_{+,-}$, D/h and E/h written at the top of the diagram are used for theoretical calculations by inputting D/h and E/h parameters from spectra in zero magnetic field and ν_+, ν_- under various external magnetic fields (Guo et al. 2022).

capture cross-section, facilitating the creation of V_B^- centers by neutron transmutation doping. After the defects have been formed, flakes are exfoliated from the ^{10}hBN crystal. These flakes are deposited over the $CrTe_2$ layer to form a $CrTe_2$-^{10}hBN heterostructure.

5.10.2 MAGNETIC IMAGING

For magnetic imaging, an ESR spectrum is recorded at each point of the photoluminescence scan of the $CrTe_2/^{10}$hBN heterostructure. The Zeeman shift of the lower ESR frequency v_- is found in this spectrum. After subtraction of the offset due to the biasing magnetic field, the Zeeman shit is given by

$$\Delta z = \gamma_e B_z \tag{5.32}$$

where γ_e is the electron gyromagnetic ratio and B_z is the projection of magnetic field along the hBN c-axis. Thus, a map of the magnetic field component B_z generated by the $CrTe_2$ flake is obtained. The sensitivity of magnetic imaging is 100 µT (Hz)$^{-1/2}$ (Kumar et al. 2022).

5.11 PLASMONIC-ENHANCED V_B^- CENTER SENSOR

5.11.1 OVERCOMING THE SHORTCOMINGS OF V_B^- DEFECTS IN hBN

Major factors restricting the magnetic field sensitivity of V_B^- defects in hBN are their low photoluminescence and contrast (= change in photoluminescence/original luminescence = ΔPL/PL) for ODMR (Gao et al. 2021a). Both these properties are improved by placing an hBN nanosheet over a microwave coplanar waveguide made of a gold film.

5.11.2 SENSOR FABRICATION

hBN nanosheets are exfoliated on a silicon wafer and subjected to He$^+$ ion implantation at low energies ~200 eV–3 keV. Spin defects are produced in the hBN nanosheet at small depths ranging from 3 to 30 nm. After ion-implantation, the hBN nanosheets having shallow spin defects are transferred over a pre-fabricated gold stripline on a sapphire substrate (Figure 5.19).

5.11.3 ODMR SETUP

An ODMR setup is used for characterizing the plasmonic effect of gold film. The V_B^- defects are excited with a 532 nm laser using an objective lens. The same lens collects the photoluminescence. The microwave signal for manipulating spins is supplied through the gold stripline. It is found that the surface plasmon of the gold film increases the photoluminescence of spin defects by 17 times and the contrast up to 46% (Gao et al. 2021a).

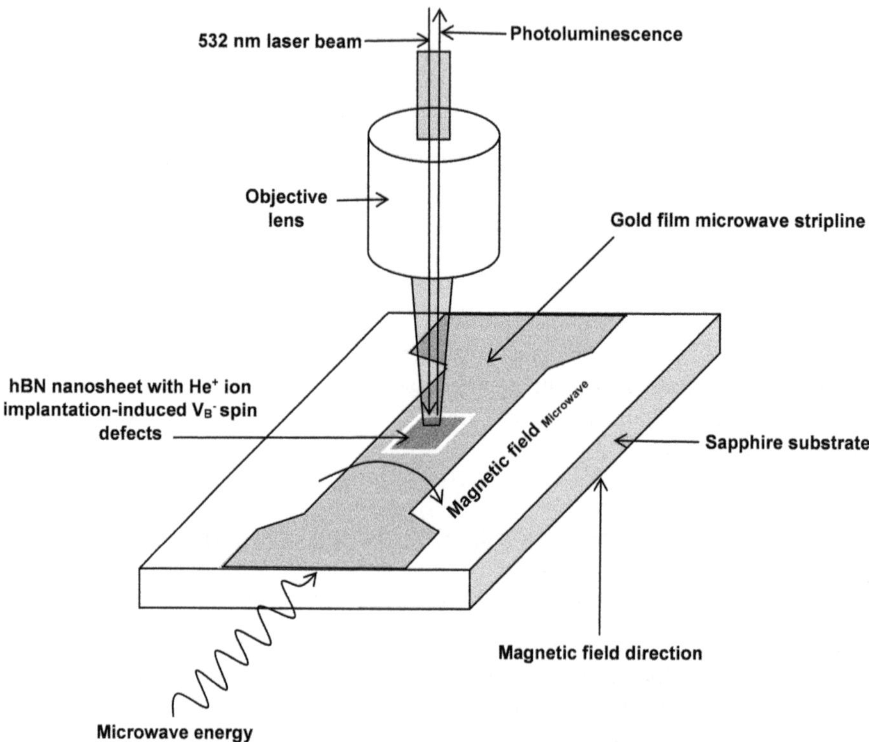

FIGURE 5.19 ODMR measurement arrangement for plasmonic-enhanced spin defects in hBN. The device consists of a sapphire substrate on which a gold film stripline is made by photolithography. Over the gold stripline, the hBN nanosheet is placed. The nanosheet has spin defects in it. Magnetic field from a permanent magnet is acting on hBN. Through the objective lens, the hBN is subjected to optical excitation with laser. At the same time, microwave energy is delivered to hBN. A photoluminescence signal is obtained in the objective lens as the outcome of simultaneous laser excitation of the hBN placed in magnetic field and spin manipulation of V_B^- in hBN by microwave power conveyed through the stripline.

5.11.4 EMISSION INTENSITY

V_B^- defects with low-loss nanopatch antennas are reported to yield comparatively higher emission intensities up to 250 times (Xu et al. 2023). The associated emission enhancement is 1685 times. Thus V_B^- spin defects coupled with nanopatch antennas work as magnetic field sensors for high-resolution measurements. The plasmonic nanocavity-enabled emission structure is illustrated in Figure 5.20.

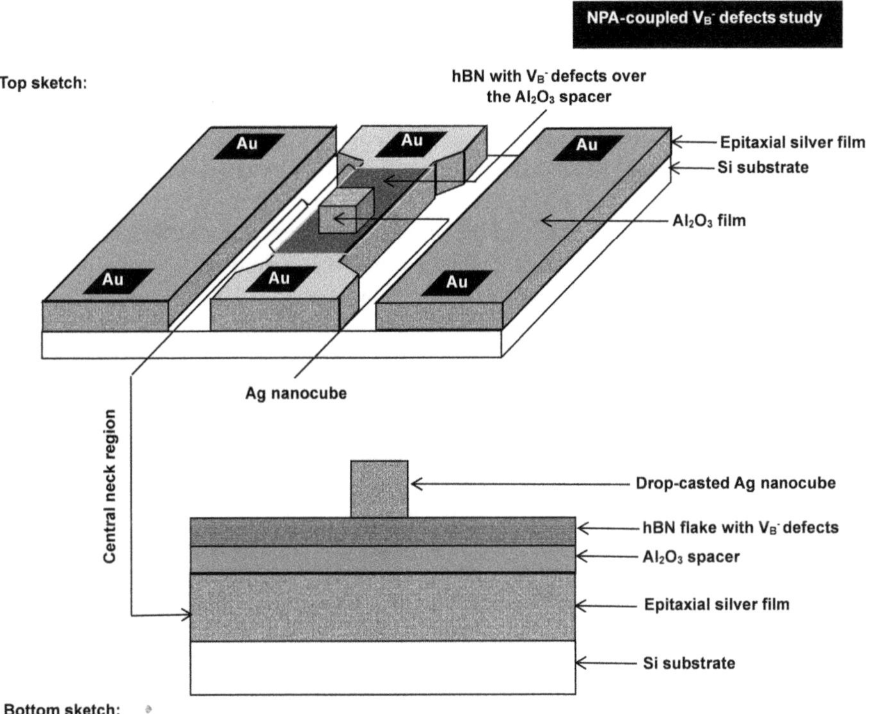

FIGURE 5.20 Structure for the study of nanopatch antenna (NPA)-coupled negatively charged boron vacancy defects. Top sketch: a coplanar microwave waveguide is made over a silicon substrate using an epitaxial silver film. On the central neck region of the waveguide, an hBN flake with V_B^- spin defects is seen. A silver nanocube is placed over the hBN flake. Au contact pads are shown by black squares. Bottom sketch: the cross-sectional diagram of the central neck region consisting of layers (from the bottom upward): silicon substrate, epitaxial silver film, Al_2O_3 spacer, hBN flake with spin defects, and drop-casted silver nanocube.

5.12 DISCUSSION AND CONCLUSIONS

The discovery of 2D magnetic materials laid a solid foundation for the physics of magnetism in two-dimensional systems. The 2D materials, chromim (III) iodide, CrI_3, a ferromagnetic insulator; and chromium germanium telluride, $Cr_2Ge_2Te_6$ (CGT), a ferromagnetic semiconductor; were the first few-layered magnetic van der Waals materials. Magnetic van der Waals materials show a magnetic ground state upon thinning to few layers or even one layer. A TMDC material vanadium diselenide (VSe_2) shows ferromagnetism in the bulk phase and as a multilayer film (Liu et al. 2018). The above materials together with iron germanium telluride, Fe_3GeTe_2 (FGT), chromium tellurosilicate, $CrSiTe_3$, which are ferromagnetic semiconductors; manganese bismuth telluride, $MnBi_2Te_4$, an intrinsic magnetic topological insulator, etc., open up new possibilities in sensor development (Wang et al. 2020, Elahi et al. 2022).

In this chapter, we pondered over several 2D material magnetic sensors. Figure 5.21 provides a rapid glance at these sensors. Research in 2D material induction coil Hall, magnetoresistive, and quantum magnetic sensors holds huge prospects and potentialities for advanced designs in forthcoming years targeting new applications.

After getting a bird's eye view of the pioneering research in physical nanosensors using 2D materials in the preceding chapters, we turn toward the sensors for chemical stimuli and take up metal ion sensors in the next chapter.

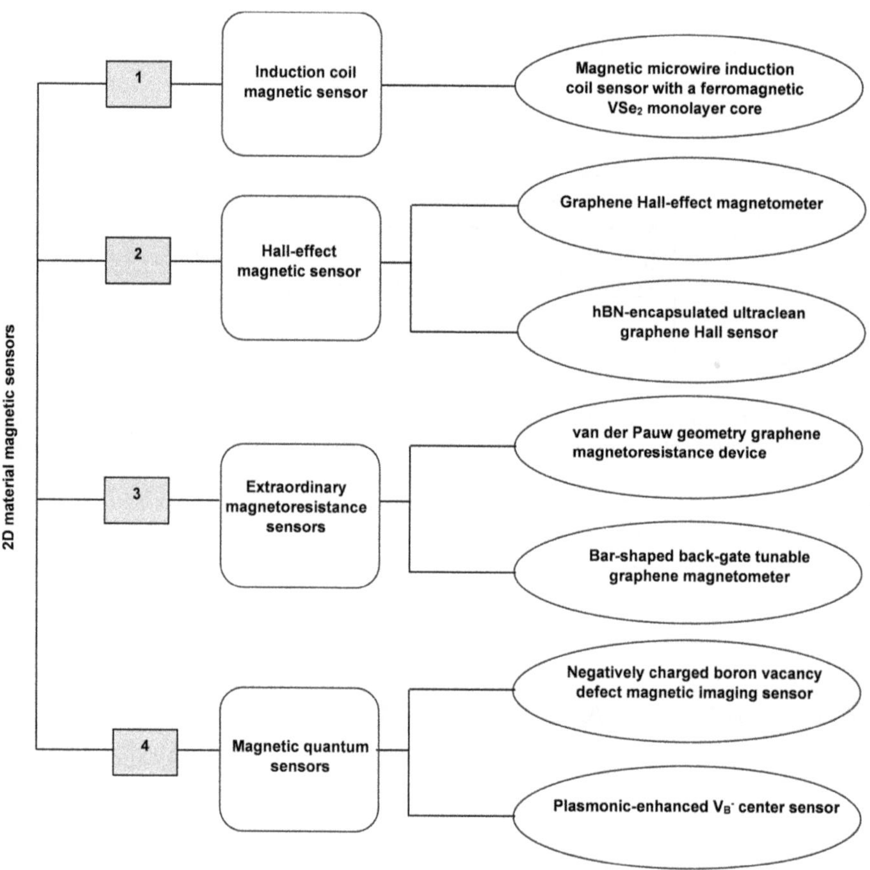

FIGURE 5.21 Visual presentation of the 2D material magnetic sensors discussed in this chapter. The sensors are sub-divided into four categories, namely, induction coil sensor with a ferromagnetic VSe_2 monolayer core (Jimenez et al. 2020), Hall-effect-sensors [graphene Hall-effect magnetometer (Izci et al. 2018) and hBN-encapsulated ultraclean graphene Hall sensor (Schaefer et al. 2020)], extraordinary magnetoresistance sensors [van der Pauw geometry graphene magnetoresistance device (Lu et al. 2011), and bar-shaped back-gate tunable graphene magnetometer (Pisana et al. 2010)], and magnetic quantum sensors [negatively charged boron vacancy defect magnetic imaging sensor (Kumar et al. 2022) and plasmonic-enhanced V_B^- center sensor (Gao et al. 2021a)].

REFERENCES

Birtwistle G. 2009 The Normal and Anomalous Zeeman Effect; The Landé g-formula, In: *The New Quantum Mechanics*, Cambridge Library Collection-Mathematics, Cambridge University Press, Cambridge, UK, pp. 14–24.

Boone T., L. Folks, J. A. Katine, E. Marinero, N. Smith and B. A. Gurney 2007 Magnetic sensitivity in mesoscopic EMR devices in I-V-I-V configuration, *2007 65th Annual Device Research Conference*, South Bend, IN, 18–20 June, pp. 247–248.

Carbonera D. 2009 Optically detected magnetic resonance (ODMR) of photoexcited triplet states, *Photosynthesis Research*, 102, pp. 403–414.

Collomb D., P. Li and S. Bending 2021 Frontiers of graphene-based Hall-effect sensors, *Journal of Physics: Condensed Matter*, 33, 243002 (22pp).

Elahi E., G. Dastgeer, G. Nazir, S. Nisar, M. Bashir, H. A. Qureshi, D.-K. Kim, J. Aziz, et al. 2022 A review on two-dimensional (2D) magnetic materials and their potential applications in spintronics and spin-caloritronic, *Computational Materials Science*, 213(2), p. 111670.

Gao S. X., B. Jiang, A.E. L. Allcca, K. Shen, M. A. Sadi, A. B. Solanki, P. Ju, Z. Xu, et al. 2021a High-contrast plasmonic-enhanced shallow spin defects in hexagonal boron nitride for quantum sensing, *Nano Letters*, 21, pp. 7708–7714.

Gao X., S. Pandey, M. Kianinia, J. Ahn, P. Ju, I. Aharonovich, N. Shivaram, and T. Li 2021b Femtosecond laser writing of spin defects in hexagonal boron nitride, *ACS Photonics*, 8(4), pp. 994–1000.

Gottscholl A., M. Kianinia, V. Soltamov, S. Orlinskii, G. Mamin, C. Bradac, C. Kasper, et al. 2020 Initialization and read-out of intrinsic spin defects in a van der Waals crystal at room temperature, *Nature Materials*, 19, pp. 540–545.

Gottscholl A., M. Diez, V. Soltamov, C. Kasper, D. Krauße, A. Sperlich, M. Kianinia, C. Bradac, I. Aharonovich, and V. Dyakonov 2021 Spin defects in hBN as promising temperature, pressure and magnetic field quantum sensors, *Nature Communications*, 12, 4480, pp. 1–8.

Guo N.-J., W. Liu, Z.-P. Li, Y.-Z. Yang, S. Yu, Y. Meng, Z.-A. Wang, X.-D. Zeng et al. 2022 Generation of spin defects by ion implantation in hexagonal boron nitride, *ACS Omega*, 7(2), pp. 1733–1739.

Hruby J., M. Gulka, M. Mongillo, I. P. Radu, M. V. Petrov, E. Bourgeois, and M. Nesladek 2022 Magnetic field sensitivity of the photoelectrically read nitrogen-vacancy centers in diamond, *Applied Physics Letters*, 120, 162402-1 to 162402-6.

Izci D., C. Dale, N. Keegan and J. Hedley 2018 The construction of a graphene Hall effect magnetometer, *IEEE Sensors Journal*, 18(23), pp. 9534–9541.

Jimenez V. O., V. Kalappattil, T. Eggers, M. Bonilla, S. Kolekar, P. T. Huy, M. Batzill and M.-H. Phan 2020 A magnetic sensor using a 2D van der Waals ferromagnetic material, *Scientific Reports*, 10, 4789, 6 pages.

Kianinia M., S. White, J. E. Fröch, C. Bradac, and I. Aharonovich 2020 Generation of spin defects in hexagonal boron nitride, *ACS Photonics*, 7(8), pp. 2147–2152.

Kumar P., F. Fabre, A. Durand, T. Clua-Provost, J. Li, J.H. Edgar, N. Rougemaille, et al. 2022 Magnetic imaging with spin defects in hexagonal boron nitride, *Physical Review Applied*, 18, L061002, pp. 1–5.

Ling S. J. 2022 Angular Momentum, In: Physics Bootcamp, Sec. 55.7, http://www. physicsbootcamp.org/sec-angular-velocity-and-angular-momentum.html

Liu W., V. Ivády, Z.-P. Li, Y.-Z. Yang, S. Yu, Y. Meng, Z.-A. Wang, N.-J. Guo, F.-F. Yan, Q. Li, et al. 2022 Coherent dynamics of multi-spin V_B^- center in hexagonal boron nitride, *Nature Communications*, 13, 5713, pp. 1–8.

Liu Z.-L., X. Wu, Y. Shao, J. Qi, Y. Cao, L. Huang, C. Liu, J.-O. Wang, et al. 2018 Epitaxially grown monolayer VSe_2: An air-stable magnetic two-dimensional material with low work function at edges, *Science Bulletin*, 63(7), pp. 419–425.

Lu J., H. Zhang, W. Shi, Z. Wang, Y. Zheng, T. Zhang, N. Wang, Z. Tang, and P. Sheng 2011 Graphene magnetoresistance device in van der Pauw geometry, *Nano Letters*, 11(7), pp. 2973–2977.

Mathur N., A. Mukherjee, X. Gao, J. Luo, B. A. McCullian, T. Li, A. N. Vamivakas and G. D. Fuchs 2022 Excited-state spin-resonance spectroscopy of V_B^- defect centers in hexagonal boron nitride, *Nature Communications*, 13, 3233, pp. 1–7.

Murzakhanov F. F., B.V. Yavkin, G. V. Mamin, S. B. Orlinskii, I. E. Mumdzhi, I. N. Gracheva, B. F. Gabbasov, et al. 2021 Creation of negatively charged boron vacancies in hexagonal boron nitride crystal by electron irradiation and mechanism of inhomogeneous broadening of boron vacancy-related spin resonance lines, *Nanomaterials*, 11, 1373, pp. 1–11.

Mu Z., H. Cai, D. Chen, J. Kenny, Z. Jiang, S. Ru, X. Lyu, T.S. Koh, X. Liu, I. Aharonovich, and W. Gao 2022 Excited-state optically detected magnetic resonance of spin defects in hexagonal boron nitride, *Physical Review Letters*, 128(21), 216402, 6 pages.

Pisana S., P. M. Braganca, E. E. Marinero, and B. A. Gurney 2010 Tunable nanoscale graphene magnetometers, *Nano Letters*, 10, pp. 341–346.

Ripka P. 2013 Security applications of magnetic sensors, *Journal of Physics: Conference Series, Sensors & their Applications XVII*, 450, 012001, pp. 1–4.

Schaefer B. T., L. Wang, A. Jarjour, K. Watanabe, T. Taniguchi, P. L. McEuen and K. C. Nowack 2020 Magnetic field detection limits for ultraclean graphene Hall sensors, *Nature Communications*, 11, 4163, pp. 1–8.

Sun J. and J. Kosel 2013 Extraordinary magnetoresistance in semiconductor/metal hybrids: A review. *Materials (Basel)*, 6(2), pp. 500–516.

Toledo J. R., D. B. de Jesus, M. Kianinia, A. S. Leal, C. Fantini, L. A. Cury, G. A. M. Sáfar, I. Aharonovich, and K. Krambrock 2018 Electron paramagnetic resonance signature of point defects in neutron-irradiated hexagonal boron nitride, *Physical Review B*, 98, p. 155203.

Turro N. J. 1969 The triplet state, *Journal of Chemical Education*, 46(1), pp. 1–6.

van den Bosc, J.C. 1957 The Zeeman Effect. In: Spektroskopie II/Spectroscopy II. *Handbuch der Physik / Encyclopedia of Physics*, 5/28, pp. 296–332.

Wang M.-C., C.-C. Huang, C.-H. Cheung, C.-Y. Chen, S. G. Tan, Y. Zhao, Y. Zhao, et al. 2020 Prospects and opportunities of 2D van der Waals magnetic systems, *Annalen der Physik (Berlin)*, 532, 1900452, pp. 1–19.

Wojciechowski A.M., P. Nakonieczna, M. Mrózek, K. Sycz, A. Kruk, M. Ficek, M. Głowacki, R. Bogdanowicz, and W. Gawlik 2019 Optical magnetometry based on nanodiamonds with nitrogen-vacancy color centers, *Materials*, 12, 2951, pp. 1–10.

Xu X., A. B. Solanki, D. Sychev, X. Gao, S. Peana, A. S. Baburin, K. Pagadala, et al. 2023 Greatly enhanced emission from spin defects in hexagonal boron nitride enabled by a low-loss plasmonic nanocavity, *Nano Letters*, 23(1), 25–33.

6 2D Materials-Based Heavy Metal Ion Sensors

Non-degradable nature together with bio-accumulation, bio-magnification, and high toxicity characteristics of heavy metal ions have exacerbated the environmental pollution reinforced by the rapid urbanization and industrialization. In this chapter, we look over the progress in 2D material chemical and biosensors and provide insights into the various research approaches followed.

6.1 EXIGENCY OF HEAVY METAL ION SENSORS

6.1.1 Heavy Metals

These are metals having:

(i) a high specific density, also termed relative density or specific gravity, and
(ii) a high atomic weight or atomic number.

A specific density >5 gcm^{-3} is generally used as a property to label a metal as a heavy one (Jaishankar et al. 2014). The main heavy metals found in water are arsenic, cadmium, chromium, mercury, lead, nickel, iron, copper, and zinc. Among these, iron, copper, and zinc are essential nutrients but toxic in large quantities. But the remaining heavy metals have severely harmful and poisonous effects on living organisms. They adversely impact the environment and are considered as lethal pollutants disturbing the ecological balance (Briffa et al. 2020).

The sources of heavy metals consist of the following:

(i) Natural sources of heavy metals are soil erosion and breaking down or dissolution of rocks and minerals known as the weathering of earth's crust.
(ii) Anthropogenic sources of these metals are mining, discharge of industrial wastes and effluents in water, release of untreated sewage sludge; pesticides such as fungicides, herbicides, and insecticides, and fertilizers used in growing crops; agricultural runoff, e.g., water from farm fields for irrigation, rain, or melted snow running into streams; urban runoff, e.g., rain and outdoor water from roofs, driveways, sidewalks, and other surfaces.

There are many adverse effects of heavy metals on human health. They cause lung cancer and impairment of gastrointestinal and neurological systems. Toxicity of heavy metals poses serious health risks for the human kidneys causing renal osteoporosis, dysfunction, and heart failure. (Alengebawy et al. 2021).

DOI: 10.1201/9781003330585-6

6.1.2 NEED FOR HEAVY METAL ION SENSORS USING 2D MATERIALS

Conventional methods for heavy metal detection require elaborate sample preparation procedures, sophisticated and expensive analytical laboratory equipment, and professionally skilled manpower (Hu et al. 2023). Therefore, rapid and real-time detection of these metals in water and food urgently needs highly sensitive, selective, and low-cost sensors. This is an area where 2D materials are touted to surpass the capabilities of traditional sensing materials for applications in agriculture, life sciences, and medical diagnostics.

6.2 ANODIC STRIPPING VOLTAMMETRY

Anodic stripping voltammetry (ASV) is a versatile electrochemical technique used for the detection of several species of trace metal ions in aqueous media, e.g., for assessing the presence of these ions in drinking water samples to monitor their quality (Wygant and Lambert 2022). It can identify and quantify concentrations of metal ions in the nanomolar range. The high sensitivity and low detection limit offered by this technique make it competitive with inductively coupled plasma mass spectrometry.

6.2.1 THE ASV SETUP

The vessel used for voltammetry is called an electrochemical cell with three electrodes: a working electrode where the electrochemical event of interest is carried out, usually a platinum or polished glassy carbon electrode, a counter electrode (Pt wire/disk or carbon-based), and a reference electrode, usually Ag/AgCl (Figure 6.1). The current flows between the working and counter electrodes while the reference electrode measures the potential applied to the working electrode with respect to a stable reference reaction (Elgrishi et al. 2018).

6.2.2 PRINCIPAL ASV STEPS

The ASV technique comprises three primary steps (Figure 6.2):

(i) Cleaning: An oxidation potential is applied for cleaning the surface of the electrode.
(ii) Cathodic deposition and stabilization: In this step, also called cathodic reduction, a reduction potential is applied to deposit the metal ions on the surface of the working electrode. The reduction potential is known as the deposition potential ($E_{\mathrm{Deposition}}$). It is several tenths of a volt more negative than reduction potential for the least easily reduced metal ion.

The step is performed in a stirred solution keeping the applied potential constant. It is a pre-concentration step because the metal ions are deposited from the large volume of the analyte solution to the small tip of the electrode.

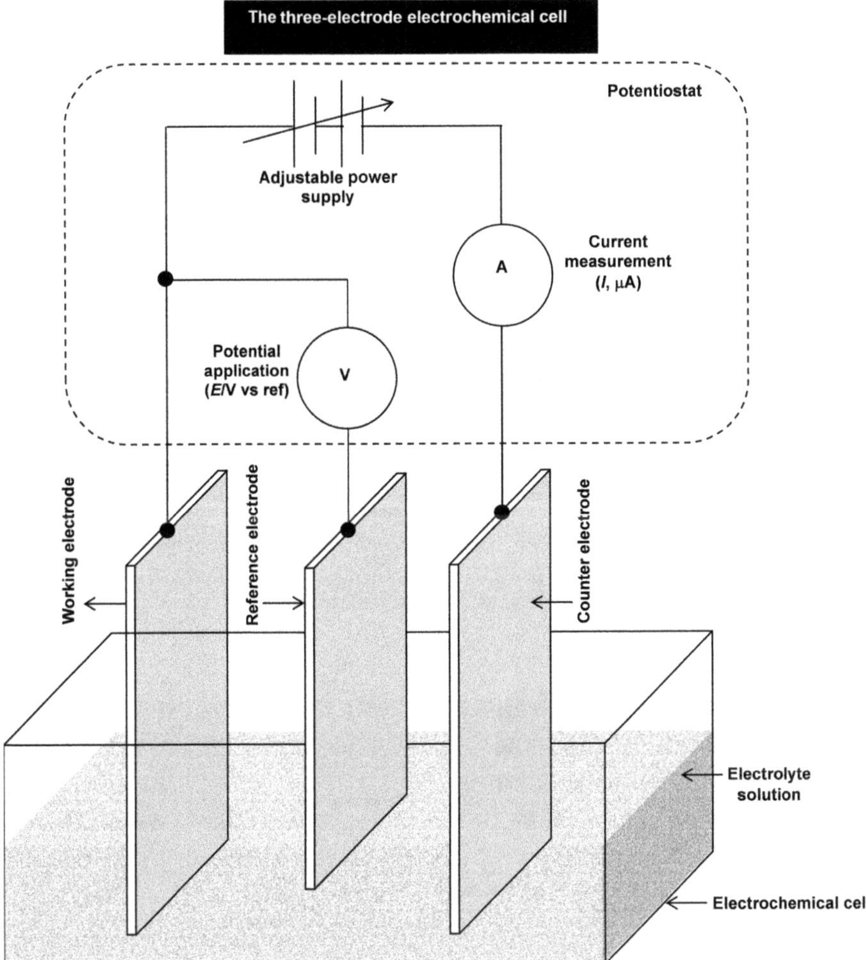

FIGURE 6.1 The electrochemical cell used in voltammetry experiments. The cell consists of a glass vessel filled with the sample electrolyte solution in which three electrodes are dipped: the working electrode where the redox reaction takes place; a reference electrode; and an auxiliary counter electrode. The potential applied to the working electrode from an adjustable power supply is measured by a voltmeter V with respect to the reference electrode, and the current flowing between the working and counter electrodes is measured by ammeter A. The potentiostat is an electronic instrument which produces/measures the potentials and currents. It controls the working electrode potential to run electroanalytical experiments.

(iii) Anodic stripping: This step, also called anodic oxidation or anodic dissolution, is done after a brief resting period for equilibration following the deposition. In this step, the deposited metals are oxidized from the working electrode into the solution by sweeping the potential toward positive side from the deposition potential $E_{\text{Deposition}}$, thereby anodically stripping the metal ionic species from the electrode surface.

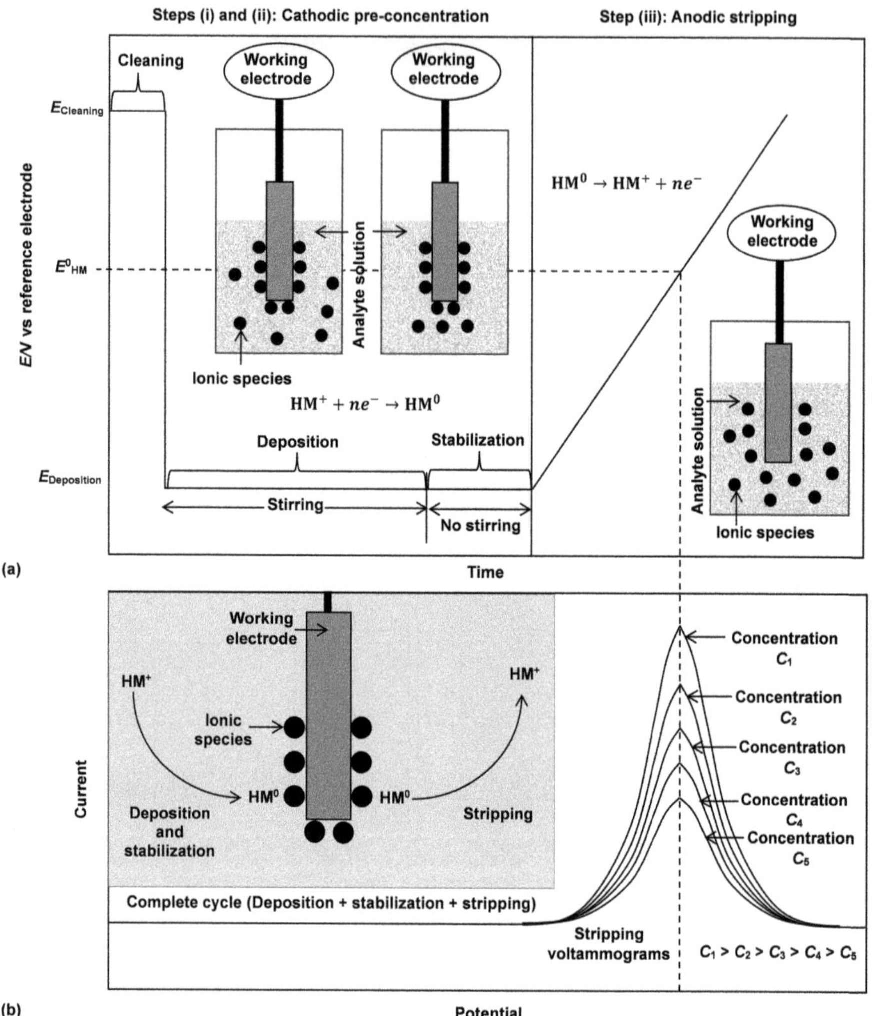

FIGURE 6.2 The standard curves of anodic stripping voltammetry in relation to the three principal steps of the analytical procedure: (a) potential-time graph and (b) current-potential curves. Part (a) shows the plot of potential E applied to the working electrode on the ordinate with respect to the reference electrode, against time on abscissa. Steps (i) and (ii) of cathodic pre-concentration involve the operations of cleaning, deposition, and stabilization. During cleaning, the potential is maintained at a high constant value $E_{Cleaning}$. The deposition and stabilization are carried out at a lower potential, the deposition potential $E_{Deposition}$. The solution is continuously stirred in the deposition period while during stabilization, no stirring is done and the solution rests undisturbed and stationary. The phenomena occurring in the analyte solution near the working electrode are shown in the insets. They indicate that the ionic species accumulate on the surface of the working electrode during deposition while they relax and settle into a stable state during equilibration. The reaction is described by the equation:

$$HM^+ + ne^- \rightarrow HM^0$$

where HM stands for heavy metal, n is a number, and e^- is the electronic charge. In step (iii) of anodic stripping, the potential applied to the working electrode is increased, and the reverse reaction:

$$HM^0 \rightarrow HM^+ + ne^-$$

takes place. The inset shows that the ionic species are released from the surface of the working electrode and they are dispersed into the analyte solution. Part (b) shows the current-potential curves called the stripping voltammograms which are generated by applying a varying potential to the working electrode relative to the reference electrode and measuring the current flowing between the working and counter electrodes. Analyte solutions of different concentrations C_1, C_2, C_3, C_4, and C_5 are taken in a decreasing order with $C_1 > C_2$, $C_2 > C_3$, ... In all cases, the current first increases with potential to reach a peak value and then falls down with the further increase in potential. For different concentrations of a given ion, the current peaks are observed at the same potential and the peak current value falls with the ion concentration. The inset shows the complete reaction cycle of the experiment. During deposition, the heavy metal ion HM^+ is deposited from the analyte solution as a neutral metal HM^0 on the working electrode. During stripping, the heavy metal HM^0 deposited on the working electrode is liberated from it into the solution in ionic form as HM^+ ions.

During stripping, the current vs. potential characteristic of the cell is measured. Peak potentials observed in the stripping curve are used for the identification of specific metal ion species. The concentration of relevant ions in solution is obtained from the integrated charge under each peak.

6.3 CHEMICALLY MODIFIED ELECTRODES

These are electrodes which are purposefully modified by controlled surface treatment for absorption, coating, or attachment of specific molecules (Bard 1983).

6.3.1 RATIONALE OF ELECTRODE MODIFICATION

The chemically modified electrodes act as selective attractive media for the particular analyte for which a suitable coating has been deposited on the electrode for helping in the detection of the relevant analyte from a solution containing a mixture of several different analytes. The conferral of this property on the electrode forms the basis of electrochemical sensors using deliberately modified electrodes to build up the concentration of a chosen trace material for facilitating its analysis. This modification imparts to them the ability to preferentially entice/trap the desired analyte from a bulk solution and concentrate it for chemical analysis.

6.3.2 Electrode Modification Methods

The electrodes are intentionally modified by:

(i)   Chemical adsorption: In chemisorption, the coating molecules are attached by forming a strong chemical bond between the electrode surface and the adsorbate through a chemical reaction.

(ii) Covalent bonding: An example is silanization of a metal electrode by treatment with an aminosilane (3-Aminopropyl) triethoxysilane (APTES) $[H_2N(CH_2)_3Si(OC_2H_5)_3]$, followed by the treatment of silanized electrode with the compound containing the desired group.

(iii) Dipping in a polymer solution: The electrode is immersed in a solution containing the selected polymer, and the solvent is allowed to evaporate leaving the desired coating on the electrode surface.

6.4   2D MATERIAL MODIFIED ELECTRODES AS SENSORS

6.4.1 Tin Oxide (SnO_2)/Reduced Graphene Oxide (rGO) Nanocomposite-Modified Glassy Carbon Electrode for Simultaneous Detection of Heavy Metal Ions

Electrochemical detection of Cd (II), Pb (II), Cu (II), and Hg (II) ions present at the same time in a solution is done by an SnO_2/rGO-modified glassy carbon electrode using square wave anodic stripping voltammetry (SWASV) (Wei et al. 2012). Figure 6.3 illustrates the procedure followed for the detection of the ions.

6.4.1.1 SnO_2/rGO Nanocomposite Preparation

Dry graphene oxide (GO) is mixed with deionized water and sonicated to obtain GO solution. This solution is added to tin (IV) chloride pentahydrate ($SnCl_4.5H_2O$) dissolved in water. After stirring and centrifuging, the mixture is sintered at 500°C for 2 h. The sintering is done in an argon atmosphere. The SnO_2/rGO nanocomposite is dispersed in alcohol.

6.4.1.2   Glassy Carbon Electrode (GCE) Modification

The GCE is polished with alumina powder and rinsed with DI water. After sonication in alcohol and DI water, it is dried in nitrogen. 5 μL aliquot of the SnO_2/rGO suspension is coated on GCE. The solvent is dried at room temperature. Inset of Figure 6.3(a) shows the SnO_2/rGO-modified electrode.

6.4.1.3   Square Wave Anodic Stripping Voltammetry Parameters

Square wave voltammetry is a type of linear potential sweep voltammetry using a staircase potential ramp modified with a series of square-wave pulses against the staircase triangular potential waveform of cyclic voltammetry (Mirceski et al. 2013).

The deposition potential of ions is −1.0V, and the deposition time is 120 s. The deposition is done by reducing Cd (II), Pb (II), Cu (II), and Hg (II) ions in 0.1 M NaAc-Hac, pH=5.0.

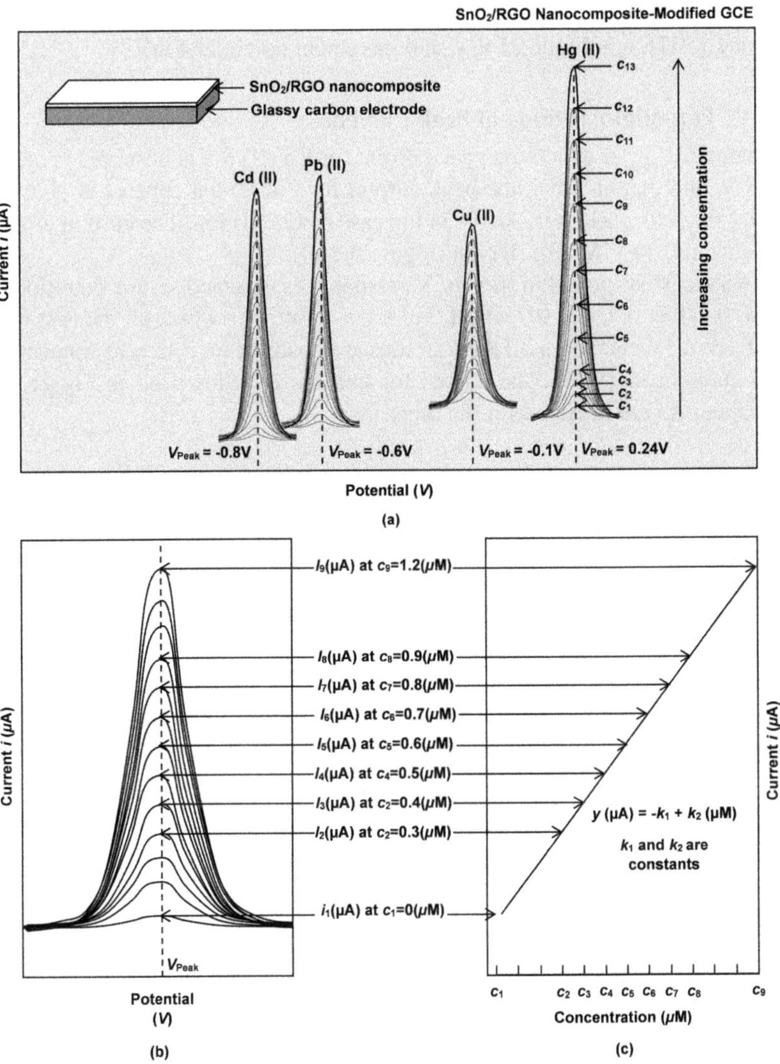

FIGURE 6.3 Detection of heavy metal ions by SnO_2/rGO nanocomposite-modified GCE: (a) SWASV current-potential curves of SnO_2/rGO-modified GCE for the simultaneous detection of Cd (II), Pb (II), Cu (II), and Hg (II) ions for a range of ionic concentrations c_1, c_2, c_3, etc. Inset shows the SnO_2/rGO-modified GCE. Peaks are seen at -0.8 V, −0.6 V, −0.1 V, and 0.24 V for Cd (II), Pb (II), Cu (II), and Hg (II) ions, respectively. (b) SWASV current-potential curves of SnO_2/rGO-modified GCE for different concentrations of a single ion. The curves are recorded for ion concentrations c_1=0, c_2=0.3, c_3=0.4, c_4=0.5, c_5=0.6, c_6=0.7, c_7=0.8, c_8=0.9, and c_9=1.2 µM. V_{Peak} is the potential at which the peak appears for this ion. The current values corresponding to the peaks of the voltage curves are i_1, i_2, i_3, i_4, i_5, i_6, i_7, i_8, and i_9 µA for concentrations c_1, c_2, c_3, c_4, c_5, c_6, c_7, c_8, and c_9 µM, respectively. The calibration graph drawn between currents at the peak potential and corresponding concentrations is a straight line represented by the equation y (µA) = $-k_1 + k_2$ (µM), where k_1 and k_2 are constants.

Anodic stripping of deposited Cd, Pb, Cu, and Hg is done at −1.0 V to 0.5 V, frequency 15 Hz, amplitude 25 mV, and increment potential 4 mV.

6.4.1.4 Potential Positions of Peak Current

The potential V_{Peak} at which the peak current for Cd (II) ion is observed is measured as −0.8 V, and that at which the peak current for Pb (II) ion appears is −0.6 V. The potential for peak current is −0.1 V in the case of Cu (II) ion. The current attains the peak value at 0.24 V for Hg (II) ion (Figure 6.3(a)).

SWASV current-potential (μA vs. V) response is obtained at concentrations of 0, 0.3, 0.4, 0.5, 0.6, 0.7, 0.8, 0.9, and 1.2 μM in a solution in which all the four ions are present. SWASV curves for all the four ions are obtained for different concentrations of ions mentioned above. The curves for one ion are illustrated in Figure 6.3(b). Identical curves are obtained for the other ions.

6.4.1.5 Linearization Equations for Simultaneous Detection of Metal Ions

The peak currents for the concentrations c_1, c_2, c_3, c_4, c_5, c_6, c_7, c_8, and c_9 are read from the SWASV curves and plotted against the respective concentrations in Figure 6.3(c) to draw the calibration graphs of the selected ion. By the same procedure, the calibration plots for all the ions, Cd (II), Pb (II), Cu (II), and Hg (II) are sketched. The calibration straight line for each ion is described by a linearization equation. The linearization equations for the four ions and their limits of detection (LoD) are (Wei et al. 2012):

$$\text{Cd}\left(\text{II}\right): i\left(\mu A\right) = -5.25 + 18.4\,c\left(\mu M\right),$$
$$\text{Correlation coefficient} = 0.995, \tag{6.1}$$
$$\text{LoD} = 1.015 \times 10^{-10}\,\text{M}$$

$$\text{Pb}\left(\text{II}\right): i\left(\mu A\right) = -4.34 + 18.6\,c\left(\mu M\right),$$
$$\text{Correlation coefficient} = 0.984, \tag{6.2}$$
$$\text{LoD} = 1.839 \times 10^{-10}\,\text{M}$$

$$\text{Cu}\left(\text{II}\right): i\left(\mu A\right) = -5.06 + 14.98\,c\left(\mu M\right),$$
$$\text{Correlation coefficient} = 0.976, \tag{6.3}$$
$$\text{LoD} = 2.269 \times 10^{-10}\,\text{M}$$

$$\text{Hg}\left(\text{II}\right): i\left(\mu A\right) = -10.5 + 28.2\,c\left(\mu M\right),$$
$$\text{Correlation coefficient} = 0.965, \tag{6.4}$$
$$\text{LoD} = 2.789 \times 10^{-10}\,\text{M}$$

6.4.2 GRAPHENE–PLATINUM NANOCOMPOSITE-MODIFIED GLASSY CARBON ELECTRODE FOR AS (III) DETECTION

Graphene with Pt nanoparticles is applied on GCE for trace level As (III) ion estimation (Kempegowda et al. 2014). Figure 6.4 presents the analytical method.

6.4.2.1 Graphene Oxide Synthesis by Modified Hummers' Method

Graphite and sodium nitrate are taken in a flask. Sulfuric acid is slowly added at 0°C. Potassium permanganate is added followed by thorough stirring. Distilled water is

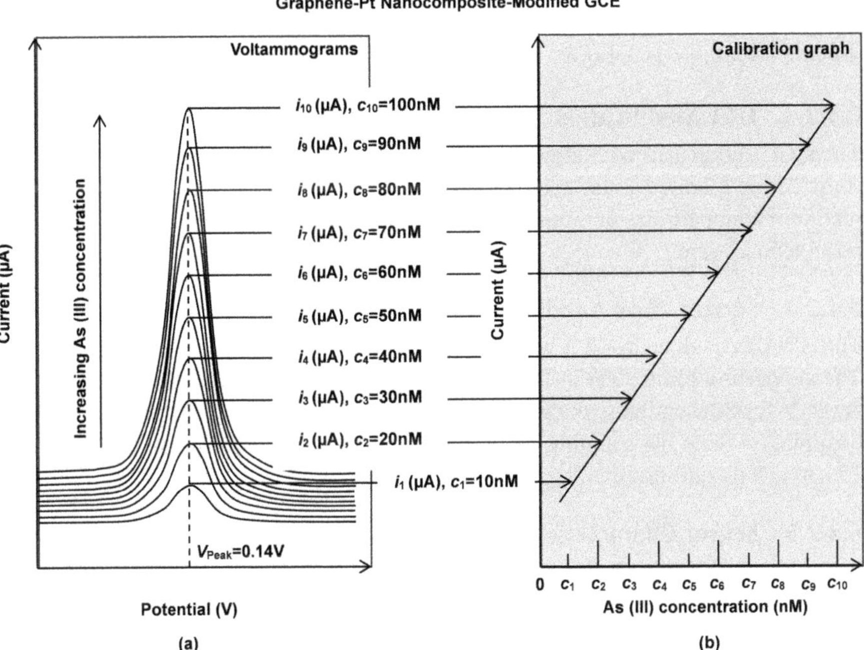

FIGURE 6.4 As (III) detection by graphene–Pt nanocomposite-modified GCE: (a) SWASV voltammograms with increasing concentrations of As (III) ion and (b) calibration graph of the modified-GCE sensor. Part (a): graphs of current in µA are plotted against potential in volt for As (III) ion concentrations c_1=10 nM, c_2 = 20 nM, c_3 = 30 nM, c_4 = 40 nM, c_5 = 50 nM, c_6= 60 nM, c_7 = 70 nM, c_8 = 80 nM, c_9 = 90 nM, and c_{10} = 100 nM. The resulting curves are rising to successively higher values of current in an ascending order of concentrations. Each curve has a peak value of current. The peaks of all the curves are in one straight line meeting the voltage axis at 0.14 V. The voltage V_{Peak} for the peak values of currents is shown as 0.14 V. The symbols i_1, i_2, i_3, i_4, i_5, i_6, i_7, i_8, i_9 and i_{10} denote the currents in µA for concentrations c_1, c_2, c_3, c_4, c_5, c_6, c_7, c_8, c_9, and c_{10} in nM. Part (b): this plot between current in µA and As (III) ion concentration in nM is generated by taking the current values i_1, i_2, i_3, etc. corresponding to the concentrations c_1, c_2, c_3, etc. from part (a). It is a straight line in the As (III) ion concentration range of 10–100 nM.

added with further stirring. Drops of hydrogen peroxide are added until evolution of gas stops. The mixture is centrifuged and the residue is separated by decantation. The residue is repeatedly washed with hydrochloric acid until the filtrate is sulfate-free. Finally, the residue is rinsed with double distilled water and vacuum dried. Thus, yellowish brown GO is obtained by exfoliation of graphite.

6.4.2.2 Graphene–Platinum Nanoparticles Composite Preparation

It is formed by one-step reduction of hexachloroplatinic acid ($H_2PtCl_6 \cdot 6H_2O$) and GO. They are mixed in a flask and a homogeneous dispersion in ethylene glycol (CH_2OHCH_2OH) is made. After reaction at 120°C with stirring and refluxing, the pale-yellow mixture turns black showing that Pt nanoparticles are formed. The resulting graphene–Pt nanoparticles composite is filtered, washed with distilled water, and dried in a vacuum desiccator.

6.4.2.3 GCE Modification

The GCE is polished with alumina slurry, washed with doubled-distilled water, and sonicated in distilled water and ethanol. 10 µL of colloidal dispersion of graphene–platinum nanoparticles composite is drop-casted over the GCE surface and dried at room temperature.

6.4.2.4 Square Wave Anodic Stripping Voltammetry Parameters

The SWASV is done at −0.2 V to 0.8 V. As (III) is taken in an electrochemical cell. The supporting electrolyte is 1M HCl. The As (III) ion is electro-reduced to As (0) at 0.4 V for 80 s on the surface of modified GCE from the bulk electrolyte solution. Stripping is done by scanning the potential in the positive direction after allowing 15 s to reach equilibrium.

6.4.2.5 Sensor Characteristics

A distinct and sharp oxidation peak is observed in the presence of As (III) in the cyclic and anodic stripping voltammograms at 0.14 V (Figure 6.4(a)). The sensor has a linear characteristic in 10–100 nM range, as shown in Figure 6.4(b). Its limit of detection is 1.1 nM (Kempegowda et al. 2014).

6.4.3 L-Cysteine/rGO-Modified GCE Sensor for Concomitantly Determining Cd (II), Pb (II), Cu (II), and Hg (II) Ions

A graphene/rGO-based organic composite GCE is made for detecting these ions (Muralikrishna et al. 2014).

6.4.3.1 Graphene Oxide Synthesis

Graphite is oxidized by a modified version of Hummers and Offmann method.

6.4.3.2 L-Cysteine-Functionalized rGO Preparation

N, N'-Dicyclohexylcarbodiimide (DCC) ($C_{13}H_{22}N_2$) is added to GO dispersed in distilled water. Stirring and sonication are done to get GO–DCC mixture. Then

L-cysteine ($C_3H_7NO_2S$) solution is made in water and its pH is adjusted to 10.4 with ammonium hydroxide. The L-cysteine solution is added to GO–DCC mixture, refluxed, and filtered. After washing with acetone, ethanol, acetonitrile, and water, it is dried in an oven at 60°C for 12 h to get L-cys-rGO.

6.4.3.3   GCE Modification

After polishing with alumina slurry, the GCE is rinsed with distilled water, sonicated in ethanol-water mixture, again rinsed with distilled water, and dried. L-cys-rGO solution in water–ethanol mixture is sonicated. 10 μL of the sonicated solution is drop-casted on the surface of GCE and dried.

6.4.3.4   Differential Pulse Anodic Stripping Voltammetry

The differential pulse anodic stripping voltammetry (DPASV) is a highly sensitive and accurate derivative of linear sweep voltammetry or staircase voltammetry in which a series of regular voltage pulses are superimposed on the potential linear sweep or stairsteps. The differential nature of DPASV reduces the influence of background charging current on the chemical analysis giving it the accuracy advantage.

The deposition potential is −0.5 V to −1.3 V vs. Ag/AgCl electrode, and the deposition time is 420 s. Sodium acetate buffer (pH=6) is used. The oxidation potential is −0. 77 V for Cd (II), −0.52 V for Pb (II), −0.053 V for Cu (II), and 0.29 V for Hg (II) vs. Ag/AgCl electrode. In the DPASV plots, the peak current rises as the concentration of metal ions increases.

Linear calibration plots are obtained for Cd (II), Cu (II), and Hg (II) ions in the range 0.4μM −2 μM while calibration characteristic for Pb (II) ion is linear in the range 0.4 μM −1.2 μM. The LoD for Cd (II) ion is 3.252 nM, for Pb (II) ion is 2.013 nM, for Cu (II) ion is 4.104 nM, and for Hg (II) ion is 5.547 nM (Muralikrishna et al. 2014).

6.4.4   Au-Reduced Graphene Oxide (rGO) Nanocomposite-Modified GCE as As (III) Sensor

A surfactant-free Au-reduced GO-modified glassy carbon electrode is able to detect As (III) in the range 0.3 ppb–20 ppb with a limit of detection ~0.1 ppb (Li et al. 2015).

6.4.4.1   Au Nanocomposite Synthesis

It is done by a one-pot reduction process of GO and $HAuCl_4$ under irradiation with ultraviolet rays. GO obtained by a modified Hummer's method is ultrasonicated in water. A yellow-brown GO solution results. After adjusting its pH to 7.0 by adding NaOH solution, 1% $HAuCl_4$ is added. The stirred mixture is loaded in a transparent vial and exposed to a high-intensity UV lamp for 20 min. while bubbling nitrogen through it for stirring, thereby ensuring uniformity of exposure to UV irradiation and preventing re-oxidation by dissolved oxygen. The Au-rGO formed is separated by centrifugation, washed with water, and redispersed in water for use.

6.4.4.2 Modified GCE Fabrication

Mirror-smooth polishing of GCE is done using alumina slurry on a wet cloth. After ethanol cleaning, water rinsing, and nitrogen drying, the GCE is coated with 5 μL Au-rGO solution and dried.

6.4.4.3 Anodic Stripping Linear Sweep Voltammetry, and Sensor Characteristics

Anodic stripping linear sweep voltammetry (ASLSV) is a simple voltammetry technique. In ASLSV, the potential between the working electrode and the reference electrode is swept linearly in time, and the corresponding current at the working electrode is measured. ASLSV involves only a single linear sweeping of potential from the lower to the upper limit, whereas in cyclic voltammetry, the potential is linearly cycled in both directions, switching backward and forward between the negative and positive potential limits.

ASLSV is done by scanning the potential from –0.3 V to 0.5 V vs. Ag/AgCl electrode in the anodic direction in 0.2M HCl solution. The resulting voltammogram is recorded. The optimum deposition potential is 0.4 V and the deposition time is 30 s. The metal is removed at 0.3 V for 30 s.

From 0.3 ppb to 20 ppb, the peak current varies linearly with As (III) concentration. The correlation coefficient is 0.993 (Li et al. 2015).

6.4.5 Alkaline-Intercalated Ti_3C_2 (Alk-Ti_3C_2)-Modified Glassy Carbon Electrode (GCE) for Simultaneous Detection of Cd (II), Pb (II), Cu (II), and Hg (II) Ions

Alkalization-intercalated Ti_3C_2 (alk-Ti_3C_2) is employed as the electrode material for concurrent electrochemical detection of four heavy metal ions (Zhu et al. 2017).

6.4.5.1 Electrode Modification Process

The process of making the modified electrode progresses as:

$$Ti_3AlC_2 \rightarrow Ti_3C_2 \rightarrow Alk - Ti_3C_2 \rightarrow Alk - Ti_3C_2 - Modified\,GCE \qquad (6.5)$$

(i) Ti_3AlC_2 powder is obtained by mixing Ti, Al, TiC, and graphite powders and pressure-less calcination of the mixture.

(ii) Multilayer Ti_3C_2 powder is made by sieving Ti_3AlC_2 powder through a 300-mesh dipped in 40wt% HF and stirring at 40°C for 48 h to etch Al, rinsing with DI water, separating by centrifuging, collecting using ethanol, and drying in a vacuum oven.

(iii) Alk-Ti_3C_2 is prepared by dispersal of multilayer Ti_3C_2 powder in 5wt% KOH solution, magnetic stirring, washing with DI water, centrifuging for separation from the supernatant, and vacuum drying. The structure of alk-Ti_3C_2 is illustrated in Figure 6.5. Its structural morphology and layer thicknesses are investigated by Yuen et al. (2021).

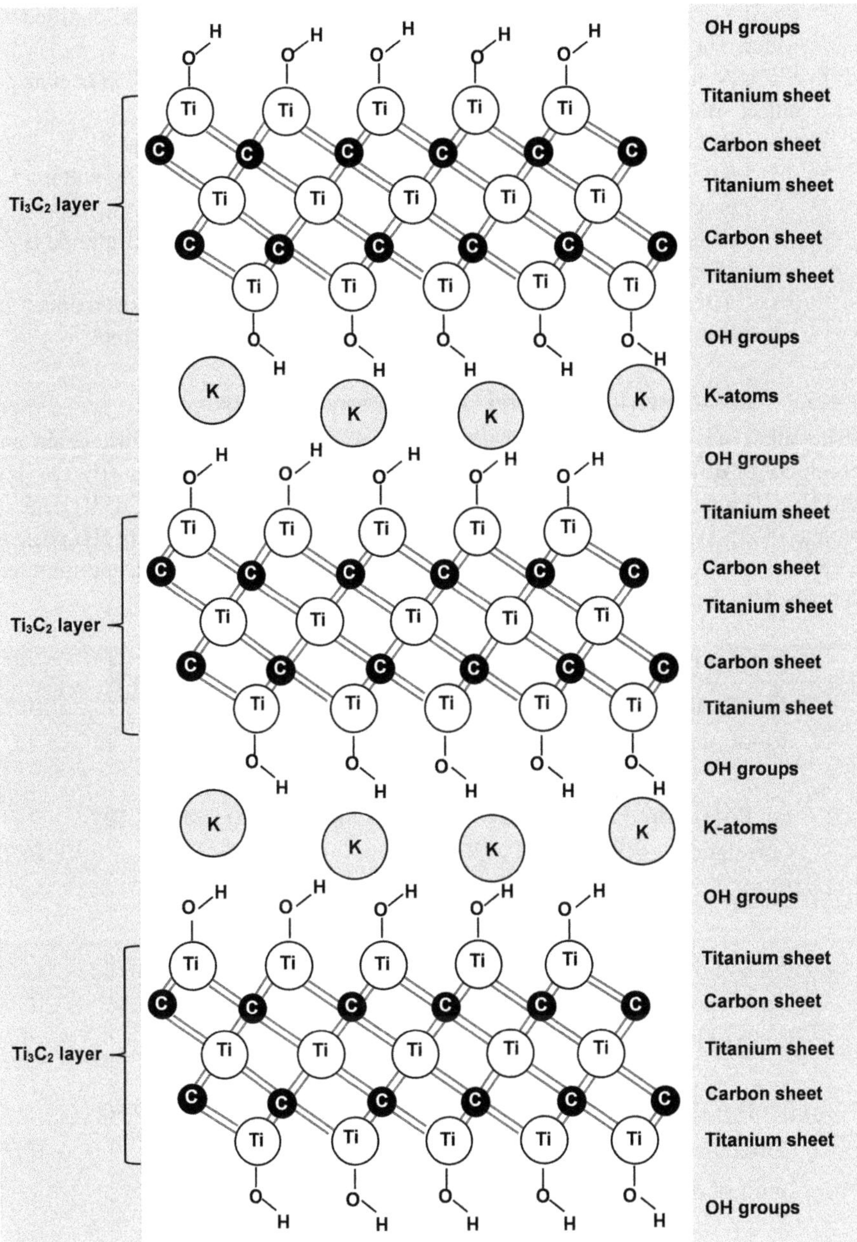

FIGURE 6.5 Alk-Ti$_3$C$_2$ structure. Three Ti$_3$C$_2$ layers are seen. The OH groups are introduced after etching in HF and the K atoms after the alkalization treatment. The unit cell of Ti$_3$C$_2$ looks hexagonal. Carbon sheets are sandwiched between titanium sheets. The thickness of a single layer of Ti$_3$C$_2$ is ~1.14–1.18 nm. The interstacking layer gap is ~2.03 nm–2.15 nm. The resulting total thickness is ~3.2 nm.

(iv) Alk-Ti$_3$C$_2$ -modified GCE is made by pipetting the sonicated suspension of alk-Ti$_3$C$_2$ in anhydrous ethanol on the GCE surface polished with alumina powder and sonicated with aqueous HNO$_3$ solution, ethanol, and distilled water. The solvents are air-dried.

(v) SWASV measurements are done using alk-Ti$_3$C$_2$-modified GCE in acetate buffer solutions containing the target heavy ions:

(a) Preconcentration is done at −1.2 V for 150 s under stirring.

(b) After waiting for 15 s to attain equilibrium, anodic stripping voltammograms are recorded in the potential range −1 V to 0.5 V. Increment potential in the squared wave potential scan is 4 mV and amplitude is 25 mV. The frequency is 15 Hz.

(c) After each anodic stripping, residues of metal ions are removed under a desorption potential of 0.8 V for 120 s with constant stirring.

6.4.5.2 Voltammetric Peaks and Linearization Equations

Using alk-Ti$_3$C$_2$ -modified GCE, distinct and well-separated voltammetric peaks are observed for different ions during simultaneous detection of Cd (II), Pb (II), Cu (II), and Hg (II) ions in the concentration range of 0.1 μM –1.5 μM. The Cd (II) peak is seen at −0.75 V, Pb (II) peak at −0.5 V, Cu (II) peak at −0.05 V, and Hg (II) peak at 0.25 V vs. Ag/AgCl electrode. The variations of peak current ion concentration are described by the linearization equations:

$$\text{For Cd(II) ion : Current}\,(\mu A) = 7.15 \times \text{Concentration}\,(\mu M) - 0.94,$$
$$\text{Correlation coefficient} = 0.9935, \tag{6.6}$$
$$\text{Limit of detection} = 0.098\,\mu M$$

$$\text{For Pb(II) ion : Current}\,(\mu A) = 32.10 \times \text{Concentration}\,(\mu M) - 1.40,$$
$$\text{Correlation coefficient} = 0.9980, \tag{6.7}$$
$$\text{Limit of detection} = 0.041\,\mu M$$

$$\text{For Cu(II) ion : Current}\,(\mu A) = 20.59 \times \text{Concentration}\,(\mu M) - 1.78,$$
$$\text{Correlation coefficient} = 0.9993, \tag{6.8}$$
$$\text{Limit of detection} = 0.032\,\mu M$$

$$\text{For Hg(II) ion : Current}\,(\mu A) = 20.47 \times \text{Concentration}\,(\mu M) - 9.80,$$
$$\text{Correlation coefficient} = 0.9694, \tag{6.9}$$
$$\text{Limit of detection} = 0.130\,\mu M$$

These results show that alk-Ti$_3$C$_2$-modified GCE offers a reliable method for selectively detecting co-existing heavy metal ions. Individual ions are also detectable (Zhu et al. 2017).

6.5 2D MATERIAL FET SENSORS

FET sensors serve as viable alternatives to electrochemical devices due to their non-dependence on electrolyte solutions and multiple electrodes. In this section, we shall come across some examples of biosensors, which will be indicated in the names of devices. The biosensors will be elucidated in Chapter 7. The biosensors are integrated bio-receptor/physico-chemical transducer devices using specific biochemical reactions mediated through biological recognition elements, e.g., deoxyribonucleic acids, enzymes, tissues, organelles, or whole cells.

6.5.1 TGA-AuNPs/rGO HYBRID-BASED FET SENSOR FOR Hg (II) IONS

An FET sensor is developed using thermally reduced GO sheets ornamented with thioglycolic acid (TGA)-functionalized gold nanoparticles (AuNPs), represented as TGA-AuNPs/rGO hybrid structure, for detecting Hg (II) ions in aqueous media (Chen et al. 2012). Figure 6.6 shows the FET sensor.

6.5.1.1 Sensor Construction and Working Principle

In this sensor (Figure 6.6), the silicon substrate is the back gate. 200 nm thermally grown SiO_2 is the gate dielectric. A Cr/Au-interdigitated electrode (IDE) pattern

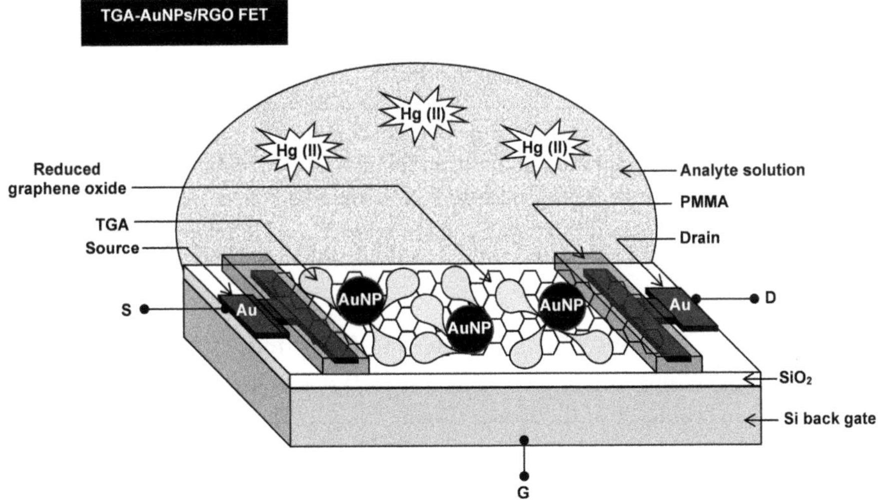

FIGURE 6.6 Schematic diagram of the TGA-AuNPs/rGO FET sensor for Hg (II) ions. The diagram shows an SiO_2/Si substrate with patterned Au electrodes labelled as the source and drain electrodes and connected to terminals S and D. The Si substrate acts as the back gate G. Between the source–drain electrodes and on the surface of SiO_2, a layer of reduced graphene oxide, AuNPs, and TGA is seen. A large portion of source/drain electrodes is coated with PMMA for protection from the solution. For measuring the sensor response, the solution containing Hg (II) ions is poured over the TGA-AuNPs/rGO composite layer.

on the SiO_2 surface provides an interdigitated source–drain geometry. Thermally reduced GO forms the conducting channel. The rGO layer is decorated with thioglycolic acid (TGA)-functionalized gold nanoparticles, which act as the recognition groups for Hg (II) ions. The TGA-AuNPs are receptors of the target Hg (II) ions, imparting sensing specificity to the device. Hg (II) ions readily adsorb on gold. TGA is a thiol, an organic compound containing the sulfhydryl group. It binds to Hg (II) ions which readily adsorb on AuNPs. As the target ions are adsorbed and bound to TGA-AuNPs, they either donate or withdraw electrons from the rGO channel. Hence, the channel conductance changes. The change in channel conductance is related to the Hg (II) ion concentration.

6.5.1.2 Making the Graphene Oxide Sheets Suspension

GO is synthesized by modified Hummers' method followed by dissolution in water, centrifugation, and exfoliation to obtain a suspension of GO sheets.

6.5.1.3 Sensor Fabrication

The sensor fabrication steps are portrayed in Figure 6.7. The fabrication process starts with an SiO_2/Si substrate (Figure 6.7(a)). An interdigitated electrode pattern is defined by electron-beam lithography on the SiO_2/Si substrate (Figure 6.7(b)). After Cr/Au metallization by electron-beam deposition, the photoresist is lifted off to form Cr/Au IDEs. A droplet of GO suspension is pipetted over the IDEs and dried. The resulting structure is heated at 300°C in flowing argon for 1 h during which GO is converted into rGO, and an intimate contact is established between rGO and Cr/Au IDEs (Figure 6.7(c)). AuNPs are deposited over the surface of rGO sheets by electrospray process using electrical dispersion and electrostatic-force-directed assembly (ESFDA) in which the charges on deposited NPs influence the congregation of subsequent particles (Figure 6.7(d)). The time for assembly of AuNPs is ~2 h. The IDE regions are protected by a PMMA covering, deposited by spin-coating, and defined by electron-beam lithography (Figure 6.7(e)). Au NPs are coated with TGA by immersion in 10 mM TGA for 24 h (Figure 6.7(f)). Extra TGA is removed by washing with DI water.

6.5.1.4 Sensor Response to Hg (II) and Interfering Ions

A droplet of the sample solution containing Hg (II) ions is dispensed over the TGA-AuNPs/rGO hybrid layer of the sensor (Figure 6.7(g)). The investigated Hg (II) ion concentration range is 2.5×10^{-8}M to 1.42×10^{-5}M. The smallest Hg (II) ion concentration sensed is 2.5×10^{-8} M. The drain current responds within <10 s to the analyte solution. Selectivity and specificity analysis showed that the sensor has a weak response toward: Na^+, Ca^{2+} ions; has a very weak response to Cd^{2+} and Zn^{2+}ions; but exhibits some sensitivity to Fe^{3+} ions. However, the limit of detection for Fe^{3+} ions (5×10^{-6} M) is much larger than that for Hg (II) ions (Chen et al. 2012).

6.5.2 DNA-Au NPs/MoS$_2$ Nanosheet Hybrid-Based FET Biosensor for Hg (II) Ions

An FET sensor for water quality monitoring is built from an MoS_2 nanosheet decorated with DNA-functionalized Au nanoparticles (DNA-AuNPs/MoS$_2$NS hybrid structure) (Zhou et al. 2016). Figure 6.8 shows the FET sensor.

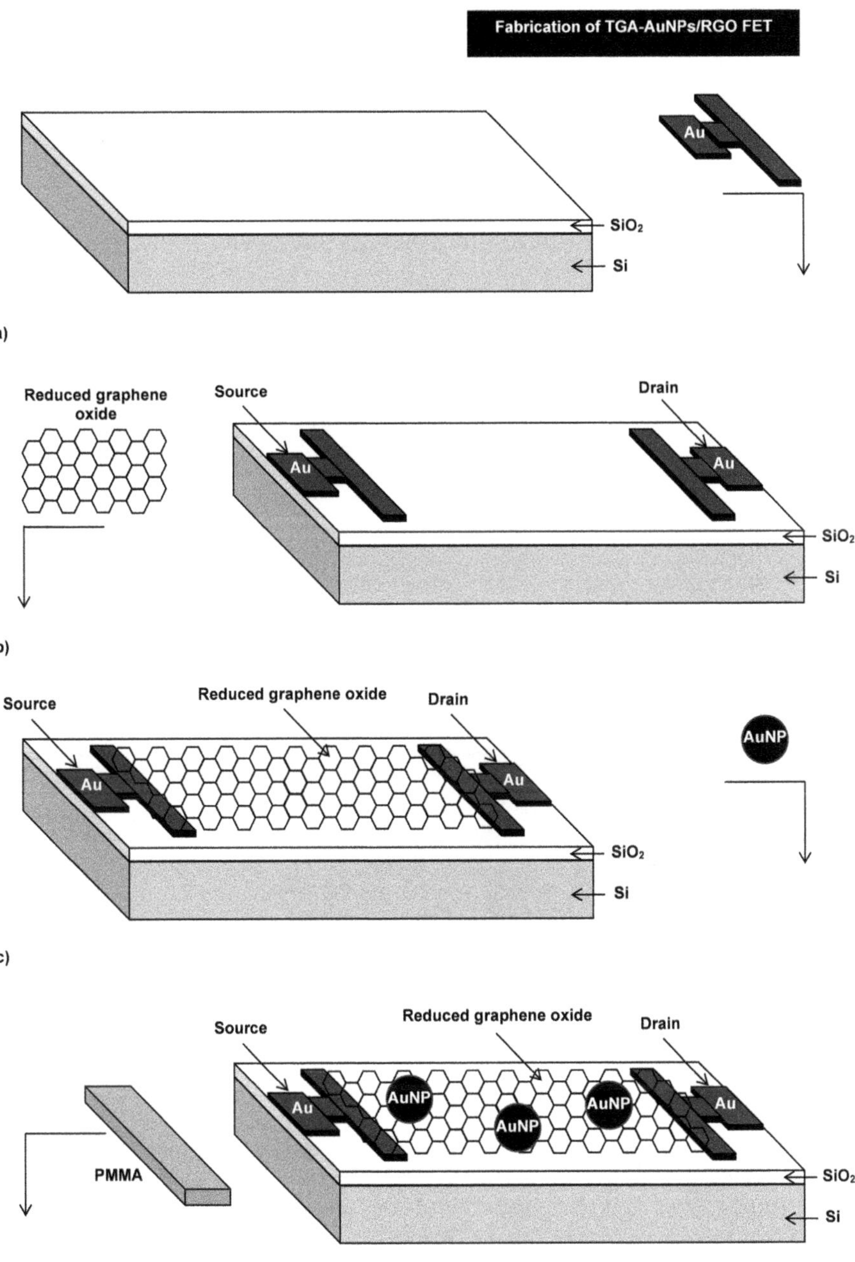

FIGURE 6.7 Fabrication of the TGA-AuNPs/rGO FET sensor, and its characterization: (a) SiO₂/Si substrate, (b) Cr/Au electrode pattern definition, (c) reduced graphene oxide transfer, (d) AuNPs deposition, (e) IDE covering by PMMA, (f) TGA coating, and (g) putting a drop of Hg (II) ion solution on the TGA-AuNPs/rGO layer. Part (a) shows the starting SiO₂/Si substrate. Part (b) shows the pattern of Au electrodes made on the SiO₂/Si substrate. Part (c) shows the reduced graphene oxide film overlaid on SiO₂ layer, touching the Au electrodes on the two sides. Part (d) shows the AuNPs deposited over the graphene film on SiO₂ layer.

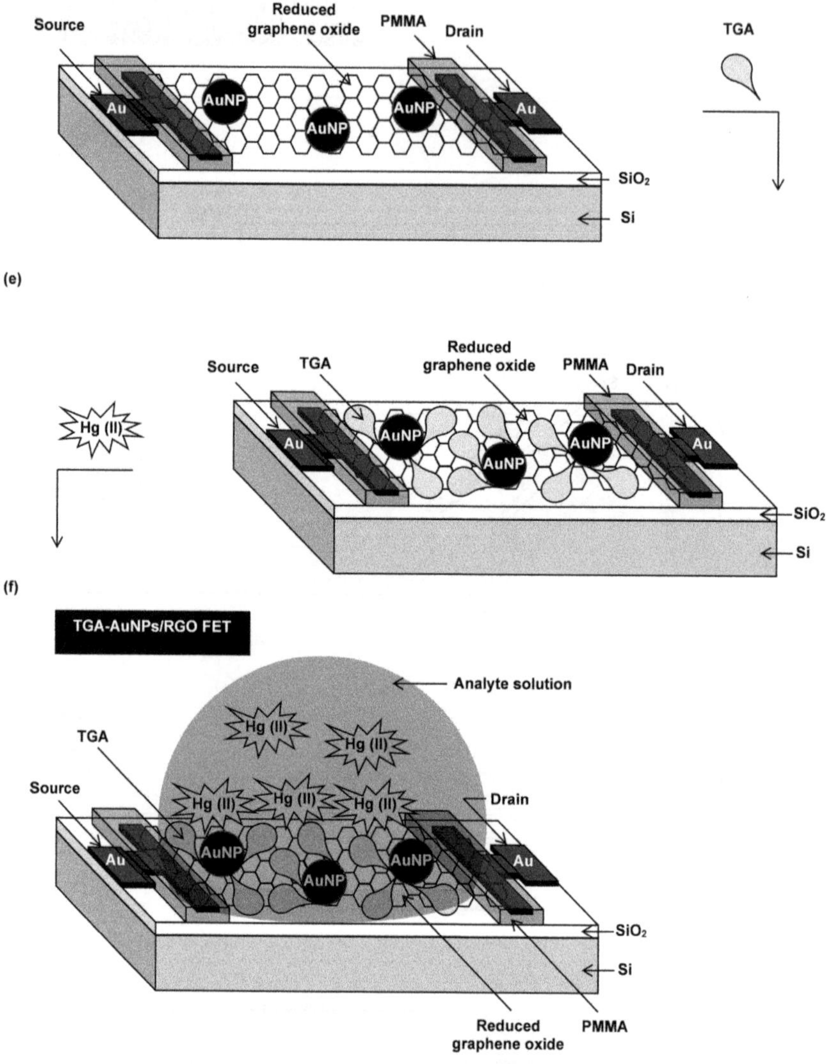

FIGURE 6.7 (Continued) Part (e) shows the portion of Au electrodes adjoining the reduced graphene oxide film encapsulated with PMMA for protection during subsequent liquid treatments. Part (f) shows the AuNPs coated with TGA. Part (g) shows the sensor covered with a droplet of analyte solution containing Hg (II) ions.

6.5.2.1 Transduction Principle

Hg (II) ions adsorb on the DNA probe molecules attached to AuNPs by reaction with thymidine of DNA molecules forming T-Hg (II)-T chelates. By this binding between Hg (II) ions and DNA probes, a positively charged layer of Hg (II) ions accumulates on AuNPs, attracting electrons from MoS_2NSs to AuNPs and creating

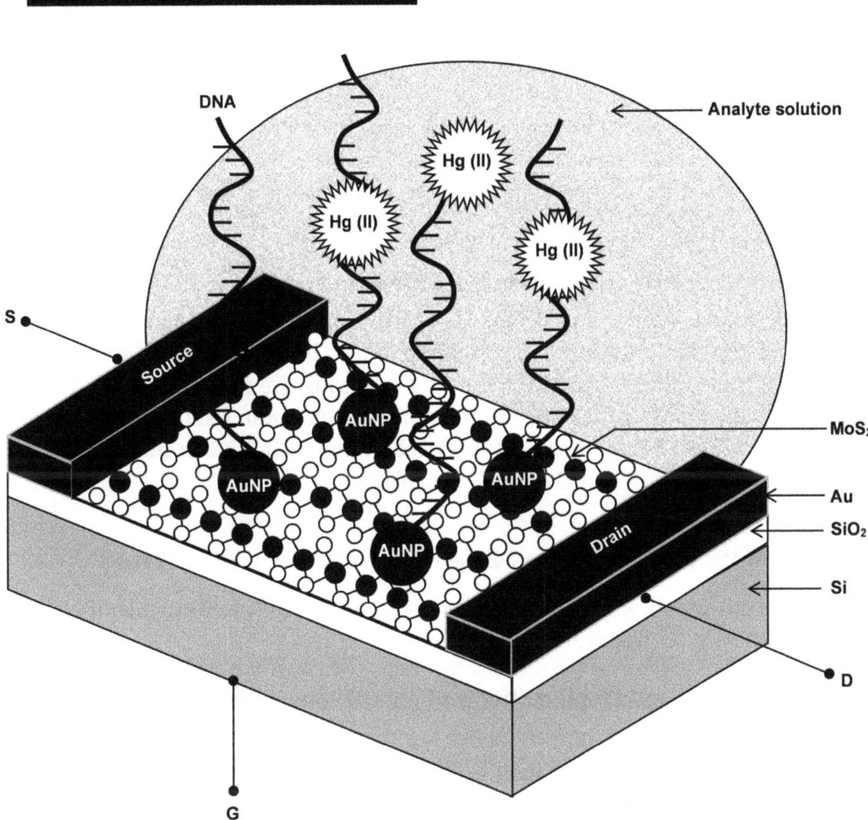

FIGURE 6.8 DNA-Au NPs/MoS$_2$ nanosheet FET sensor for Hg (II) ions. The diagram shows an SiO$_2$/Si substrate with Au source/drain electrodes on its two sides and wire terminals S, D. The Si substrate is the back gate with the wire terminal labelled as 'G'. The SiO$_2$ layer between the source/drain electrodes is covered with an MOS$_2$ nanosheet over which AuNPs are deposited. DNA is immobilized on AuNPs. A drop of Hg (II)-ion solution is poured over the DNA-AuNPs/MoS$_2$ nanosheet.

holes in MoS$_2$, thus increasing the conductivity of the P-type MoS$_2$ nanosheet and, hence, raising the drain–source current of the FET depending on the adsorbed Hg (II) ion concentration.

6.5.2.2 MoS$_2$ NSs Exfoliation from MoS$_2$ Crystals

MoS$_2$ NSs are exfoliated by lithium-ion intercalation. MoS$_2$ crystals are immersed in butyllithium (C$_4$H$_9$Li) solution in an Ar-filled flask for 7 days and centrifuged to get Li$_x$MoS$_2$, which is washed with hexane (C$_6$H$_{14}$) and subjected to ultrasonication in water. A diluted dispersion is filtered through a membrane, obtaining an MoS$_2$ film, which is delaminated and used in sensor fabrication.

6.5.2.3 Sensor Fabrication

Figure 6.9 depicts the fabrication steps of the sensor. An Au interdigitated electrode (IDE) pattern is photolithographically defined on an SiO_2 (200 nm)/Si substrate (Figure 6.9(a)). The delaminated MoS_2 film is transferred over the SiO_2 layer and the IDE pattern (Figure 6.9(b)). Thermal annealing is done at 250°C for 1 h in (Ar+H_2). A radio-frequency (RF) sputtering machine is used to deposit AuNPs on the surface of MoS_2 film (Figure 6.9(c)). The AuNPs-deposited structure is dipped in DNA solution in PBS, and after 1 h incubation at room temperature, rinsed with DI water and dried (Figure 6.9(d)).

6.5.2.4 Effect of Hg (II) Ions on FET Characteristics

The drain–source voltage V_{DS} of the FET is fixed at 0.1 V. The device is exposed to different Hg (II) ion concentrations (Figure 6.9(e)). The corresponding drain–source current I_{DS} values are recorded. The sensor shows sensitive detection of Hg (II) ions in the calibrated range of 0.1–10 nM with quick response in <2 s, and insensitivity to As^{3+}, Zn^{2+}, Fe^{3+}, Cu^{2+}, Pb^{2+}, Ca^{2+}, Cd^{2+}, etc. ions. Stability and repeatability are checked from 1 pM to 100 nM (Zhou et al. 2016).

6.5.3 MOS$_2$ DEFECTS HEALING-BASED FET SENSOR FOR HG (II) IONS

An Hg (II) ion sensor works by healing the defects in MoS_2 semiconductor channel of an FET device (Urbanos et al. 2021).

6.5.3.1 Healing and Doping Effects of Hg (II) Ions on Point Defects in MoS$_2$

Intrinsic structural defects in MoS_2 make it a reactive chemical species. Hg (II) ions have an affinity for point defects in MoS_2. They also interact with sulfur atoms in the MOS_2 lattice. Exposure to Hg (II) ions heals the defects in MoS_2 and also dopes it P-type. Hg (II) ions act as P-dopants for MoS_2. Defect healing is validated by the weakening of defect-related peaks in X-ray photoelectron spectroscopy (XPS). The binding energy of MoS_2 semiconductor shifts by 0.2 eV indicating its P-doping.

6.5.3.2 Fabrication of the MoS$_2$ FET Sensor

The sensor is fabricated by exfoliation of MoS_2 flakes on SiO_2 (285 nm)/P^{++} Si substrate, followed by laser optical lithography, 5 nm gold evaporation, and lift-off in acetone. High-vacuum annealing at 10^{-6} mbar-10^{-7} mbar at 180°C overnight removes any residual and adsorbed materials.

6.5.3.3 Variation of P-Doping Concentration of MoS$_2$ and Threshold Voltage with Hg (II) Exposure

Transfer characteristics of MoS_2 FETs exposed to Hg (II) ions confirm the P-doping effect of Hg (II) ions and a close relation between the degree of P-doping and the concentration of Hg (II) ions is established. The P-doping variation with different Hg (II) ion concentrations is used as a parameter to measure Hg (II) ion concentration, as observed through the change in threshold voltage (ΔV_{Th}) with Hg (II) ion concentration. The device is highly sensitive and selective for Hg (II) ions. It shows good reversibility. The limit of Hg (II) ion detection is 1 pM (Urbanos et al. 2021).

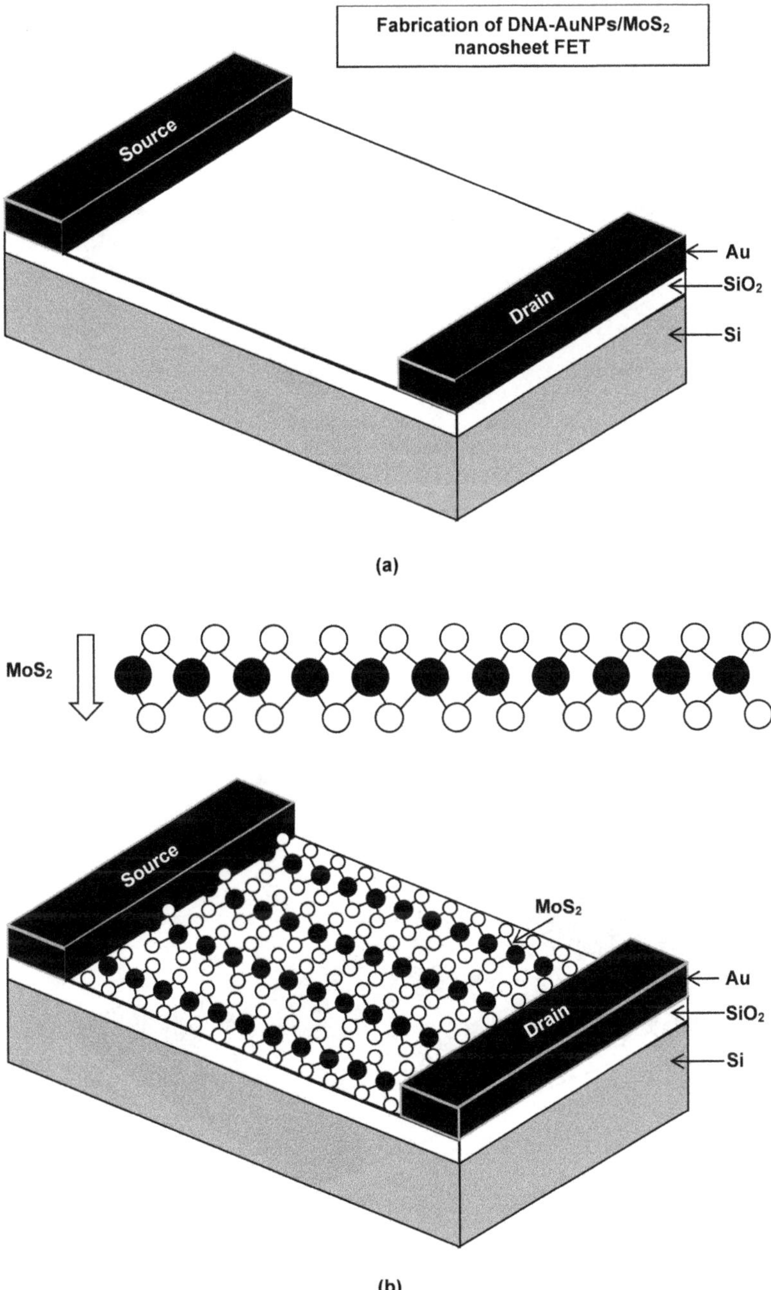

FIGURE 6.9 Fabrication and characterization of the DNA-Au NPs/MoS$_2$ nanosheet FET sensor for Hg (II) ions: (a) Au electrode patterning on an SiO$_2$/Si substrate, (b) transferring the MoS$_2$ film, (c) AuNPs sputtering, (d) dip coating in DNA solution, and (e) pouring Hg (II) ion solution. Part (a) shows Au source/drain electrodes patterned on an SiO$_2$/Si substrate. Part (b) shows the MoS$_2$ film laid over the SiO$_2$ surface between the source/drain electrodes.

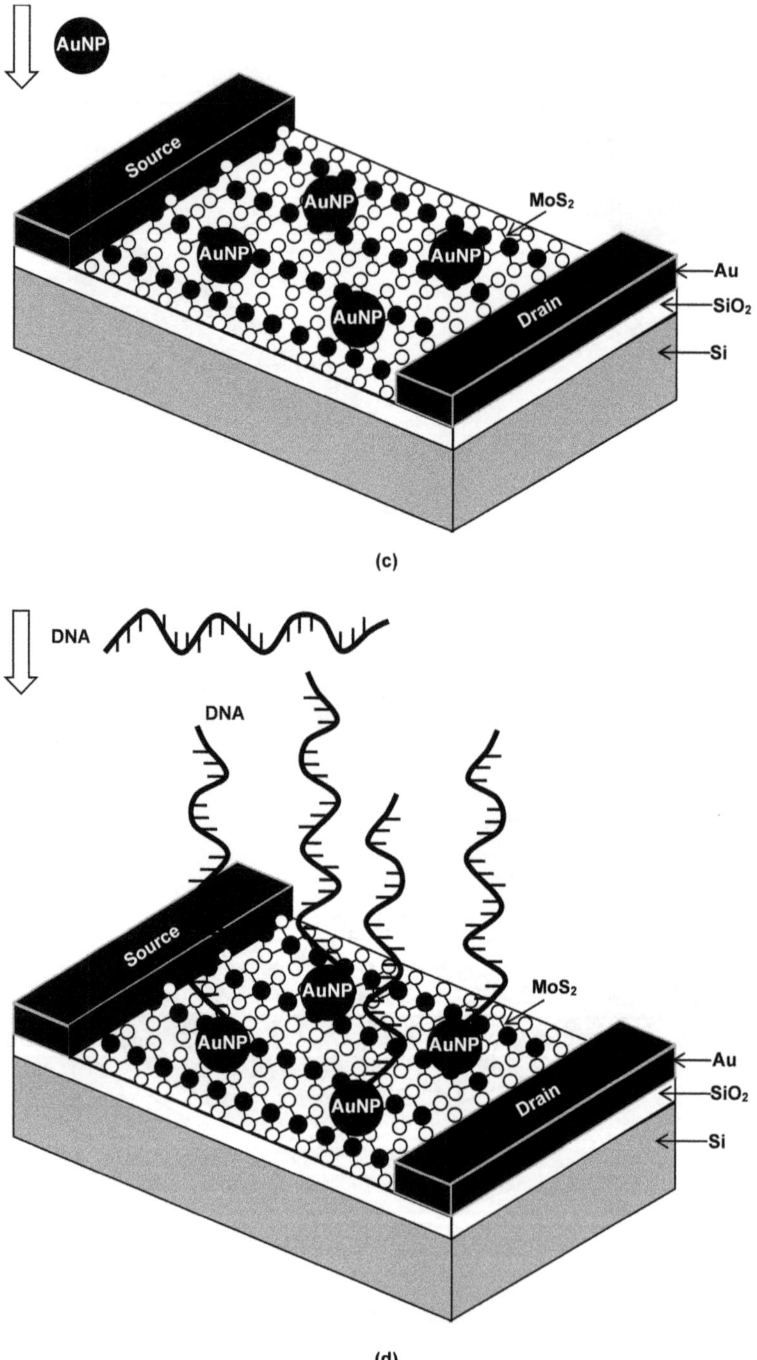

FIGURE 6.9 (Continued) Part (c) shows AuNPs sputtered over the surface of SiO_2. Part (d) shows DNA fastened to AuNPs.

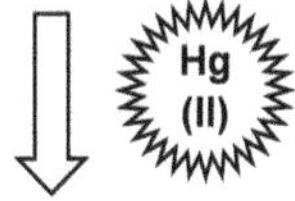

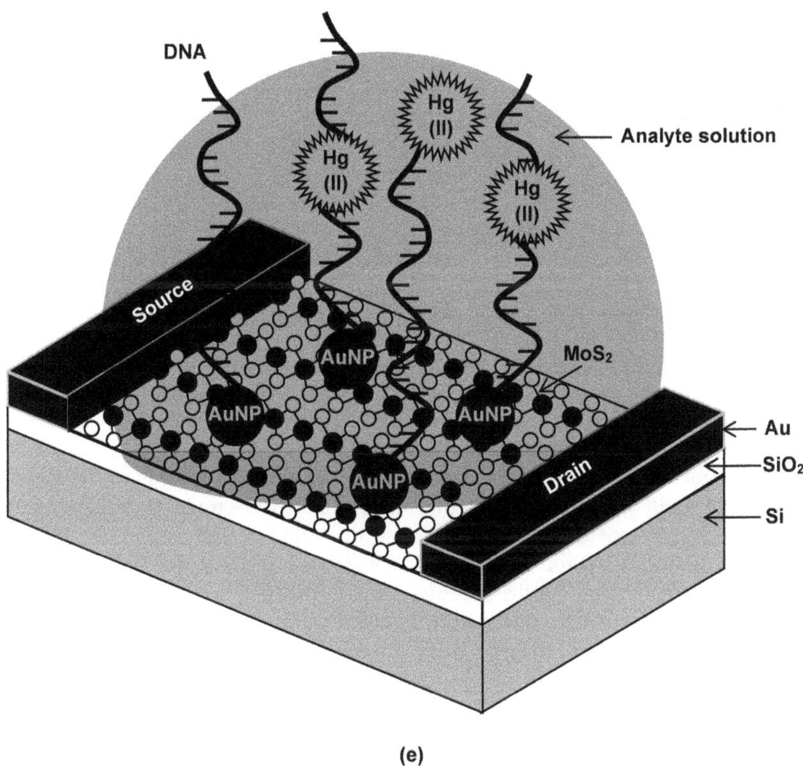

(e)

FIGURE 6.9   (Continued)   Part (e) shows a drop of Hg (II) ion solution poured on the surface of MoS_2 film.

6.5.4   DNAzyme–AuNPs/Graphene Liquid-Gated FET Biosensor for Pb (II) Ions

An FET sensor working on DNAzyme-mediated detection of Pb^{2+} ions is demonstrated. The use of DNAzyme as the recognition element for Pb^{2+} ions bestows high selectivity on the sensor (Wen et al. 2013). The sensor is shown in Figure 6.10.

6.5.4.1   Principle of Pb²⁺ Ion Detection

DNAzymes or deoxyribozymes are a family of enzymes consisting of DNA oligonucleotides taking part in catalytic activity. They are more stable but lower in

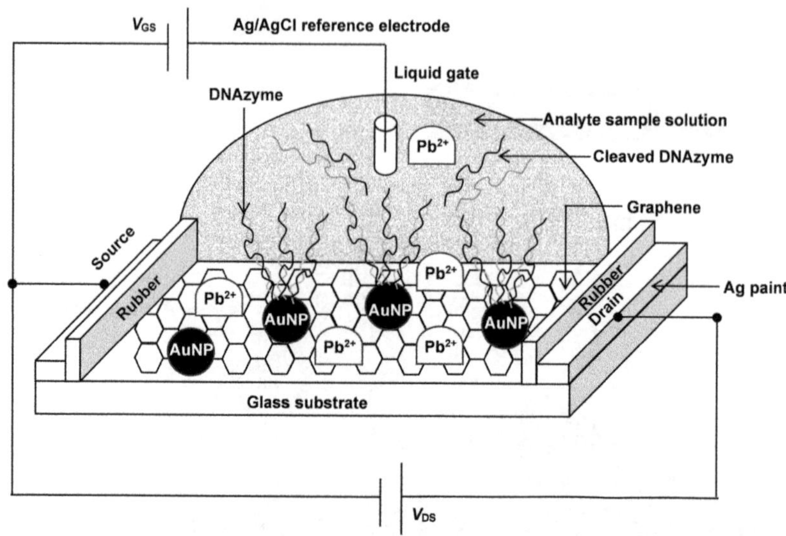

FIGURE 6.10 DNAzyme–AuNPs/graphene FET sensor for Pb (II) ions. The diagram shows a glass substrate with Ag-painted source/drain electrodes on the two edges. Rubber walls are erected adjacent to the source/drain electrodes to make an enclosed space in which liquid solutions are poured without electrically shorting the source/drain electrodes. In the space between the rubber walls, a graphene film is seen over which AuNPs are deposited. DNAzymes are immobilized on AuNPs. A drop of the Pb^{2+} ion solution is placed over the DNAzyme–AuNPs/graphene layer for carrying out the analysis. Some DNAzymes are cleaved by the Pb^{2+} ions while others are intact. An Ag/AgCl reference electrode is dipped in the solution to act as a liquid gate. The drain–source and gate–source biasings are done with batteries V_{DS} and V_{GS}.

cost than RNA and protein enzymes. The DNAzyme used here is a Pb^{2+}-dependent DNAzyme. It is a double-stranded DNA. The two strands are an enzymatic strand (17E) and a thiolated substrate strand (17S). In Pb^{2+} solution, the Pb^{2+} ions bind with the DNAzyme. Then the substrate strand is cleaved by the enzymatic strand. After this cleavage, the enzymatic strand and the unthiolated segment of the substrate strand diffuse away. The AuNP now carries the thiolated fragment of the substrate strand. Consequently, the initial coupling between the charged DNAzyme complex and graphene is altered.

The device is sensitive to Pb^{2+} ion concentration with the Dirac point (the minimum conduction point where the conduction and valence bands meet each other, $V_{g,min}$) shifting to more positive voltages with an increase in Pb^{2+} ion concentration.

6.5.4.2 Sensor Fabrication

The FET biosensor is made on a glass substrate, as shown in Figure 6.11. Graphene film is separately grown by chemical vapor deposition on a copper foil in (Ar+H_2) flow. The graphene growth is done by raising the temperature to 950°C, then passing

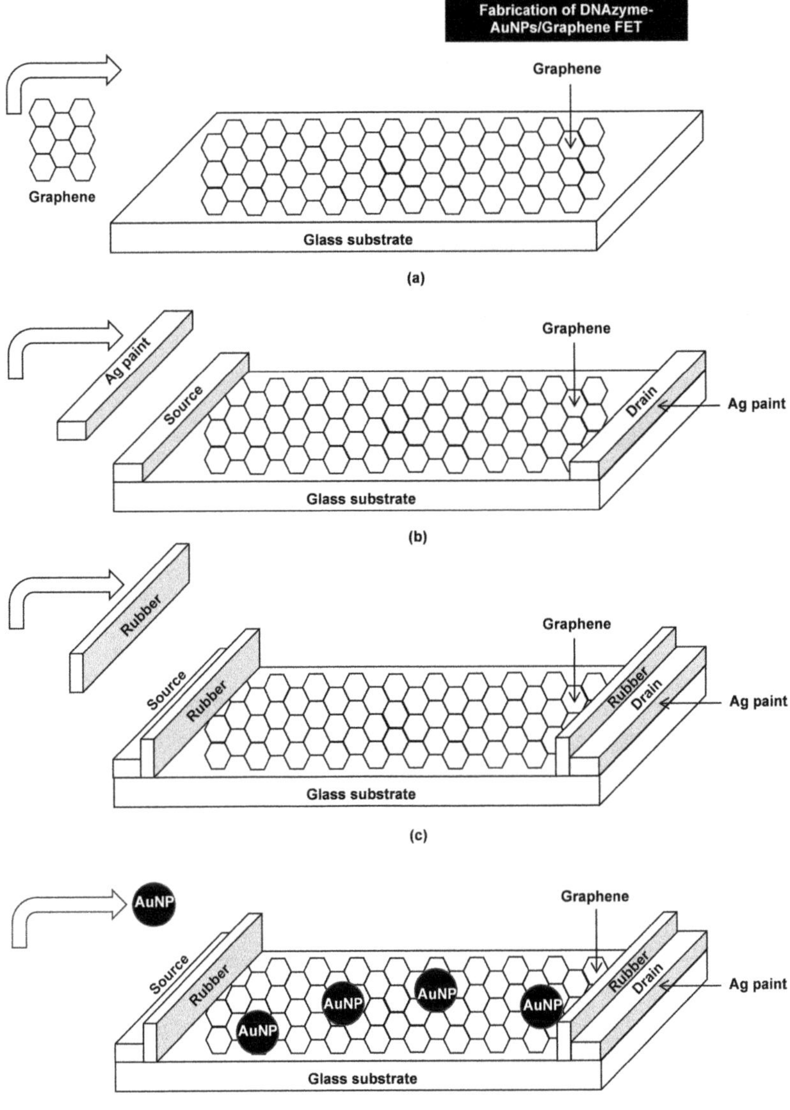

FIGURE 6.11 Fabrication process of DNAzyme–AuNPs/graphene FET sensor, and its liquid gating for measurement: (a) graphene film laying on a glass substrate, (b) silver painting for source and drain electrodes, (c) erection of silicon rubber walls, (d) AuNPs deposition, (e) DNAzyme immobilization, and (f) pouring Pb^{2+} ion sample solution. Part (a) shows graphene film laid over the SiO_2 surface of the SiO_2/Si substrate. Part (b) shows the Ag paint applied at both edges of the graphene film for making electrical contacts. Part (c) shows silicone rubber walls constructed adjoining the silver paint for source and drain electrodes to prevent any solution flowing over the paint, thereby short-circuiting the two electrodes. Part (d) shows AuNPs deposited on the graphene film by immersion in chloroauric acid solution. Note that the solution is prevented from flowing to the source/drain electrodes by the rubber walls.

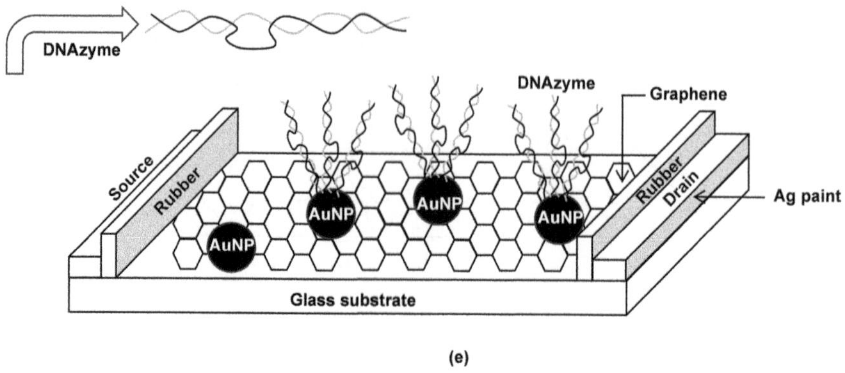

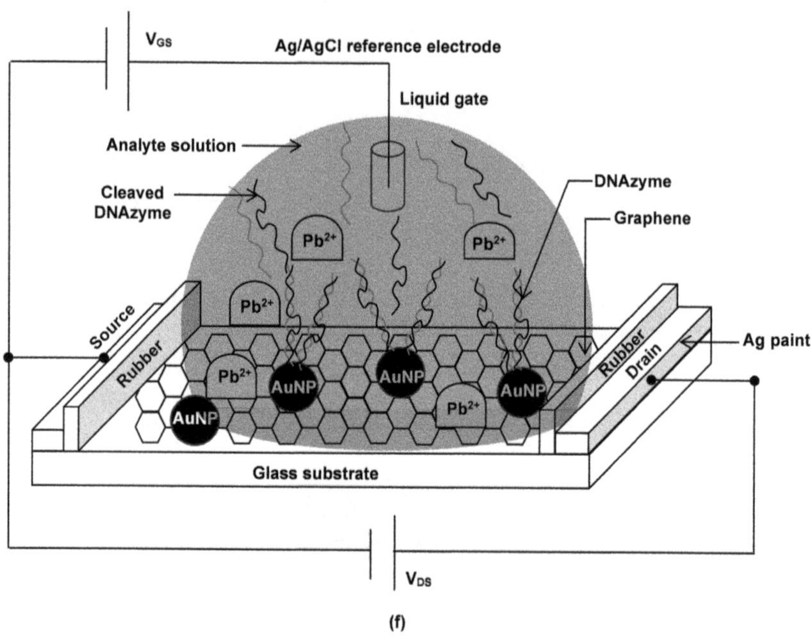

FIGURE 6.11 (Continued) Part (e) shows DNAzymes immobilized on AuNPs by incubating in DNAzyme solution. Part (f) shows the setting up of a liquid gate by immersing an Ag/AgCl reference electrode in the Pb^{2+}ion solution and connecting it to the battery applying gate-source voltage V_{GS}. Another battery is connected between source and drain electrodes for supplying the drain-source bias V_{DS}. Enzyme cleaving by Pb^{2+} ions is shown.

methane gas for 10 min, and cooling down in Ar. The graphene film on copper foil (GR/Cu) is spin-coated with PMMA and cured. The Cu foil is etched away and the graphene film is transferred over the glass substrate (Figure 6.11(a)). Silver paint is applied to make electrodes at the two ends of the graphene film (Figure 6.11(b)). The electrodes are insulated with silicone rubber forming a reaction compartment (Figure 6.11(c)). The AuNPs are deposited by immersion in $HAuCl_4$ solution to get AuNPs-coated devices (Figure 6.11(d)). On these devices, DNAzyme is immobilized by 2 h incubation with DNAzyme solution prepared by hybridizing two separately synthesized strands, 17E and thiolated-17S of DNAzyme to thiolated double-stranded DNAzyme in Tris buffer and 1M NaCl, followed by heating to 90°C for 3 min and slow cooling (Figure 6.11(e)). Rinsing in HEPEs buffer removes non-specifically adsorbed DNAzyme.

6.5.4.3 Pb²⁺ Ion Detection

The transfer characteristics of the sensor (drain–source current with respect to liquid gate voltage curves) are measured before and after reaction with various Pb^{2+} ion concentrations in the range 0.1–100 nM (Figure 6.11(f)). The shift in Dirac point ($\Delta V_{g,min}$) vs. Pb^{2+} ion concentration plot is obtained. At 0.1 nM, Pb^{2+} ion concentration $V_{g,min}$ (gate voltage of minimum conductivity) increases by 31.33 mV while at 100 nM Pb^{2+} ion concentration, the shift in $V_{g,min}$ is 150 mV. The detection limit of the sensor is 20 pM (Wen et al. 2013).

6.6 2D MATERIAL OPTICAL SENSORS

6.6.1 GRAPHENE-LIKE C_3N_4 PHOTOELECTROCHEMICAL SENSOR FOR CU (II) IONS

Photo-electrochemistry is concerned with the interaction of light with electrochemical systems. The photoelectrochemical sensor recognizes Cu (II) ions using an ITO (indium tin oxide)/graphene-like C_3N_4 electrode illuminated with light. For determining the Cu (II) ion concentration, the photocurrent is measured in an electrochemical cell under a light source (She et al. 2014). Figure 6.12 shows the photoelectrochemical sensor.

6.6.1.1 Mechanism of Cu (II) Ion Detection

Graphene-like carbon nitride (graphene-like C_3N_4) shows enhanced photocatalytic activity than bulk graphitic carbon nitride (g-C_3N_4) owing to a larger specific area (32.54 m^2 g^{-1}) than bulk form (3.3 m^2 g^{-1}) and a larger bandgap (2.79 eV) than for bulk form (2.65 eV). As a consequence, it has more photogenerated carriers and higher photocurrent making the detection of small amounts of Cu (II) ions easier.

Because the energy level of conduction band of ITO is lower than that of graphene-like C_3N_4, the photogenerated electrons move from graphene-like C_3N_4 to ITO while photogenerated holes move to the surface of graphene-like C_3N_4, leading to effective separation of the photogenerated electrons and holes and minimizing their recombination. From ITO, the photogenerated electrons move through PBS solution to the Pt electrode. Any Cu (II) ions present in PBS solution capture the electrons forming metallic Cu which is deposited on the Pt electrode. So, the Cu (II) ions help in the

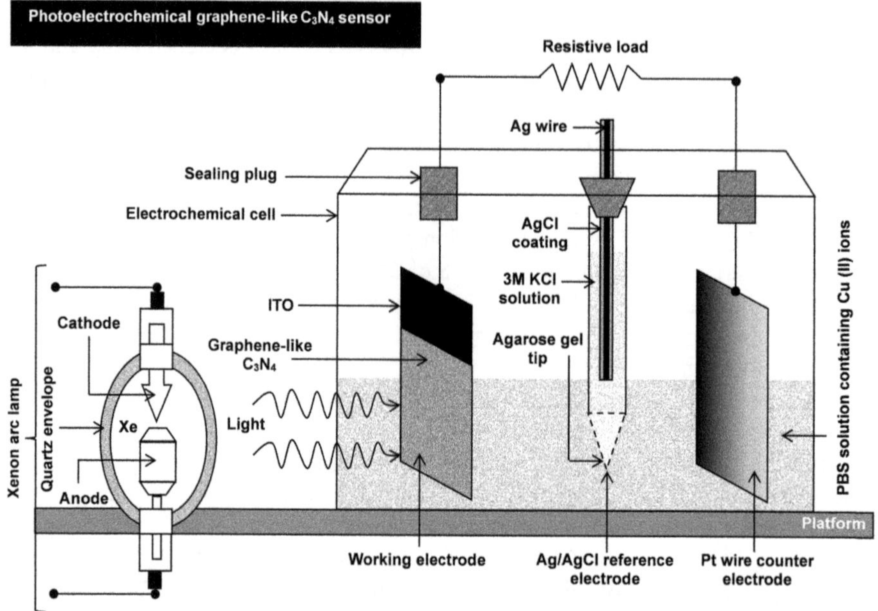

FIGURE 6.12 Photoelectrochemical sensor for Cu (II) ions. The diagram shows an electrochemical cell. In this cell, the sample of the Cu (II)-ion containing PBS solution is taken. The working electrode is made of indium tin oxide coated with graphene-like C_3N_4. This electrode is illuminated with light from a xenon arc lamp consisting of a quartz envelope filled with xenon gas, and fitted with a tungsten anode and a tungsten cathode at their opposite ends. The reference electrode is an Ag/AgCl electrode. It consists of a silver wire coated with silver chloride and dipped in a glass tube filled with 3M KCl solution. The bottom end of this glass tube is closed with a tip made of agarose gel or a similar porous material. For the counter electrode, a platinum wire is used. The current flows across a resistive load connected between the working and counter electrodes. The reference electrode is a standard stable potential against which the potential of the working electrode is determined.

separation of electrons and holes and increase the photocurrent. The photocurrent indicates the Cu (II) ion concentration.

6.6.1.2 Making the Graphene-Like C_3N_4-Modified ITO Electrode

Graphene-like C_3N_4 powder is dispersed in ethanol by ultrasonication. A small quantity of the dispersion is drop-casted on an ITO slice and dried under an IR lamp.

6.6.1.3 Preparation of Graphene-Like Carbon Nitride

Graphene-like carbon nitride (graphene-like C_3N_4) is prepared form bulk graphitic carbon nitride (g-C_3N_4) by liquid exfoliation in 1,3-butanediol (1,3-BUT) [$CH_3CH(OH)CH_2CH_2OH$]. Bulk graphitic carbon nitride is added to 1,3-BUT. The mixture is sonicated for 24 h and centrifuged. The resulting suspension is taken in a watch glass and kept in an oven at 140°C for 10 h.

6.6.1.4 Photoelectrochemical Performance of the Sensor

A 500 W Xe arc lamp is used for illumination. The photocurrent of ITO/graphene-like C_3N_4 sensor is measured for different Cu (II) concentrations in the range 0–13 µM using a three-electrode configuration consisting of graphene-like C_3N_4-modified ITO electrode as the working electrode, a platinum wire as the counter electrode and an Ag/AgCl reference electrode with PBS supporting electrolyte. A linear relationship is observed between the photocurrent and Cu (II) ion concentration, as represented by the equation between them (She et al. 2014):

$$\text{Photocurrent}\,(\mu A) = 1.4629\,\text{Cu}\,(\text{II}) - \text{Ion Concentration}\,(\mu M) + 1.2805 \quad (6.10)$$

6.6.2 2D G-C_3N_4 NANOSHEETS FLUORESCENT PROBE FOR COPPER (II) IONS

Excitation of g-C_3N_4 nanosheets at 300 nm with a UV lamp causes blue fluorescence emission at 434 nm with a quantum yield up to 10.3 %, which is >> that of bulk g-C_3N_4 (6.0 %). In the presence of a small quantity of Cu (II) ions, the intensity of emitted light decreases. This quenching of fluorescence by Cu (II) ions is applied for their detection (Guo et al. 2016).

6.6.2.1 Reason for Fluorescence Quenching

The quenching is interpreted to arise from the intense fastening affinity and rapid chelating kinetics during the reaction of Cu (II) ions with N and O atoms of the functional groups on the surfaces of g-C_3N_4 nanosheets. Furthermore, the redox potential of the $Cu^{2+}/Cu+$ couple existent between the conduction and valence bands of g-C_3N_4 nanosheets assists the PET (photo-induced electron transfer, the electron transfer between molecules in the excited and ground states), from the conduction band to Cu^{2+} ion, expediting the quenching of fluorescence.

6.6.2.2 Bulk g-C_3N_4 Synthesis by Pyrolysis of Urea

Yellow bulk graphitic carbon nitride (g-C_3N_4) powder is synthesized by 2 h-heating of urea $[CO(NH_2)_2]$ in a covered crucible at 600°C in a muffle furnace, at a heating rate 15°C min^{-1}.

6.6.2.3 Fluorescent g-C_3N_4 Nanosheets Preparation by Protonation and Exfoliation

Bulk g-C_3N_4 is added to a mixed (H_2SO_4+HNO_3) solution with DI water, ultrasonicated for 4 h, and washed with DI water. The precipitate is separated by a Buchner funnel with a pump, repeatedly washed, collected, and dried in vacuum at 60°C for 24 h. The dark yellow protonated g-C_3N_4 thus obtained is dispersed in water, ultrasonicated for 2 h, and centrifuged. The g-C_3N_4 nanosheets in the supernatant layer are taken out and stored at 4°C.

6.6.2.4 Assay Procedure and Sensor Response

The assay is carried out by the addition of Cu (II) solutions of concentrations from 0.01 to 0.4 nmol L^{-1} and measurement of fluorescence emission intensity with

$\lambda_{Excitation} = 300$ nm and $\lambda_{Emission} = 434$ nm. The fluorescence measurements are performed using a luminescence spectrometer.

The assay reveals that the intensity of fluorescence emission decreases with increasing Cu (II) ion concentration. The relation between the ratio = original emission intensity (F_0)/emission intensity (F) at a particular concentration, and the Cu (II) ion concentration is linear.

The sensor is responsive to Cu (II) ions in a broad range of concentrations with a linear range from 0.01 $nmolL^{-1}$ to 0.4 $nmolL^{-1}$. The limit of detection is 8 $pmolL^{-1}$. The detection of Cu (II) ions is not affected by the presence of interfering ions, e.g., Ag (I), Li (I), Pb (II), Zn (II), Co (II), Ca (II), Ni (II), Cd (II), Hg (II), Fe (III), and Al (III) ions (Guo et al. 2016).

6.6.3 Monolayer MoS$_2$ Chemiluminescent Sensor for Mercury (II) Ions

Monolayer MoS$_2$ is used to build a chemiluminescence sensor for Hg (II) ions working on the luminol-H$_2$O$_2$-MoS$_2$ reaction (Yang et al. 2019). Figure 6.13 shows the effect of Hg (II) ions on the chemiluminescence signal of the sensor.

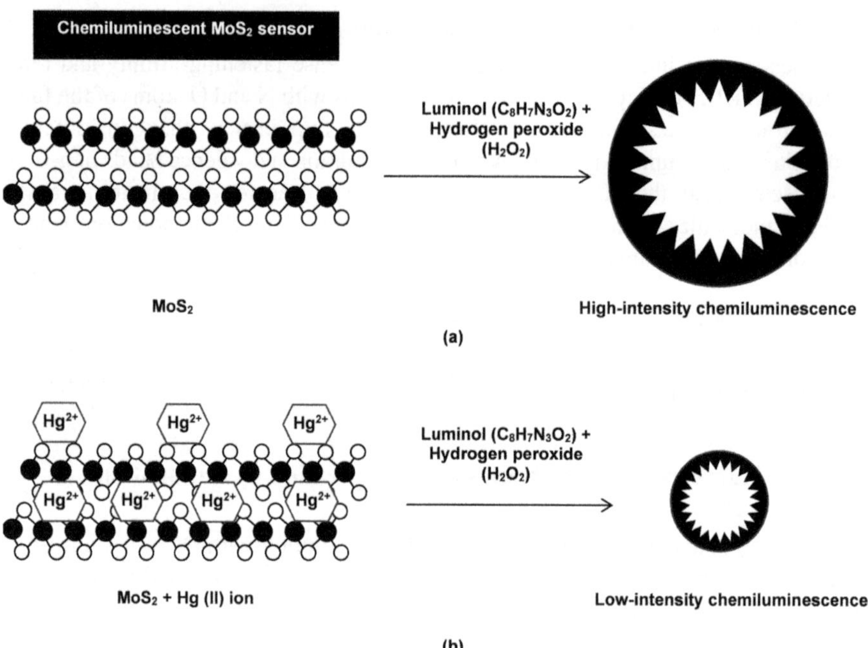

FIGURE 6.13 Chemiluminescence signals of MoS$_2$ sensor for Hg (II) ions: (a) without Hg (II) ions and (b) with Hg ions. Part (a) shows the high-intensity chemiluminescence emission produced by luminol-H$_2$O$_2$ reaction as no mercury ions are absorbed by MoS$_2$ from a Hg (II) ion-free solution. Part (b) shows the low-intensity chemiluminescence generated in (luminol+hydrogen peroxide) reaction with MoS$_2$+Hg (II) ions consequent to significant adsorption of Hg (II) ions by MoS$_2$ from a solution containing these ions.

6.6.3.1   Luminol Chemiluminescence Reaction

Luminol (3-aminophthalhydrazide or 5-amino-2,3-dihydro- 1,4-phthalazinedione) $(C_8H_7N_3O_2)$ is an organic compound, which on oxidation emits light of wavelength 425 nm. This phenomenon called chemiluminescence (CL) is used by fireflies to emit light. Glow sticks and roadside emergency lights too utilize this effect. As H_2O_2 is an oxidizing agent, the oxidation of luminol by H_2O_2 is a commonly used chemiluminescent reaction in chemical analysis.

6.6.3.2   Peroxidase-Like Activity of Monolayer MoS$_2$, and its Adsorption Ability for Hg (II) Ions

Monolayer MoS_2 displays peroxidase-like activity. In an aqueous solution, it catalyzes the oxidation of luminol by H_2O_2. The catalyzation reaction enhances the intensity of chemiluminescent emission during this reaction. Thus, the luminol-H_2O_2-MoS_2 reaction produces a strong CL emission.

Monolayer MoS_2 also shows a good adsorption ability to Hg (II) ions arising from the inherent affinity of Hg (II) ions to the sulfur atoms in MoS_2. The adsorption of Hg (II) ions by monolayer MoS_2 molecules inhibits their catalyzing activity. Hence, the intensity of luminol CL emission decreases in the presence of Hg (II) ions.

The combination of two properties of monolayer MoS_2, namely, the catalyzation of luminol oxidation and the adsorption of Hg (II) ions by it, helps in designing a sensor for Hg (II) ions which works on dependence of the quenching of the chemiluminescent emission of luminol by H_2O_2 on the concentration of Hg (II) ions. The greater the quenching of intensity, higher the Hg (II) ion concentration is. So, the concentration of Hg (II) ions in aqueous solutions is inferred from the decrease in light emission intensity of the luminol-H_2O_2-MoS_2 CL reaction.

6.6.3.3   Assay Procedure

A known quantity of MoS_2 of a specific concentration is mixed with different concentrations of mercury nitrate. After 20 min incubation, the mixtures are poured into the wells of a microplate and mixed with H_2O_2. Then luminol is injected into the wells, and the CL intensity is measured by a Fluroskan Ascent microplate reader.

6.6.3.4   Effect of Hg (II) Ion Concentration on CL Intensity of Luminol-H$_2$O$_2$-MoS$_2$ Reaction

The CL intensities of luminol-H_2O_2-MoS_2 CL reaction are measured in the absence of Hg (II) ions and in the presence of several Hg (II) ion concentrations from 0.001 μM to 40 μM. For a given Hg (II) ion concentration, the quenching efficiency is defined as

$$\text{Quenching efficieny} = \frac{\text{CL intensity in absence of Hg(II) ions} - \text{CL intensity in presence of Hg(II) ions}}{\text{CL intensity in absence of Hg(II) ions}}$$

$$(6.11)$$

A linear relationship is observed between quenching efficiency and Hg (II) ion concentration in the range 0.005–μM40 μM. The limit of detection of Hg (II) ions is estimated as 1 nM (Yang et al. 2019).

6.7 DISCUSSION AND CONCLUSIONS

2D materials-based chemical and biosensors point towards a new research direction for building a real-time, low-cost, portable, user-friendly analytical platform for rapid inline analysis to monitor environmental pollution caused by heavy metals (Table 6.1). This platform will serve as a first-sight analysis of the pollution scenario supplementing elaborate mainstream methods such as atomic absorption spectroscopy (AAS), high-performance liquid chromatography (HPLC), inductively coupled plasma optical emission spectroscopy (ICP-OES), and inductively coupled plasma mass spectroscopy (ICP-MS) (Zhou 2017). AAS uses the atomic absorption spectra of elements as their fingerprints to measure their concentrations in a sample. HPLC

TABLE 6.1

Different Types of 2D Material Sensors for Heavy Metal Ions

Sl. No.	Type of 2D Material Sensor	Sensing Device	Heavy Metal Ion (s)	Reference
1.	2D material modified electrodes	SnO_2/rGO nanocomposite-modified GCE	Cd (II), Pb (II), Cu (II) and Hg (II)	Wei et al. (2012)
		Graphene-Pt nanocomposite-modified GCE	As (III)	Kempegowda et al. (2014)
		L-Cysteine/rGO-modified GCE	Cd (II), Pb (II), Cu (II) and Hg (II)	Muralikrishna et al. (2014)
		Au-rGO nanocomposite-modified GCE	As (III)	Li et al. (2015)
		Alk-Ti_3C_2-modified GCE	Cd (II), Pb (II), Cu (II) and Hg (II)	Zhu et al. (2017)
2.	2D material FET sensors	TGA-AuNPs/rGO-FET sensor	Hg (II)	Chen et al. (2012)
		DNA-Au NPs/MoS_2 nanosheet-FET sensor	Hg (II)	Zhou et al. (2016)
		MOS_2 defects healing-FET sensor	Hg (II)	Urbanos et al. (2021)
		DNAzyme–AuNPs/graphene liquid-gated FET sensor	Pb (II)	Wen et al. (2013)
3.	2D material optical sensors	Graphene-like C_3N_4 photoelectrochemical sensor	Cu (II)	She et al. (2014)
		2D g-C_3N_4 nanosheets fluorescent probe	Cu (II)	Guo et al. (2016)
		Monolayer MoS_2 chemiluminescent sensor	Hg (II)	Yang et al. (2019)

segregates the various compounds present in a liquid sample by using high-pressure pumps and mixtures of solvents (the mobile phase) and by delivering the sample mixture into a cylindrical column containing solid adsorbents (the stationary phase). The components are then qualitatively and quantitatively analyzed. ICP-OES is an emission spectroscopy technique in which inductively coupled plasma is used to produce excited atoms and ions. These atoms/ions emit electromagnetic radiations at wavelengths characteristic of particular elements, from which they are detected. ICP-MS is a mass spectrometry method using an inductively coupled plasma for ionization of samples to produce atomic and polyatomic ions for the detection. As all these techniques require extensive instrumentation systems, the 2D material sensors will offer a simplistic, viable alternative route for detecting heavy metals.

REFERENCES

Alengebawy A., S. T. Abdelkhalek, S. R. Qureshi and M. Q. Wang 2021 Heavy metals and pesticides toxicity in agricultural soil and plants: Ecological risks and human health implications, *Toxics*, 9(3), pp. 1–33.

Bard A. J. 1983 Chemical modification of electrodes, *Journal of Chemical Education*, 60, 4, pp. 302–304.

Briffa J., E. Sinagra and R. Blundell 2020 Heavy metal pollution in the environment and their toxicological effects on humans, *Heliyon*, 6(9), pp. 1–26.

Chen K., G. Lu, J. Chang, S. Mao, K. Yu, S. Cui, and J. Chen 2012 Hg(II) ion detection using thermally reduced graphene oxide decorated with functionalized gold nanoparticles, *Analytical Chemistry*, 84(9), pp. 4057–4062.

Elgrishi N., K. J. Rountree, B. D. McCarthy, E. S. Rountree, T. T. Eisenhart, and J. L. Dempsey 2018 A practical beginner's guide to cyclic voltammetry, *Journal of Chemical Education*, 95(2), pp. 197–206.

Guo X., Y. Wang, F. Wu, Y. Ni and S. Kokot 2016 Preparation of protonated, two-dimensional graphitic carbon nitride nanosheets by exfoliation, and their application as a fluorescent probe for trace analysis of copper (II), *Microchimica Acta*, 183, pp. 773–780.

Hu T., Q. Lai, W. Fan, Y. Zhang and Z. Liu 2023 Advances in portable heavy metal ion sensors, *Sensors*, 23(8), pp. 1–17.

Jaishankar M., T. Tseten, N. Anbalagan, B. B. Mathew, and K.N. Beeregowda 2014 Toxicity, mechanism and health effects of some heavy metals, *Interdisciplinary Toxicology*, 7(2), pp. 60–72.

Kempegowda R., D. Antony and P. Malingappa 2014 Graphene–platinum nanocomposite as a sensitive and selective voltammetric sensor for trace level arsenic quantification, *International Journal of Smart and Nano Materials*, 5(1), pp. 17–32.

Li W.-W., F.-Y. Kong, J.-Y. Wang, Z.-D. Chen, H.-L. Fang, and W. Wang 2015 Facile one-pot and rapid synthesis of surfactant-free Au-reduced graphene oxide nanocomposite for trace arsenic (III) detection, *Electrochimica Acta*, 157, pp. 183–190.

Mirceski V., R. Gulaboski, M. Lovric, I. Bogeski, R. Kappl and M. Hoth 2013 Square-wave voltammetry: A review on the recent progress, *Electroanalysis*, 25(11), pp. 2411–2422.

Muralikrishna S., K. Sureshkumar, T. S. Varley, D. H. Nagaraju and T. Ramakrishnappa 2014 *In situ* reduction and functionalization of graphene oxide with L-cysteine for simultaneous electrochemical determination of cadmium (II), lead (II), copper (II), and mercury (II) ions, *Analytical Methods*, 6, pp. 8698–8705.

She X., H. Xu, Y. Xu, J. Yan, J. Xia, L. Xu, Y. Song, Y. Jiang, Q. Zhang and H. Li 2014 Exfoliated graphene-like carbon nitride in organic solvents: enhanced photocatalytic activity and highly selective and sensitive sensor for the detection of trace amounts of Cu^{2+}, *Journal of Materials Chemistry A*, 2, pp. 2563–2570.

Urbanos F. J., S. Gullace and P. Samorì 2021 Field-effect-transistor-based ion sensors: Ultrasensitive mercury (II) detection via healing MoS$_2$ defects, *Nanoscale*, 13, pp. 19682–19689.

Wei Y., C. Gao, F.-L. Meng, H.-H. Li, L. Wang, J.-H. Liu, and X.-J. Huang 2012 SnO$_2$/reduced graphene oxide nanocomposite for the simultaneous electrochemical detection of cadmium (II), lead (II), copper (II), and mercury (II): An interesting favorable mutual interference, *Journal of Physical Chemistry C*, 116, pp. 1034–1041.

Wen Y., F.Y. Li, X. Dong, J. Zhang, Q. Xiong, and P. Chen 2013 The electrical detection of lead ions using gold-nanoparticle- and DNAzyme-functionalized graphene device. *Advanced Healthcare Materials*, 2(2), pp. 271–274.

Wygant B. R., and T. N. Lambert 2022 Thin film electrodes for anodic stripping voltammetry: A mini-review, *Frontiers in Chemistry*, 9, Article 809535, pp. 1–8.

Yang H., Y. Huang, Y. Zhao and A. Fan 2019 Sensitive chemiluminescent sensing method for mercury (II) ions based on monolayer molybdenum disulfide, *Analytical Sciences*, 35, pp. 551–556.

Yuen A. C. Y., T. B. Y. Chen, B. Lin, W. Yang, I. I. Kabir, I. M. D. C. Cordeiro, A. E. Whitten, J. Mata, B. Yu, H.-D. Lu, and G. H. Yeoh 2021 Study of structure morphology and layer thickness of Ti$_3$C$_2$ MXene with small-angle neutron scattering (SANS), *Composites Part C: Open Access*, 5, p. 100155.

Zhou G. 2017 Real-time, selective detection of heavy metal ions in water using 2D nanomaterials-based field-effect transistors, Doctor of Philosophy in Engineering, The University of Wisconsin-Milwaukee, Theses and Dissertations, 1731, pp. 1–20, https://dc.uwm.edu/etd/1731

Zhou G., J. Chang, H. Pu, K. Shi, S. Mao, X. Sui, R. Ren, S. Cui and J. Chen 2016 Ultrasensitive mercury ion detection using DNA-functionalized molybdenum disulfide nanosheet/gold nanoparticle hybrid field-effect transistor device, *ACS Sensors*, 1(3), pp. 295–302.

Zhu X., B. Liu, H. Hou, Z. Huang, K. M. Zeinu, L. Huang, X. Yuan, D. Guo, J. Hu and J. Yang 2017 Alkaline intercalation of Ti$_3$C$_2$ MXene for simultaneous electrochemical detection of Cd (II), Pb (II), Cu (II) and Hg (II), *Electrochimica Acta*, 248, pp. 46–57.

7 2D Materials-Based Gas Sensors

A gas sensor or gas detector is an electronic device helping us to know whether a particular gas is present in an environment, and the concentration of the identified gas, often for the safety of personnel and for controlling atmospheric pollution, e.g., detection of the presence of combustible, flammable, explosive, or poisonous gases in factories and manufacturing plants, chemical production, oil extraction and petroleum industries, automobile exhausts, or coal mines. It is used in homes too for sounding the alarm in case of leakage of liquefied petroleum gas used for cooking, or for warning against carbon monoxide presence. It prevents fire hazards and monitors air quality for domestic and public safety. Air pollutants are implicated in causing and aggravating health problems, e.g., asthma, chronic obstructive pulmonary disease, lung cancer, leukemia, heart disease, and even untimely death.

2D materials have a larger specific surface area compared with 0D and 1D nanomaterials or bulk materials leading to higher sensitivity of 2D material gas sensors. The selectivity for gaseous adsorption varies with the symmetry of the structure and the operating temperature of the sensor, e.g., the lower symmetry of triclinic WO_3 crystal than that of monoclinic and hexagonal WO_3 crystals engenders higher sensitivity and selectivity of the triclinic WO_3 crystal for acetone molecules than monoclinic and hexagonal structures; and triclinic WO_3 crystal is more responsive to NO_2 at a lower temperature but more sensitive to acetone at a higher temperature (Ge et al. 2019).

A plethora of gas sensors have been developed utilizing the physical and chemical capabilities of 2D materials for gas adsorption, notably graphene, transition metal dichalcogenides (TMDCs), phosphorene, metal oxides, etc., A few of these devices will be discussed in this chapter.

7.1 CONFIGURATIONS OF ELECTRICAL GAS-SENSING DEVICES

7.1.1 CHEMIRESISTIVE GAS SENSOR

It works on the change in electrical resistance of a layer of 2D-sensing material upon chemical interaction with the gas. Gas concentration is measured from the change in the resistance of the 2D material before and after exposure to gas. Miniaturization and portability, ease of fabrication and upscaling for mass production, cost-effectiveness, and low-power consumption are some of the advantages of chemiresistive gas sensors.

DOI: 10.1201/9781003330585-7

7.1.2 Schottky Diode Gas Sensor

A Schottky diode is a metal–semiconductor junction. To make a Schottky diode gas sensor using 2D materials, a junction is formed between a semiconductor and a metal or a metal-like material. Among the two materials constituting the junction, either the semiconductor or the metal-like material is taken as a 2D material. The junction formation creates a Schottky barrier acting as a gate controller for current flow through the barrier. The magnitude of current flowing through the Schottky diode depends on Schottky barrier height, which in turn varies with gas adsorption. The dependence of Schottky barrier height on gas concentration tremendously improves the sensitivity of the device. Even if the adsorption of a gas produces a small change in Schottky barrier height, a large change in current takes place (Mathew and Rout 2021).

7.1.3 Heterojunction Gas Sensor

It is made of two dissimilar 2D semiconductor materials, mostly between P- and N-type materials, in contact with each other. The interface between the two participating semiconductors is the active region of the device. An interfacial potential barrier is formed due to the difference in Fermi levels of the two semiconductors. The sensor works through the change in its resistance induced by the adsorption of gas on either surface of the sensor. Redox reactions of adsorbed gas with sensing material vary the charge carrier concentration in the device thereby affecting the built-in potential barrier. As a consequence, its charge-transfer characteristic, and hence current flow behavior of the device are affected. For a rectifying heterojunction, a resistance change is observed obeying the exponential law (Yang et al. 2021).

Change in resistance of the sensor is proportional to

$$\exp\left(-\frac{\text{Electronic charge} \times \dfrac{\text{Change in barrier height at the interface of two semiconductors}}{}}{\text{Boltzmann constant} \times \text{Ambient temperature}}\right)$$

or,

$$\Delta R \propto \exp\left(-\frac{q\Delta V_{\text{Barrier}}}{k_{\text{B}}T}\right) \tag{7.1}$$

which shows that a miniscule alteration of barrier height will cause a stupendous change in the resistance of the sensor, thus greatly improving the performance of the sensor.

7.1.4 Field-Effect Transistor (FET) Gas Sensor

A 2D semiconductor film, e.g., MoS_2 monolayer, is mounted on a dielectric layer, e.g., SiO_2 on a silicon substrate. The 2D material film has metal contacts, typically Cr/Au, on both ends serving as the source and drain electrodes. The Si substrate acts as the back gate. The 2D material acts as the channel of the FET. Adsorption of a gas by the 2D material alters the channel conductance. As a result, the drain–source current registers a change. The change in drain–source current under a constant drain–source voltage is measured to determine the gas concentration. The Si gate is biased to modulate the channel conductance.

7.1.5 QUARTZ CRYSTAL MICROBALANCE GAS SENSOR

A piezoelectric quartz crystal is set into vibratory motion by an applied voltage. The resonant frequency of the crystal is measured. The resonant frequency is a function of the mass of the crystal and any mass change caused by the adsorption/desorption of gas molecules on the surface of the crystal is translated into a change in its resonant frequency. The measured change in resonant frequency is proportional to the gas concentration (Fauzi et al. 2021).

7.1.6 SURFACE ACOUSTIC WAVE-DELAY LINE/RESONATOR GAS SENSOR

A surface acoustic wave (SAW) gas sensor is made on a piezoelectric substrate (Figure 7.1). Periodic metallic electrodes called interdigitated electrodes (IDEs) are its main parts. They consist of two combs crossing each from opposite sides. They convert the radio-frequency signal to surface acoustic waves propagating on the piezoelectric surface, thus producing an electrical-to-mechanical signal transformation. They also perform the reverse signal transformation from mechanical surface acoustic wave signal to electrical signal. Hence, they can excite surface acoustic waves as well as detect them. Two configurations are used for gas sensing (Devkota et al. 2017):

(i) SAW delay line: In a two-port design, two IDTs are formed on a piezoelectric substrate at a fixed distance apart (Figure 7.1(a)). These are the input and output signal IDTs. The intervening space on the substrate separating the two IDTs is coated with a layer of the gas-sensitive 2D material which selectively adsorbs the target gas. Interaction of the gas with the 2D material layer influences the SAW velocity introducing a time-delay between the input and output signals determined by the distance between them and SAW velocity.

(ii) SAW resonator: A two-port geometrical layout consists of two IDTs with one reflection grating on the outer side of each IDT (Figure 7.1(b)). The gas-sensing 2D material film is deposited on the IDTs. A resonant cavity is formed between them.

Both the configurations have similar response and output characteristics. Upon exposure of the 2D material coating on the SAW device to gas molecules, the phase velocity of surface acoustic waves changes. The SAW phase velocity changes are detected as a shift in the resonant frequency of the SAW oscillator (Khlebarov et al. 1992).

One-port designs containing one IDT can also be used. These are known as one-port or reflective delay lines and one-port resonator.

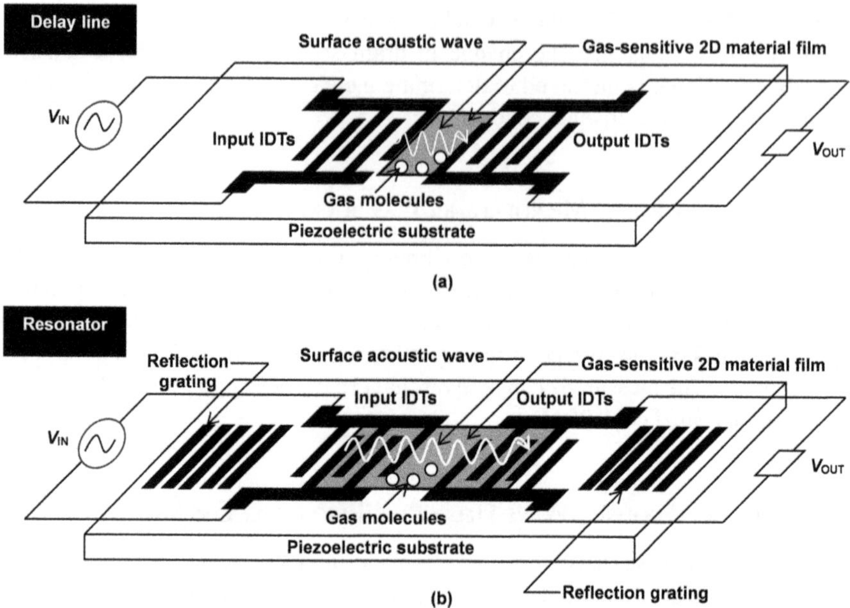

FIGURE 7.1 Surface acoustic wave gas sensor: (a) SAW delay line and (b) SAW resonator. Part (a) shows a piezoelectric substrate on the surface of which there are two pairs of inter-digitated electrodes: the input IDTs and the output IDTs. Both pairs of IDTs have contact pads at their ends. In the space between the two pairs of IDTs, the gas sensitive 2D material film is deposited. The gas molecules are adsorbed on this film. The surface acoustic wave propagates from the input IDTs to the output IDTs. The input signal V_{IN} is measured across the contact pads of the input IDTs and the output signal V_{OUT} is read across the contact pads of the output IDTs. Part (b) shows a similar construction of the SAW sensor except that there is a reflection grating on the left side of input IDTs and another reflection grating on the right side of output IDTs.

7.1.7 Optical Gas Sensor

Chemiluminescence is the emission of light in consequence of a chemical reaction. It differs from fluorescence or phosphorescence in the respect that it originates from a chemical reaction, not by absorption of photons by the molecules. Chemiluminescent reactions produce unstable electronically excited intermediate compounds which subsequently decay or break down into more stable compounds (Figure 7.2). During the formation of intermediate compounds, the molecules are raised from the ground state to the excited state by absorption of energy. During the breaking down of inter-mediate compounds, the absorbed energy is released for relaxation to the energet-ically lower ground state. The energy is liberated in the form of electromagnetic radiation in the ultraviolet, visible, or infrared portion of the spectrum.

Chemiluminescence offers a powerful light-based sensing technique for gas anal-ysis. Generally, the emission intensity is linearly dependent on the analyte

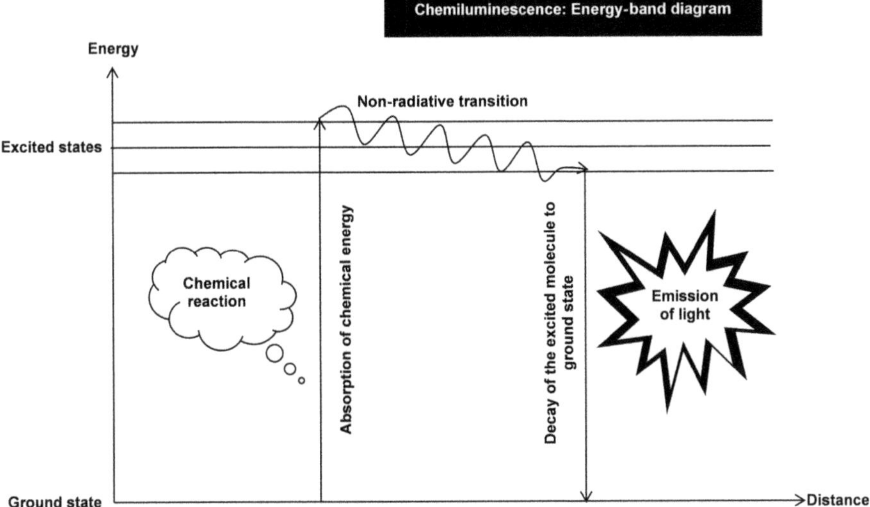

FIGURE 7.2 Energy-band diagram of chemiluminescence phenomenon. The diagram shows the plot of the energy of molecules against distance. Ground-state and excited-state energy levels are marked. By absorbing the energy released in a chemical reaction, a molecule is raised to an excited state. The excited molecule decays to the ground state by releasing the absorbed energy as a photon of light.

concentration over a wide range. High sensitivity, fast response, long-term stability, and reproducibility are feasible with little background interference using simple instrumentation. Its primary disadvantage is a weak luminescence signal. 2D nanomaterials can enhance and amplify chemiluminescence emission for improving the performance of a gas sensor.

7.2 FIGURES OF MERIT OF GAS SENSORS

7.2.1 RESPONSE AND DETECTION LIMIT

The response is defined by the equation:

$$\text{Response} = \frac{\begin{array}{c}\text{Measured variable value} \\ \text{after exposure to gas}\end{array} - \begin{array}{c}\text{Original value of measured} \\ \text{variable before gas exposure}\end{array}}{\begin{array}{c}\text{Original value of measured} \\ \text{variable before gas exposure}\end{array}} \times 100\% \quad (7.2)$$

The limit of detection (LoD) is the smallest measured variable value or the smallest sensed parameter value extracted from the measured data with adequate statistical significance or confidence level. In the context of the extracted parameter, it is the lowest concentration of target gas in a test specimen that is consistently and reliably detectable with a defined probability, typically at 95% certainty, as distinguished from background noise. It is calculated by applying the equation:

$$\text{LoD} = 3.3 \frac{\text{Standard deviation of response}}{\substack{\text{Slope of the calibration curve drawn between the} \\ \text{output signal of the sensor and the analyte concentration}}} \tag{7.3}$$

For a chemiresistive sensor,

$$\text{Response of a chemiresistor} = \frac{\substack{\text{Sensor resistance value after} \\ \text{exposure to gas}} - \substack{\text{Original value of sensor resistance} \\ \text{before gas exposure}}}{\substack{\text{Original value of sensor resistance} \\ \text{before gas exposure}}} \times 100\% \tag{7.4}$$

$$= \frac{\text{Sensor resistance in the presence of gas} - \text{Sensor resistance in air}}{\text{Sensor resistance in air}} \times 100\%$$

The LoD of a chemiresistor is measured in parts per million (ppm) or parts per billion (ppb).

7.2.2 Sensitivity and Range

The defining equation for sensitivity is

$$\text{Sensitivity} = \substack{\text{Derivative of the output signal with respect to the} \\ \text{input stimulus signal}}$$

$$= \text{Slope of the output} - \text{input characteristic curve of the sensor}$$

$$= \frac{\text{Change in output signal}}{\text{Change in input signal}} \tag{7.5}$$

$$= \frac{\substack{\text{Measured variable value} \\ \text{after exposure to gas}} - \substack{\text{Original value of measured} \\ \text{variable before gas exposure}}}{\text{Concentation of gas}}$$

Sensitivity is also defined as a ratio of measured variable values in the target gas and reference gas, usually air:

$$\text{Sensitivity} = \frac{\text{Measured variable value in the target gas}}{\text{Measured variable value in reference gas} \left(\text{air}\right)} \tag{7.6}$$

For a chemiresistive sensor,

$$\text{Sensitivity}_{\text{Oxidizing gas}} = \frac{\text{Resistance of the gas sensor in target gas} \left(R_{\text{Gas}}\right)}{\text{Resistance of the gas sensor in air} \left(R_{\text{Air}}\right)} \tag{7.7}$$

for an oxidizing gas, and

$$\text{Sensitivity}_{\text{Reducing gas}} = \frac{\text{Resistance of the gas sensor in air}\left(R_{\text{Air}}\right)}{\text{Resistance of the gas sensor in target gas}\left(R_{\text{Gas}}\right)} \tag{7.8}$$

for a reducing gas.

The range of a sensor is the difference between maximum and minimum values of the sensed parameter that can be measured. Dynamic range DR of a sensor specifies the range of possible values that a continuously changing signal acquires. It is expressed as a ratio given by

$$DR = \frac{\text{Largest value of the measurand}}{\text{Smallest value of the measurand}} = \frac{\text{Strongest output signal}}{\text{Weakest output signal}} \tag{7.9}$$

7.2.3 Selectivity

It is the ability of a sensor to discriminate the target gas from interfering gases.

7.2.4 Response Time

It is the time in which the output signal of the sensor changes from its initial value to 90% of the final equilibrium value upon exposure to gas, i.e., 90% of its ultimate saturated response.

7.2.5 Recovery Time

It is the time taken by the sensor output signal to fall to 90% of the final equilibrium value after the step withdrawal of the gas.

7.3 GAS SENSORS FABRICATED USING GRAPHENE AND GRAPHENE OXIDE

7.3.1 B- and N-Codoped Graphene (BNGr) Nanosheets Sensor for NO_2

An ultrasensitive NO_2 gas sensor is designed and demonstrated by applying the concept of co-doping of heteroatoms in graphene to form BNGr nanosheets (Srivastava et al. 2022). A heteroatom is the atom of any element other than carbon or hydrogen. The purpose of co-doping graphene with boron and nitrogen is to improve the response of the sensor to NO_2. The sensor is resistive in behavior showing changes in resistance with variation in concentration of NO_2. Figure 7.3 displays the fabrication process sequence of the sensor.

7.3.1.1 Synthesis of BNGr

For BNGr synthesis, a copper foil is taken (Figure 7.3(a)). BNGr nanosheet is synthesized on the copper foil (Figure 7.3(b)) by low-pressure chemical vapor deposition at 950°C and ~31.9 Torr using methane as a carbon precursor and borazine (H_3NBH_3) as the precursor for doping B and N. It is spin coated with PMMA and air

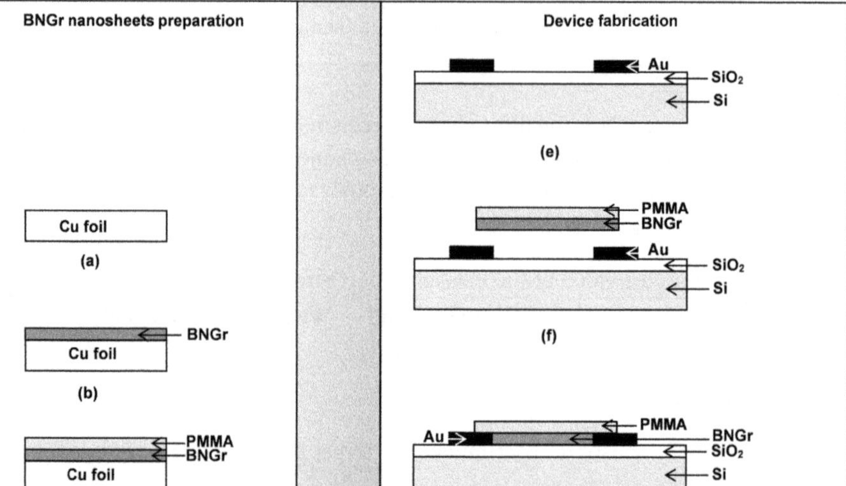

FIGURE 7.3 Fabrication of BNGr nanosheets sensor: (a) starting with a copper foil, (b) synthesizing BNGr on the copper foil, (c) spin coating PMMA on BNGr, (d) etching copper, (e) thermal evaporation and lithography for making gold electrodes on SiO_2/Si substrate, (f) lowering PMMA/BNGr over prepatterned gold electrodes, (g) pressing PMMA/BNGr over gold electrodes, and (h) gas sensor formation after PMMA removal. Parts (a), (b), (c), and (d) show a copper foil, BNGr grown over the copper foil, PMMA/BNGr/Cu foil, and PMMA/BNGr, respectively. Part (e) shows the pattern of gold electrodes over an SiO_2/Si substrate. Part (f) shows bringing PMMA/BNGr over Au electrodes/ SiO_2/Si substrate. Part (g) shows aligning and pressing together PMMA/BNGr against Au electrodes/SiO_2/Si substrate. Part (h) is the picture of the sensor in which BNGr film is deposited over a pair of interdigitated gold electrodes formed on an SiO_2/Si substrate.

dried to get: PMMA/BNGr/Cu foil (Figure 7.3(c)) from which the Cu foil is etched in 3:1 water:nitric acid solution obtaining: PMMA/BNGr (Figure 7.3(d)).

7.3.1.2 Transfer of PMMA/BNGr Over Electrode Pattern

Prepatterned gold electrodes on an SiO_2/Si substrate are taken (Figure 7.3(e)). The PMMA/BNGr is transferred over patterned gold electrodes formed on the SiO_2 film over an Si substrate to form: PMMA/BNGr/Au electrode pattern/SiO_2/Si substrate

(Figure 7.3(f)). PMMA is removed in acetone when we get: BNGr/Au electrode pattern/SiO$_2$/Si substrate (Figure 7.3(g), (h)).

7.3.1.3 Sensor Response

The sensor is characterized by measuring the relative change in resistance = (change in resistance on exposure to NO$_2$/original resistance in air) × 100%. It shows a response of 0.05% for 1ppb NO$_2$. The response increases to 3.29% for 5 ppm NO$_2$. For 1 ppb NO$_2$, the response time is 177 s and the recovery time is 392 s. It is also found that response of a sensor made with pure graphene film is very feeble as compared to that of BNGr sensor indicating the improvement in response caused by co-doping graphene with B and N (Srivastava et al. 2022).

7.3.2 UV-Illuminated Graphene–Silicon Schottky Diode Sensor for NO$_2$ and THF

A graphene–Si Schottky diode gas sensor enhanced by UV irradiation is demonstrated (Drozdowska et al. 2023). Figure 7.4 shows the Schottky diode gas sensor.

7.3.2.1 Sensor Fabrication

A 90-nm thick silicon dioxide film thermally grown on an N-type silicon wafer of resistivity 1 Ω-cm –10 Ω-cm, is selectively etched, and CVD graphene monolayer on a Cu foil is transferred over the resulting SiO$_2$/Si substrate by electrochemical delamination. After patterning graphene by selective etching using reactive ion etching in an oxygen plasma, Cr (10 nm)/Au (100 nm) contacts are formed by thermal evaporation. For making contact to Si, a portion of SiO$_2$ is etched and a Cr film is deposited. A graphene/N-doped silicon Schottky junction diode is formed. It has a barrier height of 0.7 eV –0.8 eV at room temperature in the dark condition.

7.3.2.2 Sensor Performance Assessment

To assess the response of Schottky diode toward gases, it is tested in 1 ppm –3 ppm NO$_2$, an oxidizing gas under 275 nm UV illumination. The resistive part of the forward current–voltage characteristic of the diode dominated by resistance of graphene is affected by NO$_2$. The junction current rises. The LoD is 75 ppb NO$_2$ at 4 V bias.

Tetrahydrofuran (THF), a reducing gas affects the forward and reverse characteristics of the diode, notably their exponential parts, increasing the Schottky barrier height up to tens of meV, and degrading the detection limit of THF to 31 ppm (Drozdowska et al. 2023).

7.3.3 Graphene-FET Sensor for SO$_2$

7.3.3.1 Working Principle

A graphene FET sensor for detecting sulfur dioxide works on the principle of shift of Dirac point of graphene toward the positive side on exposing the graphene FET to sulfur dioxide. This shift of Dirac point occurs due to the strong P-doping of graphene in the presence of SO$_2$ (Ren et al. 2012).

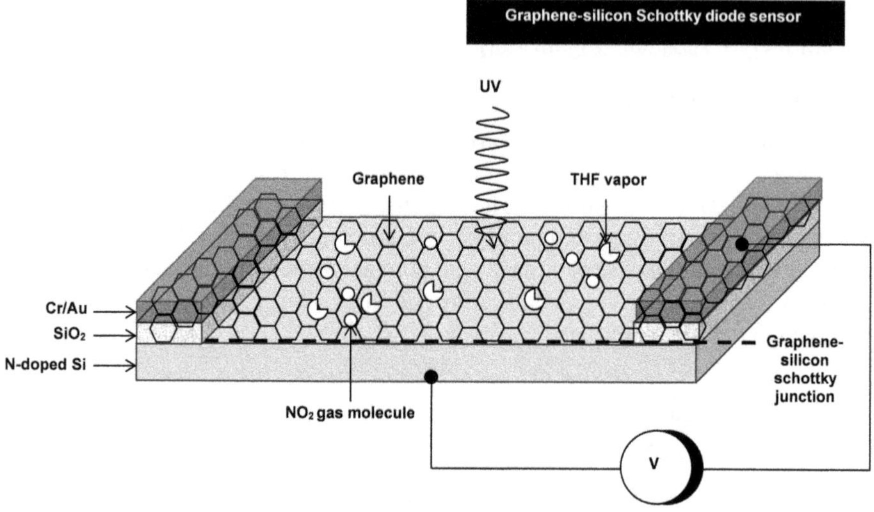

FIGURE 7.4 Graphene–silicon Schottky diode gas sensor for NO_2 gas and THF vapors. The diagram shows an N-doped silicon substrate at both edges of which silicon dioxide strips are patterned. A graphene film is overlaid over the silicon surface. It also covers the silicon dioxide strips. Cr/Au electrodes are formed over the portions of the graphene film covering the silicon dioxide strips. The graphene film is illuminated with UV radiation. It is exposed to NO_2 gas molecules and THF vapors. A voltmeter is connected between the N-doped Si and the metallic contact to the graphene film.

7.3.3.2 Fabrication of the Graphene FET

Step (a): Graphene synthesis on Cu foil: large-area graphene is grown over a 25 μm thick Cu foil by CVD in a cold-wall furnace by:

   (i)   annealing the Cu foil by directly heating with 35 A current up to 880°C in 1×10^{-2} Torr vacuum, and

   (ii)   exposing it to methane at 880°C for 10 min.

Step (b): Graphene transfer over an SiO_2/Si substrate: a fish-out method is used to avoid any polydimethylsiloxane (PDMS) or poly-methyl methacrylate (PMMA) residues. In this method, the Cu foil is removed by etching in an aqueous iron nitrate solution. Each etching process yields two exfoliated graphene films. After dissolution of the Cu foil, the SiO_2/Si substrate is brought in contact with the graphene film. The graphene film is pulled away from the solution onto the substrate obtaining clean monolayer graphene without PMMA contamination.

Step (c): Gold electrode deposition over the graphene on SiO_2/Si substrate: two Au electrodes are deposited on graphene-on-SiO_2 (280 nm)/Si substrate by thermal evaporation through a shadow mask. Thus, an FET with graphene channel and silicon back gate is realized.

7.3.3.3 Response of the Graphene FET to SO_2

A drain–source voltage of 2.53 V and a gate bias −50 V is applied. At room temperature for 50 ppm SO_2, the Dirac point shifts by 0.678 Vppm^{-1} toward the positive side and the resistance decreases ~ 60%. At 100°C for 50 ppm SO_2, the shift in Dirac point is 1.09 $Vppm^{-1}$ toward the positive side while the resistance decreases over 100%. The considerably long time taken by the sensor for resetting is shortened at a high temperature to ~10 min at 100°C. The sensor shows a linear increase in sensitivity between room temperature and 100°C (Ren et al. 2012).

7.3.4   GRAPHENE-COATED QCM SENSOR FOR VOLATILE ORGANIC COMPOUNDS

7.3.4.1   Sauerbrey Equation

Quartz crystal microbalance (QCM) consisting of a quartz disc sandwiched between metal electrodes is a device which is highly sensitive to changes in mass (Figure 7.5). Its working principle is explained in Figure 7.5(a). When an alternating voltage is applied across the metal electrodes, the piezoelectric quartz disc undergoes periodic mechanical deformation and oscillates up and down in synchronization with the applied voltage. Any changes in the mass of the device (Δm) such as by adsorption of gas molecules on a film coated on the crystal surface cause changes in the resonant frequency of the device (Δf) given by the well-known Sauerbrey equation:

$$\Delta f = -\frac{n\Delta m}{C} \tag{7.10}$$

where n is a number representing the particular harmonic of vibration =1, 3, 5, …, and

$$C = \text{mass sensitivity constant} = 17.7 \text{ng} \left(\text{cm}^2 - \text{Hz} \right)^{-1} \text{for a } 5 - \text{MHz crystal} \tag{7.11}$$

As the QCM is able to detect small changes in mass to the extent of mass of a molecule, it is suitable for gas detection. The use of graphene as a gas-sensitive film is receiving increasing attention. Figure 7.5(b) shows a graphene-coated QCM. Large surface area and extremely low mass of graphene monolayer make it appropriate for fabricating a gas-sensitive graphene resonator showing high sensitivity and fast recovery (Quang et al. 2014).

7.3.4.2   Graphene Growth and Transfer

Graphene is grown on copper foils by CVD at 1000°C for 10 min in (CH_4+H_2) at a total pressure of 40 Pa. The Cu foils are cut into circular discs of diameter 12 mm. The discs are coated with poly-methyl methacrylate (PMMA) and the Cu foils are removed by etching in iron (III) chloride ($FeCl_3$) solution in water followed by thorough DI water rinsing and placing over the circular gold film on the upper surface of a 5 MHz AT-cut quartz crystal plate (25.4 mm diameter) with pre-deposited gold electrodes on both its surfaces. For PMMA removal, the PMMA/graphene/QCM is immersed in acetone and rinsed with water.

7.3.4.3   Resonant Frequency Shifts in VOC Vapors

The sensor is exposed to target gases, and the change in its resonant frequency is measured using a frequency counter. The device shows sensitivity to different VOCs such

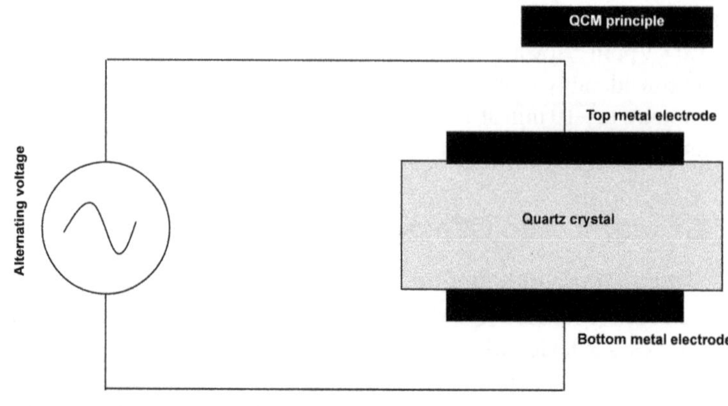

$$\text{Normalized frequency shift} = -\frac{2(\text{Resonant frequency of fundamental mode of the quartz crystal})^2 \times \text{Change in mass of the crystal}}{\text{Piezoelectrically active area of crystal}\sqrt{\text{Density of quartz} \times \text{Shear modulus of quartz}}}$$

(a)

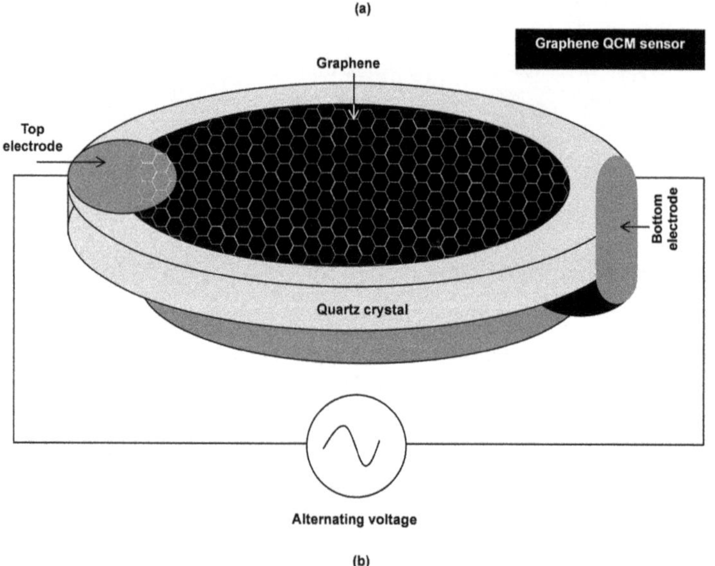

(b)

FIGURE 7.5 Operating principle of quartz crystal microbalance and graphene-coated QCM for the detection of VOCs: (a) principle of operation of QCM and (b) the QCM sensor for VOCs. Part (a) shows that a quartz crystal with metal electrodes on both sides is connected to an alternating voltage supply. The Sauerbrey equation giving the linear relationship between the normalized frequency shift of the quartz crystal and the change in mass of the crystal is stated. The normalized frequency shift means the observed frequency shift of the mode of vibration of the crystal divided by the mode number. Besides the change in mass, the frequency shift depends on the square of resonant frequency of the fundamental mode of the crystal, the actual crystal area covered by the electrode, and properties of quartz, namely, the square root of the product of its density and modulus of rigidity. Part (b) shows the graphene-based gas sensor with a CVD graphene film transferred to the circular gold disc over the quartz substrate of a QC sensor. The top and bottom electrodes and the AC voltage source are seen.

as ethanol, butanol, isopropanol, and acetone. It is most sensitive to ethanol vapors in which the frequency changes are 19.4 Hz at 90 ppm and 34.2 Hz at 180 ppm. The frequency shifts in acetone vapors are 14.8 Hz and 32.3 Hz at 100 ppm and 200 ppm, respectively (Quang et al. 2014).

7.3.5 Fe_2O_3–Graphene Nanosheets Chemiluminescence H_2S Sensor

Fe_2O_3–graphene nanosheets can detect H_2S because ferric oxide is an efficient detector of hydrogen sulfide through the chemiluminescence signal. This signal is generated from intermediate H_2S oxidation. Recall from Section 7.1.7, chemiluminescence is the phenomenon of emission of light by decay of molecules from excited to ground states ensuing their promotion to the excited states via a chemical reaction. A chemiluminescence sensor for H_2S has been designed using this property of Fe_2O_3–graphene nanosheets (Jiang et al. 2014).

7.3.5.1 Making Vertically Arranged Fe_2O_3/Graphene Nanosheets

Fe_2O_3 is a magnetic material. Apart from its role in chemiluminescence, the additional advantage of ferric oxide comes from its magnetic properties, helping in magnetically controlled alignment of Fe_2O_3/graphene nanosheets. Magnetic properties of Fe_2O_3 allow to align Fe_2O_3–graphene nanosheets vertically or horizontally, thereby producing vertically arranged Fe_2O_3/graphene nanosheets (VAFe/GN) and horizontally arranged Fe_2O_3/graphene nanosheets (HAFe/GN). Two types of nanosheets are prepared, and their H_2S responses are evaluated. Figure 7.6 illustrates the process for making VAFE/GN, which is implemented in two steps:

(i) Preparation of Fe_2O_3/graphene nanosheets powder: black powder of Fe_2O_3/graphene nanosheets is prepared by mixing Fe $(NO_3)_3.9H_2O$ [Ferric nitrate nonahydrate, or Iron (III) nitrate nonahydrate] with ethanol solution of graphene oxide (GO); the GO is made by Hummer's method. The mixture is ultrasonicated, transferred to a stainless-steel vessel and pressurized with supercritical CO_2 up to 5 MPa at 0°C (Figure 7.6(a)). The vessel is sealed (Figure 7.6(b), kept in an oven at 120°C for 2 h, then kept in a salt-bath furnace at 350°C for 2 h, and, finally, cooled and suddenly depressurized when the powder is obtained (Figure 7.6(c)).

(ii) Alignment of Fe_2O_3/graphene nanosheets: the powder is dispersed in water, ultrasonicated, and subjected to a magnetic field (Figure 7.6(d)) followed by vacuum filtration of the dispersion through an Anodisc membrane filter under a vertically directed magnetic field. Thus, 4–6 μm thick VAFE/GN are obtained (Figure 7.6(e)).

HAFE/GN are made by a similar process.

7.3.5.2 CL Emission

The CL emission of VAFE/GN is 450 absorption units for 15 ppm H_2S at 190°C. The response time of the sensor is 500 μs while its recovery time is <30 s. The detection limit of VAFe/GN is < 10 ppm H_2S at 130°C. The VAFE/GN sensor is

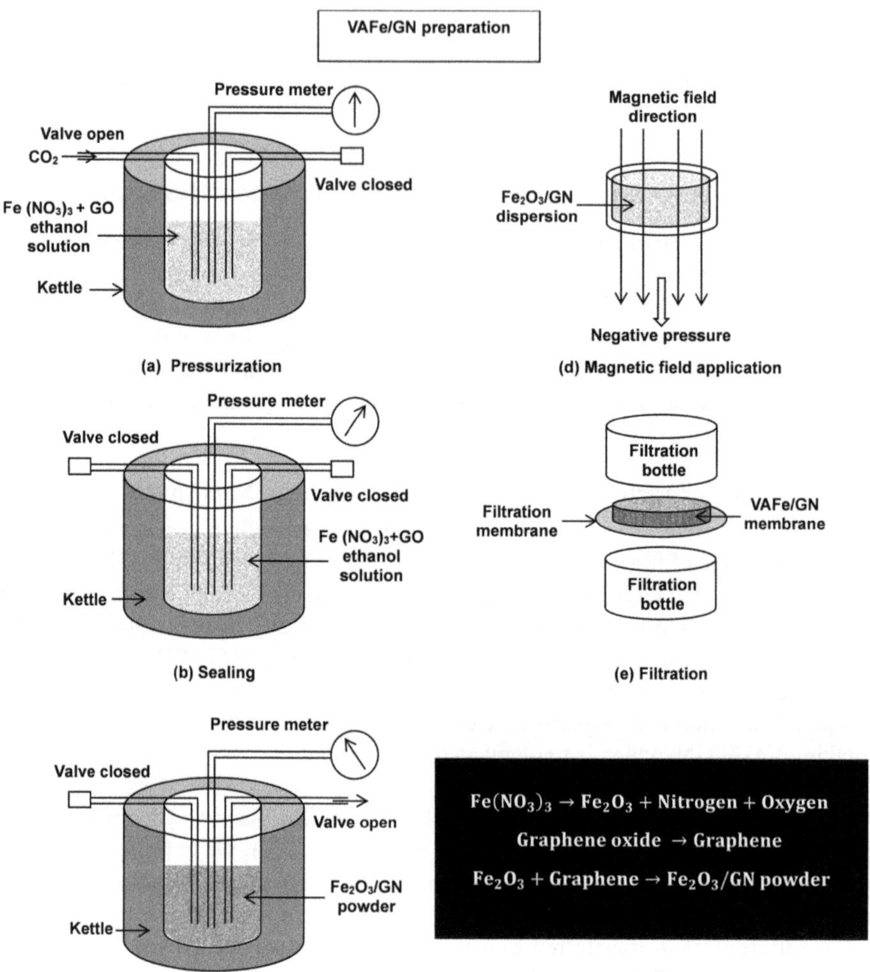

FIGURE 7.6 Preparation of VAFE/GN: (a) pressurization, (b) sealing, (c) depressurization by sudden exhausting, (d) exposure of Fe_2O_3/GN to magnetic field and directed flow for vertical alignment of nanosheets, and (e) showing the end stage of step (d) after filtration. Parts (a), (b), and (c) show a kettle with a central hollow portion containing $Fe(NO_3)_3$ mixed with graphene oxide solution in ethanol in parts (a), (b), and Fe_2O_3/GN powder in part (c). In part (a), the kettle is pressurized with supercritical CO_2. In part (b), the kettle is sealed. Part (c) shows the kettle after it has been kept in a furnace and depressurized during which the Fe_2O_3/GN powder is formed. Part (d) shows the application of a vertical magnetic field with directed flow. Part (e) shows the end result of part (d) after vacuum filtration.

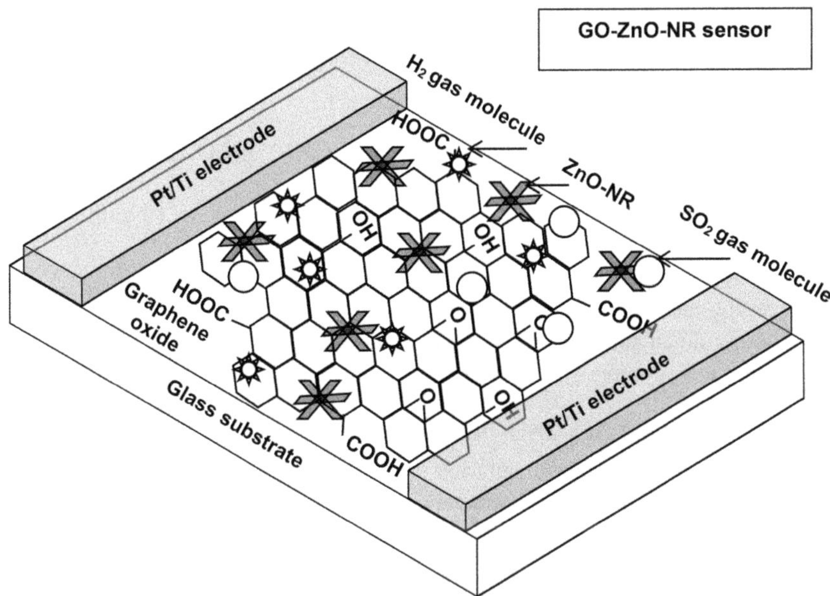

FIGURE 7.7 GO–ZnO–NR composite sensor for SO_2 and H_2. The diagram shows Pt/Ti electrodes on two edges of a glass substrate. Over the glass substrate, a nanocomposite is deposited. It is made of ZnO nanorods and graphene oxide (GO) with attached oxygen-containing functional groups, e.g., =O, -OH, -O-, and -COOH. The sulfur dioxide and hydrogen gas molecules are adsorbed on the surface of the GO and ZnO nanorods composite.

insensitive to CO_2, ethanol, benzene, or toluene. Moreover, the VAFE/GN sensor shows a higher chemiluminescence sensitivity to H_2S than the HAFe/GN sensor. CL emission of HAFE/GN sensor is 350 absorption units for 23 ppm H_2S at 190°C (Jiang et al. 2014).

7.3.6 GRAPHENE OXIDE–ZnO NANOROD SENSOR FOR SO_2 AND H_2

Room temperature sensing of SO_2 and H_2 at sub-100 ppm levels is demonstrated using a GO–ZnO nanorod (GO–ZnO–NR) composite (Dhingra et al. 2020). In this composite, the GO constituent provides a large specific area. Figure 7.7 shows the GO–ZnO–NR composite sensor. The fabrication and response behavior of the sensor are described below.

7.3.6.1 Pt/Ti Interdigitated Electrodes

The Pt (90 nm)/Ti (10 nm) IDEs are patterned on glass substrates by photolithography.

7.3.6.2 GO Synthesis

GO synthesis is done by the protocol of Hummers' method.

7.3.6.3   Hydrothermal growth of GO–ZnO–NR Composite

The GO and ZnO nanorods composite (GO–ZnO–NR) is grown by a hydrothermal method consisting of:

(i)   dissolution of zinc nitrate hexahydrate, Zn $(NO_3)_2.6H_2O$ in DI water with constant stirring;

(ii)   drop-by-drop addition of GO solution, adjusting the pH to 12 with NaOH solution;

(iii)   transference of the mixture to a Teflon hydrothermal container;

(iv)   keeping the mixture at 200°C for 4 h;

(v)   cooling down and washing with DI water; and

(vi)   drying to get composite powder.

7.3.6.4   Making the GO–ZnO–NR Gas Sensor

The GO–ZnO–NR composite solution in DI water is drop-coated on the pre-fabricated Pt/Ti electrodes. The contact pads are protected from the coating solution. The coated substrate is kept on a hot plate at 50°C for ½ h and then at 70°C for ½ h.

7.3.6.5   Gas-Sensing Response

The response of the sensor defined as: sensor resistance in target gas/sensor resistance in air, increases from 2.97 at 25 ppm to 5.45 at 100 ppm for SO_2. The response time is 80 s and the recovery time is 75 s. For H_2, it rises from 3.12 at 25 ppm to 5.82 at 100 ppm. The response time is 30 s and the recovery time is 40 s. The response characteristics are linear from 25 to 100 ppm.

7.3.6.6   Gas-Sensing Mechanism

(a)   Sulfur dioxide: exposure to SO_2 leads to the formation of an overlayer of zinc sulfite, $ZnSO_3$ (solid) on ZnO nanorods. The gain of O_2^- species by ZnO nanorods during the conversion of ZnO into $ZnSO_3$ is associated with removal of electrons from the GO–ZnO–NR composite, thereby increasing the resistance of the device.

(b)   Hydrogen: although H_2 is a reducing gas, a rise in resistance is observed on exposure to H_2, which is interpreted to be due to the dissociation of H_2 molecules into $2H^+$, and their adsorption on $Zn^{2+}O^{2-}$ pairs from ZnO forming Zn–H and O–H species. As before, the gain of O_2^- species by H^+ is accompanied by the withdrawal of electrons from the GO–ZnO–NR composite. Hence, the sensor shows an increase in resistance (Dhingra et al. 2020).

7.4   GAS SENSORS FABRICATED USING TMDCs

7.4.1   RESISTIVE CO GAS SENSOR USING SnO₂ NANOPARTICLES/MoSe₂ NANOFLOWERS NANOCOMPOSITE

A high-sensitivity, humidity-resistant CO gas sensor is developed by designing a nanocomposite from tin oxide nanoparticles and molybdenum diselenide nanoflowers (Yang et al. 2020). The sensor is sketched in Figure 7.8.

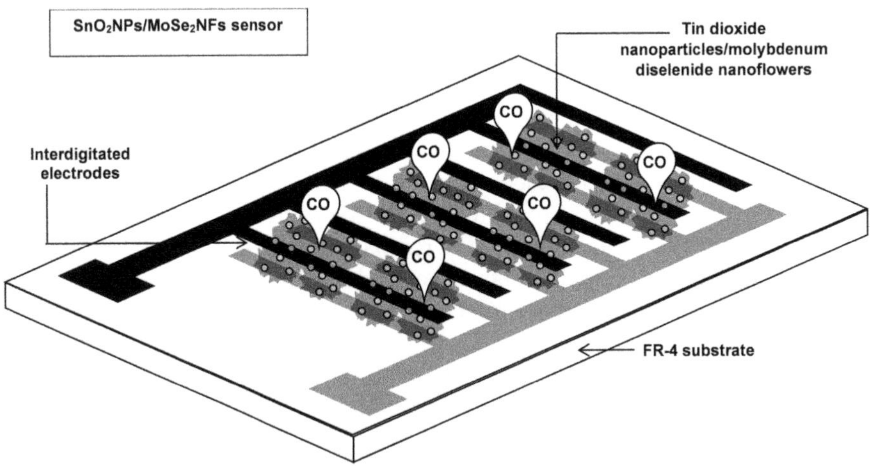

FIGURE 7.8 CO gas sensor using a gas-sensing film made from metal-organic frameworks-derived tin dioxide nanoparticles and molybdenum diselenide nanoflowers. The diagram shows an FR-4 substrate on which an interdigitated electrode pair is formed. The electrode pair is coated with a nanocomposite film containing SnO_2 nanoparticles and $MoSe_2$ nanoflowers. CO gas molecules are adsorbed on the surface of the $SnO_2NPs/MoSe_2$ nanoflowers film.

7.4.1.1 SnO₂ Nanoparticles Preparation

$SnSO_4$ (tin sulfate) solution is added to a dispersion of H_2BDC (terephthalic acid) and NaOH (sodium hydroxide) in DI water. Sn-based metal-organic framework (Sn-MOF) is formed after a 5 h reaction. Thermal treatment of Sn-MOF at 400°C for 3 h yields SnO_2 NPs by its decomposition.

7.4.1.2 MoSe₂ Nanoflowers Preparation

A solvothermal route is adopted. $Na_2MoO4 \cdot 2H_2O$ (sodium molybdate dihydrate) and $NaBH_4$ (sodium borohydride) are dispersed in a (DI water+ ethanol) solution. Hydrazine hydrate ($NH_2NH_2.H_2O$) solution containing Se powder is added to the dispersion. After 48 h heating at 200°C, rinsing with DI water and ethanol, and overnight drying, 2 h calcining at 700°C in nitrogen gives pure $MoSe_2$.

7.4.1.3 SnO₂ NPs/MoSe₂ NFs Nanocomposite Preparation

The nanocomposite is made by mixing SnO_2 powder with $MoSe_2$ powder in DI water, stirring, and drying.

7.4.1.4 Sensor Fabrication

Dispersion of SnO_2 NPs/ $MoSe_2$ NFs nanocomposite in DI water is sonicated to form a suspension. The suspension is drop-coated on pre-printed Cu/Ni IDEs on an FR-4 substrate and dried at 60°C. FR= Flame retardant, and the number '4' shows that the substrate is a laminate made of woven glass fabric impregnated with epoxy resin binder for reinforcement.

7.4.1.5 Sensor Response

The gas response is characterized by exposing the sensor to different CO concentrations and measuring the device resistance. The response % is calculated by subtracting the resistance in CO from that in air, dividing by the resistance in air and multiplying the result by 100. At 1 ppm CO, the response is 3.61, at 5 ppm, it is 5.28, and at 10 ppm, it is 6.98. The responses at 50, 100, and 500 ppm are 8.02, 9.25, and 12.21, respectively. The responses of the $SnO_2/MoSe_2$ film are much better than those of individual SnO_2 and $MoSe_2$ films. The response of the $SnO_2/MoSe_2$ film remains constant over a period of 30 days showing a good stability. The device is selective to CO detection without interference from CH_4, H_2, CO_2, SO_2, and H_2S.

7.4.1.6 CO Gas-Sensing Mechanism

Both SnO_2 and $MoSe_2$ are N-type semiconductors. In the presence of air, oxygen molecules are adsorbed on the surface of the $SnO_2/MoSe_2$-sensing film. These adsorbed oxygen molecules annex electrons from the underlying $SnO_2/MoSe_2$ film. Thus, oxygen anions are created on the surface of the sensing film. CO molecules act as electron donors. CO reacts with adsorbed oxygen anions forming carbon dioxide and releasing electrons to the $SnO_2/MoSe_2$ film:

$$2CO\left(Gas\right) + O_2^-\left(Adsorbed\right) \rightarrow 2CO_2\left(Gas\right) + e^-\left(Electron\right) \tag{7.12}$$

This gain of electrons by the N-type $SnO_2/MoSe_2$ film increases the concentration of electrons. The increased electron concentration is observed as an increase in conductance and decrease in resistance of the film. The superior response of $SnO_2/MoSe_2$ film as compared to separate SnO_2 and $MoSe_2$ films is ascribed to the N–N junctions formed at the interface between N-type SnO_2 and N-type $MoSe_2$ films (Yang et al. 2020).

7.4.2 Self-Powered P-WSe₂/N-WS₂ and N-MoS₂/P-WSe₂ Heterostructure Gas Sensors for NO₂ and NH₃ Detection

The photovoltaic effect of 2D TMD heterostructures forms the basis of gas sensors for detecting NO_2 and NH_3 (Kim et al. 2020). Changes in photocurrent with gas concentration are recorded. Two heterostructures are examined, P-WSe₂/N-WS₂ and N-MoS₂/P-WSe₂ for sensor development. These are shown in Figure 7.9.

7.4.2.1 Self-Powering Capability

The self-powering capability of the sensor is derived from photovoltaic effect. Both the heterostructures, WSe₂/WS₂ and MoS₂/WSe₂ show zero current in the dark. Under illumination, electrons and holes are generated. The electrons diffuse toward the N-type WS_2 or MoS_2 layer. The holes move to the P-type WSe_2 layers. This separation of photogenerated carriers produces a built-in electric field and therefore an open-circuit voltage. This field acts as a driving force for gas sensing

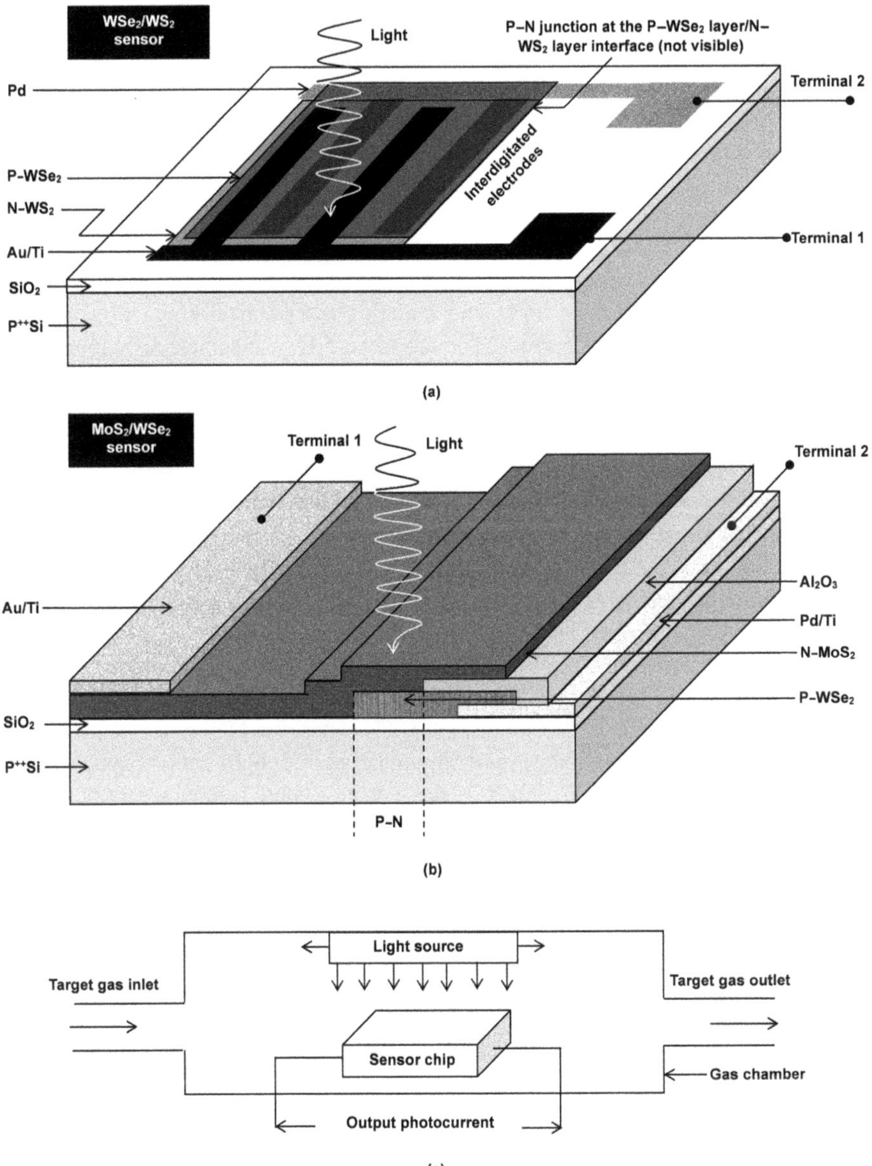

FIGURE 7.9 Photovoltaic gas sensors made of heterostructures and their exposure to target gas: (a) WSe$_2$/WS$_2$ sensor, (b) MoS$_2$/WSe$_2$ sensor and (c) gas exposure setup. Parts (a) and (b) show heterostructures fabricated on SiO$_2$/P^{++} Si substrates. In both (a) and (b), the incident light and electrical terminals 1 and 2 are indicated. The layers seen in part (a) are: Au/Ti bottom electrodes and contact pad, WS$_2$ layer, WSe$_2$ layer, and Pd top electrodes and contact pad while those in part (b) are WSe$_2$ layer, Pd/Ti contact pad, MoS$_2$ layer, and Au/Ti contact pad. The MOS$_2$ film is separated from the Pd/Ti contact pad by an insulating Al$_2$O$_3$ film. Part (c) shows the gas chamber with an inlet and an outlet for target gas, and a source of light at the roof. Light falls on the sensor chip below, and the output photocurrent is measured across the terminals of the chip.

7.4.2.2　Gas-Sensing Mechanisms

The gas-sensing mechanism relies on the physisorption of electron acceptor NO_2 and electron donor NH_3 gas molecules.

In the WSe_2/WS_2 sensor, the adsorbed NO_2 molecules on P-type WSe_2 layer act as electron acceptors creating holes in the P-type WSe_2 layer. The consequent increase in hole concentration raises the output photocurrent. The adsorbed NH_3 molecules on P-type WSe_2 layer act as electron donors neutralizing holes and decreasing hole concentration. The decrease in hole concentration lowers the output photocurrent.

In the MoS_2/WSe_2 sensor, adsorbed NO_2 gas molecules on N-type MoS_2 act as electron acceptors producing holes and reducing the electron population, thereby decreasing the output photocurrent. The adsorbed NH_3 molecules on N-type MoS_2 act as electron donors furnishing additional electrons. Hence, NH_3 adsorption causes an increase of output photocurrent.

7.4.2.3　P-Type WSe_2/N-Type WS_2 Sensor Fabrication and Response

Fabrication: P-type WSe_2/N-type WS_2 sensor (Figure 7.9(a)) is fabricated by *in situ* synthesis of WSe_2/WS_2. A tube furnace is heated to the growth temperature ~700°C. WS_2 is made from tungsten hexachloride (WCl_6) and hydrogen sulfide, and WSe_2 from tungsten hexachloride and diethylselenide, $(CH_3)_2Se$. Tungsten hexachloride is flowed with Ar carrier gas. Hydrogen is continuously flowed in the tube to prevent contamination by carbon. The WSe_2/WS_2 heterostructure is transferred over the Au/Ti bottom electrode. Top electrode over WSe_2 is made by evaporating palladium.

Response: To examine the sensor response, it is exposed to NO_2 gas (electron acceptor) and NH_3 gas (electron donor) under illumination with light without applying any external voltage (Figure 7.9(c)). For NO_2, the sensor response percentage = (increase in current in target gas relative to air/original current in air) ×100% is positive, increasing from 178% at 10 ppm to 322% at 500 ppm, while for NH_3, the sensor response is negative, decreasing from −19% at 10 ppm to −23% at 500 ppm. This sensor has P-type gas-sensing surface with a characteristic showing high selectivity to NO_2. The response time for NO_2 detection is 387 s, while its recovery time is 728 s.

7.4.2.4　N-Type 1L MoS_2/P-Type 3L WSe_2 Sensor Fabrication and Response

Fabrication: N-type 1L MoS_2/P-type 3L WSe_2 sensor (Figure 7.9(b)) is fabricated by *ex situ* synthesis of MoS_2/WSe_2. The 3L WSe_2 is made by a self-limiting layer synthesis at 700°C using 150 cycles of atomic layer deposition with a precursor-purge-reactant-purge sequence in which the precursor is tungsten hexachloride, heated to 90°C in a canister, and the reactant is diethylselenide with hydrogen injected during reaction time. The precursor is flowed with Ar carrier gas. The WSe_2 film pattern is defined with reactive ion etching. Aluminum is evaporated, oxidized to form aluminum oxide, and patterned. Then MoS_2 is deposited at 700°C using molybdenum (V) chloride ($MoCl_5$) heated to 90°C as a precursor in Ar carrier gas, and dimethyl sulfide, $(CH_3)_2S$, as a reactant. H_2 gas is flowed and 10 mg NaCl is added to hot zone of the reactor at 700°C. MoS_2 is patterned by reactive ion etching. Au/Ti electrode is formed by evaporation.

Response: For NH_3, the sensing response is positive, increasing from 62% at 10 ppm to 165% at 500 ppm, while for NO_2, the sensor response is negative, decreasing from −15% at 50 ppm to −36% at 500 ppm. This sensor shows N-type gas-sensing surface with a characteristic selective to NH_3. The response time for NH_3 detection is 205 s. Its recovery time is 263 s (Kim et al. 2020).

7.4.3 G-C₃N₄-Functionalized SnSe₂ Nanorods Composite Sensor for SO₂

Selective SO_2 detection at ppm level is reported using a composite gas-sensing nanomaterial made from $SnSe_2$, an N-type TMDC material functionalized with $g\text{-}C_3N_4$. Both $SnSe_2$ and $g\text{-}C_3N_4$ are 2D materials and act collaboratively to enable a high sensitivity, selectivity, and reversibility for SO_2 detection at ppm concentration levels (Zhang et al. 2022a). Figure 7.10 shows the sensor made with $g\text{-}C_3N_4/SnSe_2$ composite.

7.4.3.1 Fabrication Process of the Sensor

(i) $SnSe_2$ powder preparation by hydrothermal reduction: Tin chloride dihydrate ($SnCl_2.2H_2O$) and selenium dioxide (SeO_2) are mixed in DI water and stirred for ½ h after which hydrazine hydrate ($N_2H_4.H_2O$) is added to the mixture. The resulting solution is transferred to a Teflon-lined stainless-steel autoclave. After hydrothermal treatment at 180°C for 24 h, the precipitate is filtered, washed, and dried to obtain a black $SnSe_2$ powder.

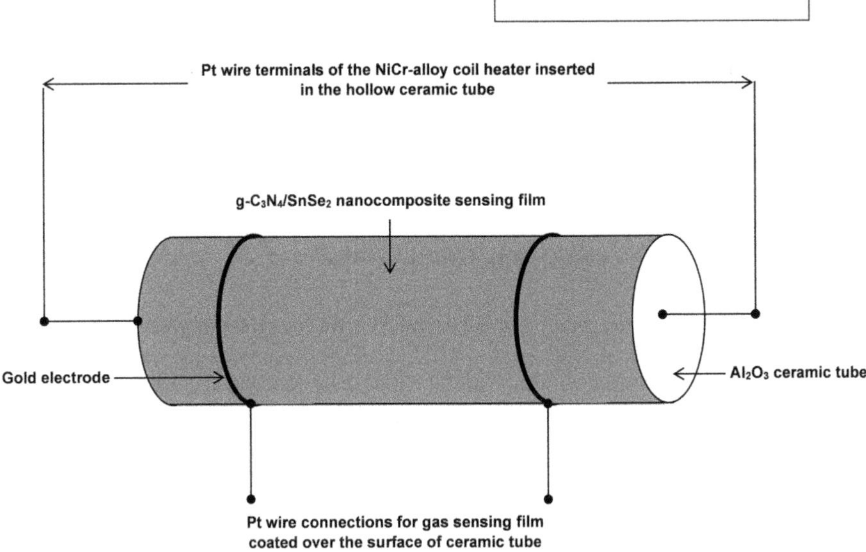

FIGURE 7.10 The $g\text{-}C_3N_4/SnSe_2$ composite sensor for SO_2. The diagram shows an Al_2O_3 ceramic tube with gold electrodes and platinum connecting wires. The surface of the ceramic tube is coated with a $g\text{-}C_3N_4/SnSe_2$ nanocomposite film. An NiCr-alloy heater is inserted inside the hollow ceramic tube and platinum wire terminals are taken out for supplying current.

(ii) g-C_3N_4/$SnSe_2$ composite nanomaterial synthesis: The $SnSe_2$ and g-C_3N_4 powders are dissolved in DI water, blended together by stirring for 1 h to ensure anchorage of g-C_3N_4 on the surface of $SnSe_2$, and dried at 60°C overnight to form the g-C_3N_4/$SnSe_2$ composite.

(iii) Making the ceramic tube-based sensor: An alumina ceramic tube with pre-formed gold electrodes and attached platinum wires has a NiCr heating coil inside. The g-C_3N_4/$SnSe_2$ composite is coated on the surface of the ceramic tube. It is dried at 60°C for 6 h and aged at 200°C for 24 h.

7.4.3.2 Gas Sensitivity

For 20 ppm SO_2, the gas response defined as [{(resistance of the sensor in air–resistance of the sensor in SO_2 ambience)/resistance of the sensor in air}×100%] for the 30% g-C_3N_4/$SnSe_2$ sensor shows the highest achieved value of 28.9% at 200°C. Pristine g-C_3N_4 and $SnSe_2$ sensors show a much inferior response at their optimum operating temperatures.

7.4.3.3 Sensing Mechanism of SO_2

(i) Sensor in air: Oxygen molecules in air adsorb on the sensor surface forming adsorbed oxygen molecules:

$$O_2\left(\text{air}\right) \rightarrow O_2\left(\text{adsorbed}\right) \tag{7.13}$$

Adsorbed oxygen molecules extract electrons from the conduction band of $SnSe_2$ (in g-C_3N_4/$SnSe_2$ composite) to form adsorbed O^- ions:

$$O_2\left(\text{adsorbed}\right) + 2e^- \rightarrow 2O^-\left(\text{adsorbed}\right) \tag{7.14}$$

resulting in the creation of a depletion region on the sensor surface.

(ii) Sensor in SO_2: The SO_2 molecules adsorb on the sensor surface forming adsorbed SO_2 molecules:

$$SO_2\left(\text{air}\right) \rightarrow SO_2\left(\text{adsorbed}\right) \tag{7.15}$$

SO_2 being a reducing gas, the adsorbed SO_2 molecules lose electrons to the conduction band of $SnSe_2$ (in g-C_3N_4/$SnSe_2$ composite) by reacting with the adsorbed O^- ions residing on the surface of the sensor, and SO_2 is itself oxidized to SO_3:

$$SO_2\left(\text{adsorbed}\right) + O^-\left(\text{adsorbed}\right) \rightarrow SO_3 + e^- \tag{7.16}$$

Recalling that $SnSe_2$ is an N-type material, the electrons received by it on adsorption of SO_2 cause an increase in the free carrier concentration, raising its conductivity and thereby lowering the resistance of the sensor.

7.4.3.4 Effects of Decoration of SnSe$_2$ with g-C$_3$N$_4$

There are two effects:

(i) Increase of specific surface area: Being a nanomaterial, g-C$_3$N$_4$ increases the specific surface area of the sensor, meaning that the sensor has more active sites. As a consequence, it has more adsorbed oxygen molecules, and therefore more adsorbed oxygen ions on its surface than a sensor without g-C$_3$N$_4$.

(ii) N–N heterojunction formation at the g-C$_3$N$_4$/SnSe$_2$ interface: Both C$_3$N$_4$ and SnSe$_2$ are N-type semiconductors. The bandgap of SnSe$_2$ is 1.3 eV while that of g-C$_3$N$_4$ is 2.7 eV. The work function of SnSe$_2$ is 4.3 eV while that of g-C$_3$N$_4$ is 4.67 eV. When put in contact with other, the difference of Fermi levels between the two semiconductors causes electron transference from SnSe$_2$ to g-C$_3$N$_4$ until the Fermi levels are equalized and equilibrium is reached. SnSe$_2$ loses electrons becoming positively charged, and g-C$_3$N$_4$ gains electrons becoming negatively charged. A built-in potential along with a depletion region is produced at the interface of the two semiconductors. Upon adsorption of SO$_2$, the electron concentration in the depletion region increases, and the width of depletion region decreases. As a result, the resistance of the sensor becomes smaller. The built-in potential across the g-C$_3$N$_4$/SnSe$_2$ interface helps in electron transfer. So, the sensitivity is still further increased. Thus, the g-C$_3$N$_4$/SnSe$_2$ heterojunction between the two 2D nanomaterials greatly boosts the sensor response (Zhang et al. 2022a).

7.5 GAS SENSORS FABRICATED USING OTHER 2D MATERIALS

7.5.1 BLACK PHOSPHOROUS NANOSHEET FET SENSOR FOR NO$_2$

7.5.1.1 Operating Principle

A sensor for NO$_2$ gas is made by utilizing the variation in electrical conductivity of a black phosphorous nanosheet (BPNS) with NO$_2$ concentration (Cui et al. 2015). The black phosphorous nanosheet is a P-type semiconductor in which holes are majority carriers and electrons are minority carriers. NO$_2$ is an oxidizing gas behaving as an electron acceptor. Therefore, adsorption of NO$_2$ gas molecules on the surface of the black phosphorous nanosheet is accompanied by the withdrawal of electrons from the nanosheet and simultaneous formation of holes in the electron-vacated sites. Consequently, the hole concentration in the P-type BPNS rises, and its electrical conductivity increases. The sensor made with black phosphorous nanosheet is shown in Figure 7.11.

7.5.1.2 Sensor Fabrication

For device fabrication, black phosphorous nanosheet exfoliated from black phosphorous crystal by the Scotch tape is transferred over 300 nm thick SiO$_2$ thermally grown over <100> orientation heavily-doped Si substrate. Degenerative doping of

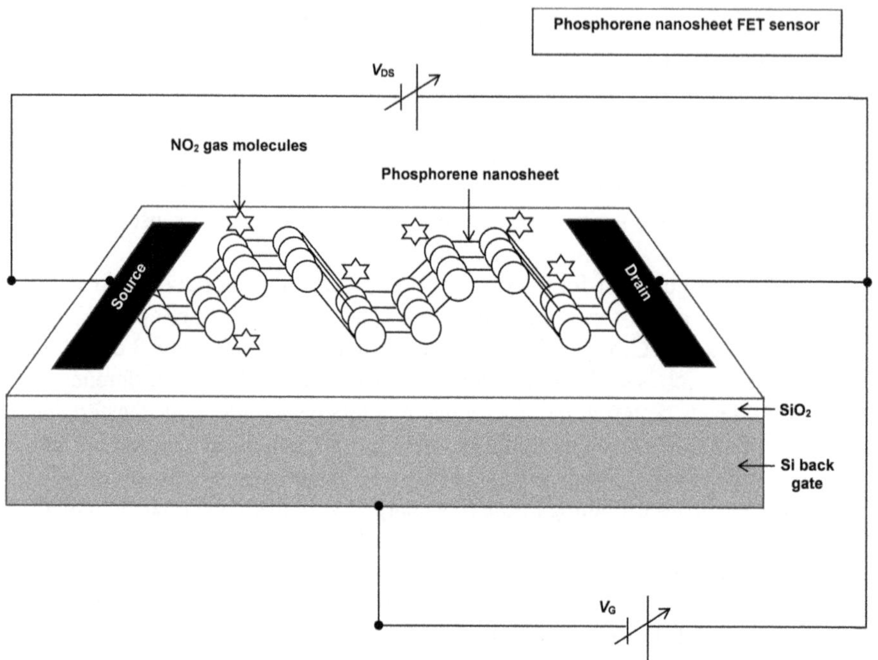

FIGURE 7.11 Phosphorene nanosheet sensor for NO_2. The diagram shows a silicon substrate acting as the back gate of the FET sensor. Over the silicon substrate is the silicon dioxide gate dielectric layer. A black phosphorous nanosheet is laid over the silicon dioxide film. The nanosheet acts as the semiconductor channel. The Ni/Au metal contacts are formed on both ends of the nanosheet constituting the source and drain terminals of the FET. NO_2 gas molecules are adsorbed on the surface of the black phosphorous film. Batteries are connected to supply drain–source voltage V_{DS} and gate bias V_G.

Si imparts to it a metallic character. After electron-beam lithography and pattern definition, contacts are made by (20 nm Ni/40 nm Au) metallization. The 4.8-nm thick black phosphorous nanosheet forms a conducting channel across the 2 μm gaps between adjacent gold fingers. Annealing in Ar at 200°C improves the BPNS/gold interface.

7.5.1.3 Sensor Response

For evaluating the sensor performance, a constant voltage of 0.6 V is applied between its electrodes and the conductance of the sensor is measured after exposing it to known concentrations of NO_2. A measurement cycle is composed of three sequential steps: dry air injection into the chamber, injection of target gas balanced in dry air, and sensor recovery in dry air. The gas sensitivity is defined as a difference between the relative conductance change (change in conductance (ΔG)/original conductance (G_0)) in air = 0 at the first cycle and that in the target gas at the end of switching off the gas for each concentration, i.e., the difference between $\Delta G/G_0 = 0$ in air (first cycle) and $\Delta G/G_0$ at the end of NO_2 cutting-off for each NO_2 concentration. The sensitivity is found to be 190% at 20 parts per billion (ppb) NO_2 at room temperature. The response of the sensor is stable in dry air over 0–5 days. The sensor shows insensitivity toward H_2S, H_2, and CO gases (Cui et al. 2015).

7.5.2 THE 2D G-C_3N_4@TiO_2 NANOPLATE/LANGASITE SAW NO_2 GAS SENSOR

An enhanced performance NO_2 sensor is developed using a 2D g-C_3N_4@TiO_2 nanocomposite as a chemical interface. The sensor has a highly selective response toward NO_2 against interfering gases, e.g., H_2S, NO, NH_3, and CO at room temperature. It exhibits a low detection limit and high sensitivity toward NO_2, apart from fast response/recovery times (Pasupuleti et al. 2022).

7.5.2.1 Piezoelectric Substrate

The piezoelectric substrate for the SAW sensor is langasite (lanthanum gallium silicate) abbreviated as LGS, with the chemical formula: $La_3Ga_5SiO_{14}$, an admired material for SAW sensors, widely preferred for high accuracy and long-term stability of frequency. Other merits favoring langasite for SAW sensor are its high-quality factor reaching up to 25000, excellent piezoelectric properties, intermediate between those of quartz and lithium tantalate, up to temperatures >1000°C, and high melting point (1470°C) with no phase transitions up to the melting point.

7.5.2.2 Gas Adsorption Film

The gas-sensitive film is laid over the SAW sensor fabricated on the LGS substrate. It is a hybrid nanocomposite made by the integration of 2D g-C_3N_4 with TiO_2 nanoplates of {001} facets. The 2D g-C_3N_4 is synthesized by thermal polymerization.

7.5.2.3 Frequency Shift and the Mass Loading Enhancement Factors

Target gas is adsorbed in this nanocomposite film of the SAW sensor producing the response signal. The sensor shows a negative shift of frequency on exposure to NO_2. The frequency shift ($\Delta\nu$) is~19.8 kHz for 100 ppm NO_2 at room temperature. It is 2.4 times that of pristine TiO_2 nanoplate/langasite SAW sensor without 2D g-C_3N_4.

The high sensitivity results from an increased mass loading effect. The loading effect is aided by the large surface area, oxygen vacancies, hydroxyl (OH), and amine groups (organic functional groups in which the N-atom has a lone electron pair, and three bonds) of the N–N hybrid heterojunction of 2D g-C_3N_4@TiO_2 nanoplates through the provision of plentiful active sites, facilitating the adsorption and diffusion of NO_2 molecules (Pasupuleti et al. 2022).

7.5.3 LAYERED-DOUBLE HYDROXIDE VOC GAS SENSOR

An LDH co-precipitation method is applied to fabricate sensors for VOC detection (Vigna et al. 2021). The sensor is depicted in Figure 7.12.

7.5.3.1 Synthesis of LDHs

Four LDHs (ZnAl–Cl, ZnFe–Cl, ZnAl– NO_3, and MgAl–NO_3) are synthesized by dropwise addition of 2M sodium hydroxide solution to a solution of suitable

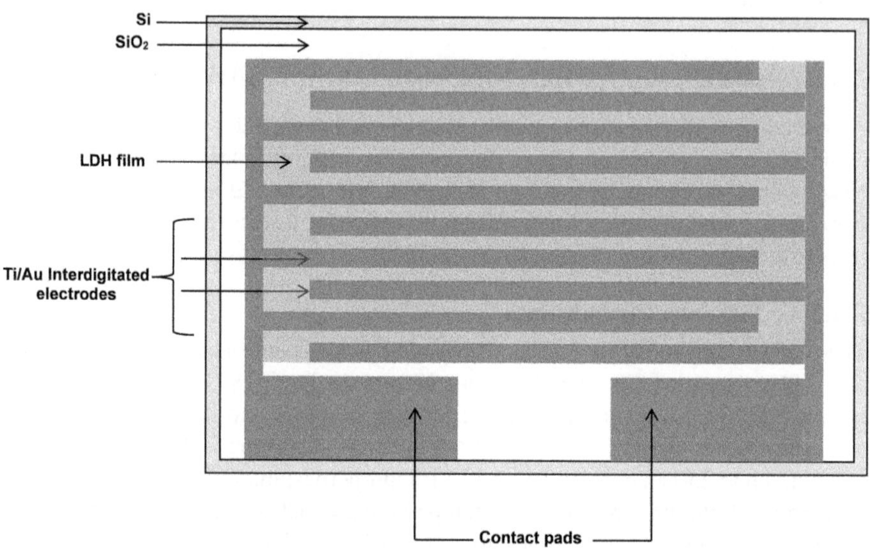

FIGURE 7.12 Geometrical layout of the LDH gas sensor for VOCs. The diagram shows the SiO$_2$/Si substrate over which lie Ti/Au interdigitated electrodes and contact pads. The electrodes are coated with the LDH film by drop-casting using a PDMS mold for defining the boundaries.

quantities of two salts in decarbonated distilled water as follows until reaching a pH = 10:

> Salts used for ZnAl–Cl are: ZnCl$_2$ 6H$_2$O and AlCl$_3$·xH$_2$O (zinc chloride hexahydrate and aluminum chloride hydrate)
> Salts used for ZnFe–Cl are: ZnCl$_2$·6H$_2$O and FeCl$_3$·6H$_2$O (zinc chloride hexahydrate and iron (III) chloride hexahydrate)
> Salts used for ZnAl–NO$_3$ are: Zn (NO$_3$)$_2$·6H$_2$O and Al (NO$_3$)$_3$·9H$_2$O (zinc nitrate hexahydrate and aluminum nitrate nonahydrate)
> Salts used for MgAl–NO$_3$ are: Mg (NO$_3$)$_2$·6H$_2$O and Al (NO$_3$)$_3$·9H$_2$O (magnesium nitrate hexahydrate and aluminum nitrate nonahydrate)

The pH is maintained at 10.00 by continuously adding NaOH solution, and the temperature is kept at 70°C (except for LDH ZnFe–Cl, which is kept at room temperature) for 24 h under flowing nitrogen. The precipitates are collected by centrifugation. They are repeatedly washed in distilled water and dried in a vacuum to get LDH powders.

7.5.3.2 Chemoresistive Sensor Fabrication

Four sets of devices are made, one for each synthesized LDH composition. On a SiO$_2$/Si wafer, contact pads and IDEs are photolithographically patterned followed by Ti/Au evaporation and lift-off process to delineate the electrode fingers and pads. Dispersions of LDH powders in EtOH are drop-coated on the active areas of the devices using a micropipette. The dispersions are protected from spreading over the inactive areas by a PDMS mold. The devices are heated on a hotplate at 65°C for 10 min to evaporate the solvent.

7.5.3.3 Gas-Sensing Characteristics

The sensor responses to different gases (ammonia, acetone and chlorine, and ethanol vapors) are determined by measuring the baseline resistance R_{Baseline} of each device, and resistance R_{Gas} after exposure to a known concentration of analyte gas. The response S is calculated by the formula:

$$S = \left(1 - \frac{R_{Baseline}}{R_{Gas}}\right) \times 100\% \tag{7.17}$$

All four types of LDH-based sensors could reversibly detect the analyte vapors at room temperature, showing sensing response values of up to 6%. The sensitivity and selectivity of the LDH sensor to a particular gas are tailored by altering the composition of the LDH film. Several sensors can be integrated in an array to construct an electronic nose (Vigna et al. 2021).

7.5.4 OPTICALLY TRANSPARENT 2D MoO_{3-x} NANOSHEETS-BASED MOISTURE SENSOR

A moisture sensor with optical transmittance of 86% is fabricated by liquid-phase exfoliation of 2D MoO_{3-x} nanosheets from α-MoO_3 powder (Zhang et al. 2022b). Figure 7.13 shows the 2D MoO_{3-x} nanosheets sensor.

7.5.4.1 Exfoliation of MoO_{3-x} Nanosheets

For liquid-phase exfoliation of MoO_{3-x} nanosheets, α-MoO_3 powder is grinded for ½ h, mixed with ethyl cellulose ($C_{20}H_{38}O_{11}$) in ethanol (CH_3CH_2OH), and probe sonicated in ice-water bath for 1h. The process yields a light blue-colored dispersion of exfoliated MoO_{3-x} nanosheets. The nanosheets are prevented from re-aggregating by ethyl cellulose.

7.5.4.2 Sensor Fabrication

The sensor is made on a glass substrate coated with a conducting FTO (fluorine-doped tin oxide) film. The FTO film is laser-etched to form stripe electrodes. The substrate is cleaned ultrasonically in DI water and ethanol. A small quantity of PEDOT: PSS is added to a diluted dispersion of MoO_{3-x} nanosheets in ethanol to prepare a mixed precursor. The PEDOT: PSS is a polymer mixture of two ionomers: poly(3,4-ethylenedioxythiophene) and polystyrene sulfonate.

The PEDOT: PSS plus MoO_{3-x} nanosheets dispersion-precursor is deposited by spin coating over the electrodes, keeping the contact areas protected. The coated substrate is baked at 100°C for ½ h.

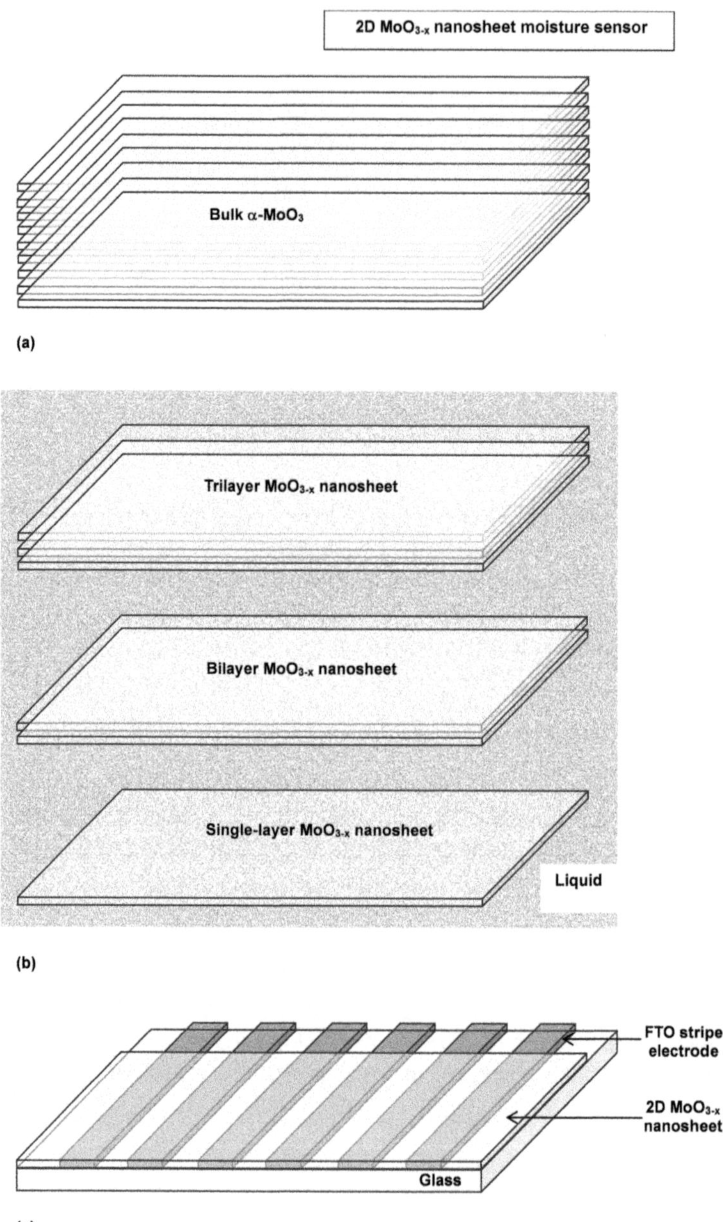

FIGURE 7.13 Transparent 2D MoO_{3-x} nanosheets moisture sensor: (a) bulk α-MoO_3, (b) liquid-phase exfoliated MoO_{3-x} nanosheets, and (c) the fabricated device. Diagram (a) shows the layered structure of α-MoO_3 comprising the stacked MoO_{3-x} nanosheets. Diagram (b) shows the exfoliated stacks consisting of three, two, and single-layer MoO_{3-x} nanosheets dispersed in liquid phase. Diagram (c) shows a glass substrate over which there is a laser-etched pattern of FTO stripe electrodes coated with MoO_{3-x} nanosheet except for the contact regions.

7.5.4.3 Humidity Sensitivity of the Device

On applying a constant voltage of 5 V across the sensor stripe electrodes, a monotonic change in current is observed in proportion to ambient relative humidity from 11% to 95%, giving a sensitivity of 582. The response time of the sensor is 0.12 s. Its recovery time is 0.53 s. The sensor is used to develop an expiration monitoring system by which the moisture response signal is converted into the brightness of a light-emitting diode (Zhang et al. 2022b).

7.6 DISCUSSION AND CONCLUSIONS

Table 7.1 compiles the 2D material gas sensors discussed in this chapter. A competing technology for 2D material gas sensors is the highly sensitive, long-lasting, low-cost semiconductor metal-oxide sensors using SnO_2, TiO_2, WO_3, ZnO, and other oxides. These sensors are extensively used for the detection of pollutant gases.

TABLE 7.1

Types of 2D Material Sensors Gas Sensors

S. No.	2D Material (s) Used	Sensing Device	Gas (es) Detected	Reference
1.	Graphene and graphene oxide	UV-illuminated graphene–silicon Schottky diode sensor	NO_2 and THF	Drozdowska et al. (2023)
		Graphene-FET sensor	SO_2	Ren et al. (2012)
		Graphene-coated QCM sensor	Volatile organic compounds (VOCs)	Quang et al. (2014)
		Fe_2O_3-graphene nanosheets chemiluminescence sensor	H_2S	Jiang et al. (2014)
		Graphene oxide–ZnO nanorod sensor	SO_2 and H_2	Dhingra et al. (2020)
2.	TMDCs	SnO_2 nanoparticles/$MoSe_2$ nanoflowers-resistive sensor	CO	Yang et al. (2020)
		P-WSe_2/N-WS_2 and N-MoS_2/P-WSe_2 heterostructure gas sensors	NO_2 and NH_3	Kim et al. (2020)
		g-C_3N_4 functionalized-$SnSe_2$ nanorods composite sensor	SO_2	Zhang et al. (2022a)
3.	Miscellaneous 2D materials	Black phosphorous nanosheet FET sensor	NO_2	Cui et al. (2015)
		2D g-C_3N_4@TiO_2 nanoplate/langasite SAW sensor	NO_2	Pasupuleti et al. (2022)
		Layered-double hydroxide sensor	Volatile organic compounds	Vigna et al. (2021)
		2D MoO_{3-x} nanosheets sensor	Moisture	Zhang et al. (2022b)

However, they require high operating temperatures ~400°C leading to a huge power consumption. 2D materials provide a viable substitute to this technology allowing operation at room temperature or at a comparatively lower temperature than metal-oxide sensors. The substitution meets the industrial demand for minimizing power wastage. In addition, such sensors can be made on flexible substrates used in wearable electronics to make wearable gas sensors.

The drawbacks of 2D material gas sensors are the lack of a steady state on exposure to the gas, and the slow recovery after exposure. These limitations hinder proper calibration. A remedy for this problem is to analyze the time differential of the signal output (TDSO) instead of the signal output. This approach has been found to be more convenient for sensor calibration and comparison (Ricciardella et al. 2020).

Reliable sensors for monitoring low concentrations of 'criteria air pollutants' are most sought after. These pollutants include ozone (O_3), nitrogen dioxide (NO_2), sulfur dioxide (SO_2), carbon monoxide (CO), and particulate matter (PM10 and PM2.5). A small chunk of reported devices in this category were studied in this chapter. After chemical sensors, we move to sensors for detecting biological analytes in the next chapter.

REFERENCES

Cui S., H. Pu, S. A. Wells, Z. Wen, S. Mao, J. Chang, M. C. Hersam and J. Chen 2015 Ultrahigh sensitivity and layer-dependent sensing performance of phosphorene-based gas sensors, *Nature Communications*, 6(8632), pp. 1–9.

Devkota J., P. R. Ohodnicki, and D. W. Greve 2017 SAW sensors for chemical vapors and gases. *Sensors (Basel)*, 17(4), pp. 1–28.

Dhingra V., S. Kumar, R. Kumar, A. Garg and A. Chowdhuri 2020 Room temperature SO_2 and H_2 gas sensing using hydrothermally grown GO–ZnO nanorod composite films, *Materials Research Express*, 7(6), pp. 1–11.

Drozdowska K., A. Rehman, J. Smulko, S. Rumyantsev, B. Stonio, A. Krajewska, M. Słowikowski, M. Filipiak, et al. 2023 Enhanced gas sensing by graphene-silicon Schottky diodes under UV irradiation, *Sensors and Actuators B: Chemical*, 396(134586), pp. 1–9.

Fauzi F., A. Rianjanu, I. Santoso, and K. Triyana 2021 Gas and humidity sensing with quartz crystal microbalance (QCM) coated with graphene-based materials – A mini review, *Sensors and Actuators A: Physical*, 330, p. 112837.

Ge L., X. Mu, G. Tian, Q. Huang, J. Ahmed, and Z. Hu 2019 Current applications of gas sensor based on 2-D nanomaterial: A mini review, *Frontiers in Chemistry*, 7, pp. 1–7.

Jiang Z., J. Li, H. Aslan, Q. Li, Y. Li, M. Chen, Y. Huang, J. P. Froning, et al. 2014 A high efficiency H_2S gas sensor material: paper like Fe_2O_3/graphene nanosheets and structural alignment dependency of device efficiency, *Journal of Materials Chemistry A*, 2, pp. 6714–6717.

Khlebarov Z. P., A. I. Stoyanova and D.I. Topalova 1992 Surface acoustic wave gas sensors, *Sensors and Actuators B: Chemical*, 8(1), pp. 33–40.

Kim Y., S. Lee, J.-G. Song, K.Y. Ko, W. J. Woo, S.W. Lee, M. Park, H. Lee, Z. Lee, H. Cho, et al. 2020 2D Transition metal dichalcogenide heterostructures for p- and n-type photovoltaic self-powered gas sensor, *Advanced Functional Materials*, 30(43), 2003360, pp. 1–11.

Mathew M. and C. S. Rout 2021 Schottky diodes based on 2D materials for environmental gas monitoring: A review on emerging trends, recent developments and future perspectives, *Journal of Materials Chemistry C*, 9, pp. 395–416.

Pasupuleti K. S., M. Reddeppa, S.S. Chougule, N. H. Bak, D. J. Nam, N. Jung, H. D. Cho, S. G. Kim, and M. D. Kim 2022 High performance langasite based SAW NO_2 gas sensor using 2D g-C_3N_4@TiO_2 hybrid nanocomposite, *Journal of Hazardous Materials*, 427, p. 128174.

Quang V. V., V. N. Hung, L. A. Tuan, V. N. Phan, T. Q. Huy, and N. V. Quy 2014 Graphene-coated quartz crystal microbalance for detection of volatile organic compounds at room temperature, *Thin Solid Films*, 568, pp. 6–12.

Ren Y., C. Zhu, W. Cai, H. Li, H. Ji, I. Kholmanov, Y. Wu, R. D. Piner, and R. S. Ruoff 2012 Detection of sulfur dioxide gas with graphene field effect transistor, *Applied Physics Letters*, 100, 163114-1–163114-4.

Ricciardella F., K. Lee, T. Stelz, O. Hartwig, M. Prechtl, M. McCrystall, N. McEvoy, and G. S. Duesberg 2020 Calibration of nonstationary gas sensors based on two-dimensional materials, *ACS Omega*, 5(11), pp. 5959–5963.

Srivastava S., P. Pal, D. K. Sharma, S. Kumar, T. D. Senguttuvan, and B. K. Gupta 2022 Ultrasensitive boron–nitrogen-codoped CVD graphene-derived NO_2 gas sensor, *ACS Materials Au*, 2, pp. 356–366.

Vigna L., A. Nigro, A. Verna, I. V. Ferrari, S. L. Marasso, S. Bocchini, M. Fontana, A. Chiodoni, C.F. Pirri, and M. Cocuzza 2021 Layered double hydroxide-based gas sensors for VOC detection at room temperature, *ACS Omega*, 6, pp. 20205–20217.

Yang S., G. Lei, H. Xu, Z. Lan, Z. Wang, and H. Gu 2021 Metal oxide-based heterojunctions for gas sensors: A review, *Nanomaterials (Basel)*, 11(4), pp. 1–26.

Yang Z., D. Zhang, and D. Wang 2020 Carbon monoxide gas sensing properties of metal-organic frameworks-derived tin dioxide nanoparticles/molybdenum diselenide nano-flowers, *Sensors and Actuators B: Chemical*, 304, p. 127369.

Zhang H., Q. Pan, Y. Zhang, Y. Zhang and D. Zhang 2022a High-performance sulfur dioxide gas sensor based on graphite-phase carbon-nitride-functionalized tin diselenide nanorods composite, *Chemosensors*, 10(401), pp. 1–13.

Zhang Y., H. Ma, S. Wu, H. Yu, L. Wu, W. Li, J.-L. Sun, H. Wang, and H. Fang 2022b Transparent humidity sensor with high sensitivity via a facile and scalable way based on liquid-phase exfoliated MoO_{3-x} nanosheets, *Sensors and Actuators Reports*, 4(100092), pp. 1–7.

8 2D Materials-Based Biological Sensors

8.1 EMERGING APPLICATIONS OF BIOLOGICAL SENSORS MADE FROM 2D MATERIALS

Research on biological sensors has witnessed an explosive growth in recent years. 2D materials-based biological sensors are at the forefront of research. Research on 2D material-based biological sensors holds enormous potential for developing cost-effective, point-of-care diagnostic tools providing quick medical test reports, either at the patient bedside or in the physician's clinic about disease-causing micro-organisms and disease-indicating markers in bodily fluids such as blood, urine, saliva, and sweat (Filice et al. 2021). Besides healthcare, these biological sensors are likely to be used for on-site, *in situ*, and online measurements of critical parameters in food industry (for the detection of pathogens and pesticide residues and for analysis of yogurt, wine, and beer), agriculture (for assessing toxins in crops, judging quality and ripening of fruits, testing for crop diseases, and carrying out veterinary screening tests of farm animals), and environmental monitoring (for the detection of pollutants in air, water, and soil). This chapter examines the advances in biological nanosensors made from 2D materials for biomedical, food, environmental, and biodefense applications.

8.2 THE 2D MATERIAL BIOLOGICAL SENSORS

8.2.1 BIOLOGICAL SENSOR

Biological sensors are defined as a generic class of sensors using or involving biomolecules, the molecules produced by living organisms (carbohydrates, proteins, lipids, nucleic acids, and others); and biochemicals, the chemicals existing in or obtained from living organisms, or those used in processes dealing with living matter. As an important example of biochemicals, mention may be made of the various drugs and pharmaceuticals used in the diagnosis, treatment, or prevention of diseases. A biological sensor need not always contain a biomolecule as an essential component but must be a sensor of biological interest, being related to medicine, food, and environment. A major subcategory of biological sensors is the biosensor, which requires a special attention due to its overwhelming significance, and so constitutes a complete discipline in itself.

8.2.2 BIOSENSOR AND ITS MAIN COMPONENTS

Like any other type of biosensor, a 2D material biosensor is a self-sufficient, integrated, easy-to-use analytical device using biomolecules or living organisms for

DOI: 10.1201/9781003330585-8

the detection of chemical substances (Figure 8.1). It is a three-component device consisting of (Bhalla et al. 2016):

(i) Bioreceptor: It is a biological recognition element, usually biologically derived or biomimetic materials, e.g., enzymes, tissues or whole cells, micro-organisms, organelles, nucleic acids (DNA, RNA), aptamers, proteins, peptides, molecularly imprinted polymers (MIPs), and antibodies, having specific affinities to the analytes of interest. It selectively identifies or binds with the target analyte.

(ii) Physicochemical transducer: It converts the biochemical information produced by the interaction between the bioreceptor and the analyte into an electrical or optical signal. Common types of transducers work on mechanical, piezoelectric, pyroelectric, gravimetric, electrochemical, thermal, photoelectric, fluorescent, chemiluminescent, and magnetic principles. The 2D material used in the construction of this component is the core material that interfaces with the analyte and therefore determines the sensor response.

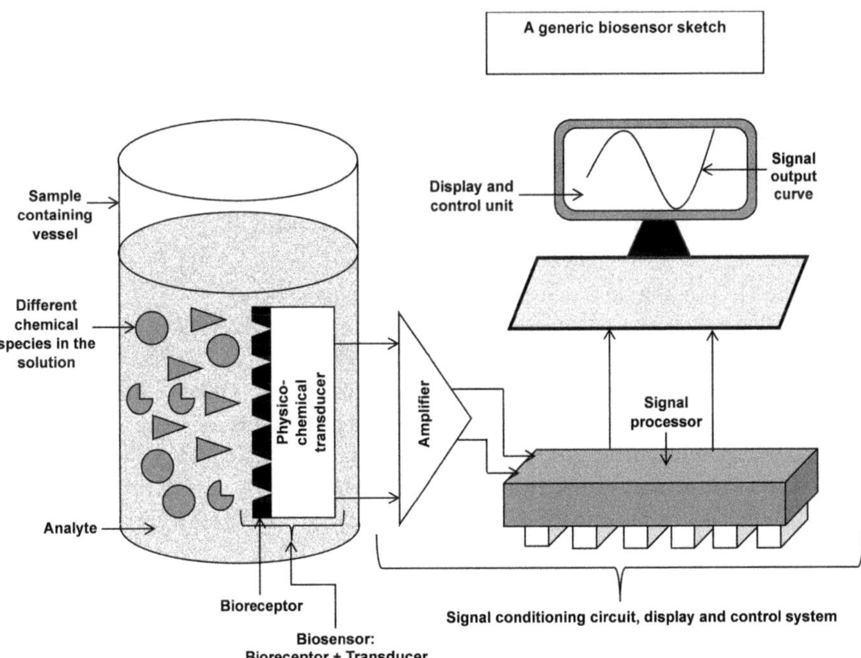

FIGURE 8.1 Generic components of a biosensor. The diagram shows a bioreceptor on the surface of a physico-chemical transducer. The bioreceptor plus physico-chemical transducer constitutes the biosensor and is enclosed within curly brackets for clarity. The signal produced by the biosensor is fed to a signal conditioning circuit and control system. The control system comprises an amplifier, a signal processor, and a display and control unit. The biosensor is immersed in the specimen analyte solution containing different chemical species.

(iii) Signal conditioning circuit, display, and control system: It filters out the noise or interference in the signal, amplifies it, performs linearization, and transforms it into a compatible format for data acquisition, display, and machine control.

Biological and biosensors are studied from the consideration of the bio-recognition element, e.g., antibodies, enzymes, and DNA, or from the viewpoint of the transduction method implemented. We shall follow the latter approach. In the background of the translation of generated response into an electrical/optical signal, the 2D material biological and biosensors are classified into modified-electrode electrochemical sensors, FET-based sensors, and optical sensors.

8.3 2D MATERIAL-MODIFIED ELECTRODE ELECTROCHEMICAL SENSORS FOR BIOLOGICAL APPLICATIONS

8.3.1 Au/MXene Nanocomposite–GCE Enzymatic Bioelectrode Sensor for Glucose

A GO_x/AuNPs/MXeneNSs/Nafion/GCE linear amperometric biosensor is fabricated by immobilizing glucose oxidase (GO_x) enzyme on a Nafion-solubilized gold nanoparticles (AuNPs)-incorporated MXene nanosheets (MXeneNSs) composite over a glassy carbon electrode (GCE) (Rakhi et al. 2016). The fabrication of the sensor is described in Figure 8.2.

In this multilayer electrode, each layer has a defined role, as elaborated below.

8.3.1.1 Roles of Different Layers in the Bioelectrode

 (i) Go_x: It acts as a catalyst in the oxidation of glucose to glucono-d-lactone and H_2O_2 in the presence of oxygen in the atmosphere.

 (ii) MXeneNSs: The MXenes have high metallic conductivity. The high in-plane conductivity of MXene nanosheets (MXeneNSs) makes them suitable as an enzyme immobilization matrix helping to improve the electron transfer kinetics between the active redox centers of the enzyme. Moreover, the hydraulic surface of MXene nanosheets provides stability in an aqueous environment. Another advantage of using MXene is that it is reliably produced in large quantities by wet chemical synthesis offering scalability of the process.

(iii) AuNPs: The anchoring of highly conductive biocompatible Au nanoparticles (AuNPs) on the surfaces of MXene nanosheets elevates their electrical conductivity facilitating the electron transfer process between GO_x and GCE, and thus lowering the high potential required for the oxidation/reduction of H_2O_2 over GCE. Effectively, the electrocatalytic activity of AuNPs/MXene composite surpasses that of bare MXene matrix for enzyme immobilization.

(iv) Nafion: It is used for the proper dispersal of the AuNPs/MXene composite to make a well-dispersed, uniform, and stable film over GCE, furnishing better adhesion of enzyme with GCE. Nafion also reduces the permeability of negatively charged enzyme substrates because it is a negatively charged polyelectrolyte. In this way, it suppresses the interference signals and improves the selectivity of the device.

(v)   GCE: It is a mechanically stable support for enzyme immobilization. It is impermeable to liquids and gases.

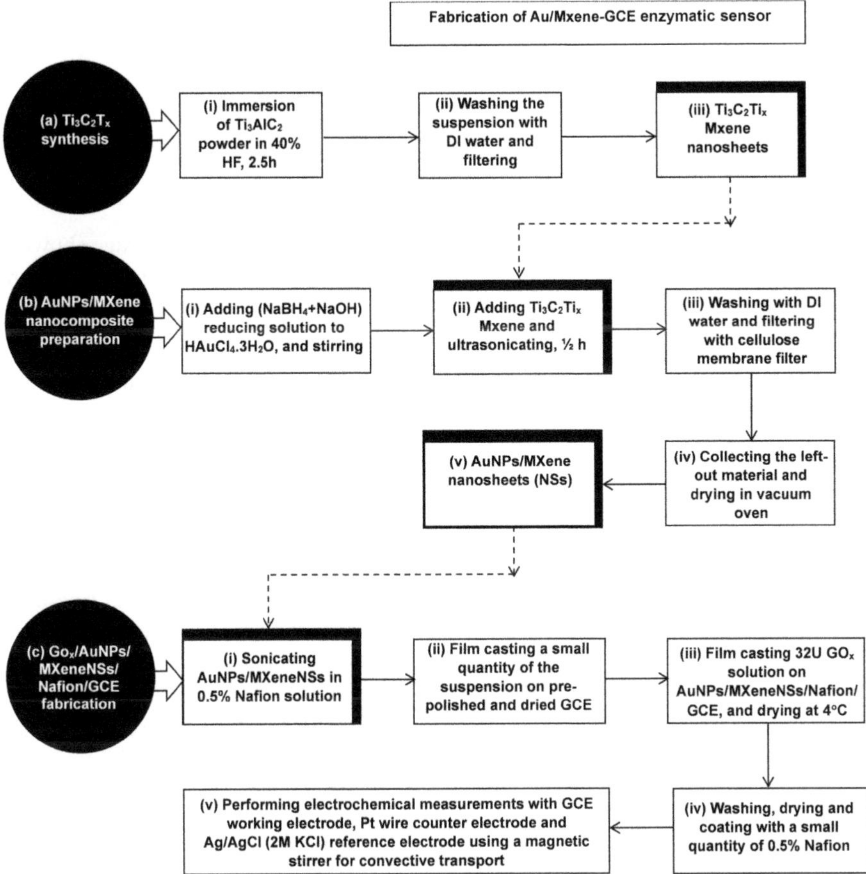

FIGURE 8.2 Flow charts of fabrication process stages of an Au/MXene nanocomposite–GCE enzymatic bioelectrode glucose sensor. The stages are concerned with making: (a) Ti₃C₂Tₓ, (b) AuNPs/MXene nanocomposite, and (c) Goₓ/AuNPs/MXeneNSs/Nafion/GCE. Part (a) shows Ti₃C₂Tₓ synthesis consisting of the steps: immersing Ti₃AlC₂ powder in HF (2.5 h), washing with DI water, and filtering to get Ti₃C₂Tiₓ MXene nanosheets. Part (b) shows AuNPs/MXene nanocomposite preparation consisting of the steps: adding (NaBH₄+NaOH) solution to HAuCl₄.3H₂O, stirring, adding Ti₃C₂Tiₓ MXene to the mixture, ultrasonicating (½ h), washing with DI water, filtering, collecting the left-out material, and drying in vacuum to get AuNPs/MXene nanosheets (NSs). Part (c) shows the fabrication and use of Goₓ/AuNPs/MXeneNSs/Nafion/GCE consisting of the steps: sonicating AuNPs/MXeneNSs in 0.5% Nafion solution and film casting on GCE; film casting 32U GOₓ solution on AuNPs/MXeneNSs/Nafion/GCE and drying (4°C); washing, drying, and coating with 0.5% Nafion; and electrochemical measurements.

8.3.1.2   Cyclic Voltammetric (CV) Responses

In cyclic voltammetry (CV), a pair of redox peaks are observed at -0.49V for GO_x/MXeneNSs/Nafion/GCE bioelectrode. The same appear at a lower potential in GO_x/AuNPs/MXeneNSs/Nafion/GCE bioelectrode because of the relatively faster charge transport in the AuNPs-enriched bioelectrode. Approximately three-fold lower reduction current is achieved for the GO_x/MXeneNSs/Nafion/GCE bioelectrode than that of GO_x/AuNPs/MXeneNSs/Nafion/GCE bioelectrode. For a particular glucose concentration, the time taken by the current to stabilize is 20 s for the former but 10 s for the latter electrode.

For GO_x/MXeneNSs/Nafion/GCE biosensor, the linear range of glucose detection is 0.5 mM –6mM, LoD = 100 μM. For the GO_x/AuNPs/MXeneNSs/Nafion/GCE biosensor, the linear range is 0.1 mM –18mM, LoD = 5.9 μM, and sensitivity = $4.2\,\mu AmM^{-1}cm^{-2}$.

The biosensor retains 93% of its starting response after 2 months storage showing its suitability for Diabetes Mellitus monitoring over an extended period (Rakhi et al. 2016).

8.3.2   Au NPs-MoS$_2$ Nanosheet-GCE Sensor for Ascorbic Acid, Dopamine, and Uric Acid

This is an electrochemical sensor using a chemically modified electrode for electro-catalytic oxidation of three analytes (ascorbic acid, dopamine, and uric acid) displaying clearly decisive peaks (Sun et al. 2014). The electrode used is a 3 mm-diameter glassy carbon electrode (GCE). It is the working electrode of a 3-electrode voltametric cell consisting of a Pt wire counter electrode and a saturated calomel reference electrode with a 0.1 M phosphate-buffered solution as the supporting electrolyte.

8.3.2.1   MoS$_2$ Nanosheets Preparation by Intercalation Exfoliation

Intercalated MoS_2 is formed with n-butyllithium (C_4H_9Li) solution in Ar ambience for 2 days at room temperature. After removing the superfluous n-butyllithium solution, and leftover solvent by flowing argon, Li-intercalated MoS_2 is obtained by adding deoxidation water, followed by 1 h-sonication, and, finally, centrifugation of water-dispersed MoS_2 nanosheets for getting rid of LiOH and dissolved impurities.

8.3.2.2   Decoration of GCE with AuNPs@MoS$_2$ Nanocomposite by AuNPs Electrodeposition

This is done by dropping the MoS_2 solution on the GCE surface, air-drying for 16 h, immersing it in chloroauric acid ($HAuCl_4$) solution, and electrodepositing AuNPs by pulse voltammetry.

8.3.2.3   Cyclic Voltammetry

In voltammograms (current at the AuNPs@MoS$_2$/GCE against applied voltage vs. SCE), three distinct oxidation peaks are noticed at -0.046 V, 0.105 V, and 0.242 V corresponding to ascorbic acid (AA), dopamine (DA), and uric acid (UA), respectively. The peak-to-peak separation potentials are:

$$\Delta V_{DA-AA} = 0.105 - (-0.046) = 0.151\,V = 151\,mV, \tag{8.1}$$

$$\Delta V_{UA-DA} = 0.242 - 0.105 = 0.137\,V = 137\,mV, \tag{8.2}$$

and

$$\Delta V_{UA-AA} = 0.242 - (-0.046) = 0.288\,V = 288\,mV, \tag{8.3}$$

so that the peaks are easily distinguishable for detecting these analytes individually as well as simultaneously.

8.3.2.4 Differential Pulse Voltammetry

Differential pulse voltammetry (DPV) profiles are measured in a mixture of AA, AA, and UA. Plots of oxidative peak currents vs. concentrations of these analytes are linear: AA from 1 mM to 70 mM, limit of detection (LoD) = 100μM; DA from 50 nM to 4 mM, LoD = 50 nM; and UA from 10 μM to 7 mM, LoD = 10 μM. Dopamine measurements in 1% human serum confirm the usefulness of the sensor for real bio-specimens (Sun et al. 2014).

8.3.3 THE 2D-CONDUCTIVE MOF NANOSHEETS-BASED SENSOR FOR CANCER BIOMARKER H_2O_2 IN LIVE CELLS

A high-density AuNPs-decorated 2D-conductive MOF nanosheets-based nanozymatic electrochemical biosensor is fabricated (Huang et al. 2022). Here, conductive metal–organic framework (C-MOF) nanosheets (NSs) = C-MOF NSs = Cu-HHTP [HHTP = 2,3,6,7,10,11-hexahydroxytriphenylene)-NSs].

The C-MOF NSs offer a large surface area. Additionally, they have plentiful Cu–O_4 open metal sites. So, they show good catalytic activity towards H_2O_2. The copious exposed oxygen atoms act as fastening sites for AuNPs.

8.3.3.1 Synthesis of AuNPs/Cu-HHTP-NSs

To synthesize AuNPs/Cu-HHTP-NSs, [Cupric acetate monohydrate or Cu $(OAc)_2 \cdot H_2O$ + sodium dodecyl sulfate (SFS) or $CH_3(CH_2)_{11}OSO_3Na$] is added to aqueous (HHTP + NaOH) solution, stirred, sonicated for 1 h and left overnight. After centrifuging, the precipitate is washed with methanol. Then it is redispersed in water and ethanol in an ultrasonic ice bath for 1 h. The dispersion is left unperturbed for 12 h. Upper colloidal solution is taken out and freeze-dried. The process is continued until exfoliated Cu-HHTP-NSs are obtained. These are redispersed in ethanol. After adding hydrogen tetrachloroaurate (III) trihydrate ($HAuCl_4 \cdot 3H_2O$) into Cu-HHTP-NSs dispersion, stirring for ½ h, and heating at 65 °C for 5 h, the precipitates are collected through centrifugation at 14,000 RPM, washed thrice with ethanol and water, and freeze-dried. The GCE is modified with Au-NPs/Cu-HHTP-NSs to get Au-NPs/Cu-HHTP-NSs/GCE.

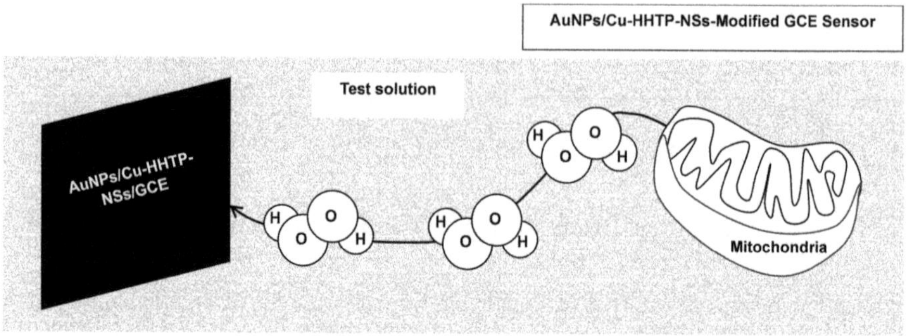

FIGURE 8.3 Using Au-NPs/Cu-HHTP-NSs-modified glassy carbon electrode to track H_2O_2 molecules released from mitochondria in human colon cells. The diagram shows the modified GCE, and H_2O_2 molecules moving in the test solution after liberation from mitochondria.

8.3.3.2 Electrochemical Impedance Spectroscopy

The nanohybrid-modified GCE is characterized by electrochemical impedance spectroscopy (EIS) using ferrocyanide/ferricyanide [Fe $(CN)_6$]$^{3-/4-}$ as redox species. Its electrocatalytic performance toward H_2O_2 is studied by CV showing a reduction peak at -0.6 V. The reduction current increases linearly with H_2O_2 concentration. The modified-glassy carbon electrode (GCE) exhibits a high sensitivity for H_2O_2 ~ 188.1 μA cm^{-2} mM^{-1}, and a very low LoD ~5.6 nM, enabling real-time monitoring of mitochondria-liberated H_2O_2 from human colon cells to distinguish colon cancer cells from normal cells for clinical diagnosis.

8.3.3.3 Revealing Cancer Cells

A biosensing experiment is carried out to illustrate how the cancer cells are revealed in a test specimen (Figure 8.3).

N-Formyl-methionyl-leucyl-phenylalanine (fMLP), $C_{21}H_{31}N_3O_5S$, is added in a test solution containing NCM-460 (a normal human colon mucosal epithelial cell line), SW-48 (human colon adenocarcinoma cell line), and HCT-116 (a human colorectal carcinoma cell line), as a stimulator to initiate live cells to liberate H_2O_2. Current changes of 0.22, 0.64, and 0.52 μA occur for NCM-460, SW-48, and HCT-116, respectively. On the other hand, no detectable signal is seen with control specimens having no cells. Greater amount of extracellular H_2O_2 production by human colon cancer cells than normal cells is therefore evident implying the effective revelation of cancer cells by this sensor (Huang et al. 2022).

8.3.4 GRAPHENE/NAFION-MODIFIED ELECTRODE FOR PARACETAMOL, ASPIRIN, AND CAFFEINE

A graphene-Nafion (GR-NF)-modified glassy carbon electrode (GR-NF/GCE) shown in Figure 8.4 is used to simultaneously measure paracetamol (PAR), an analgesic and antipyretic agent; aspirin (ASA), a nonsteroidal anti-inflammatory drug; and caffeine (CAF), a central nervous system stimulant for reducing fatigue and drowsiness (Yiğit et al. 2016).

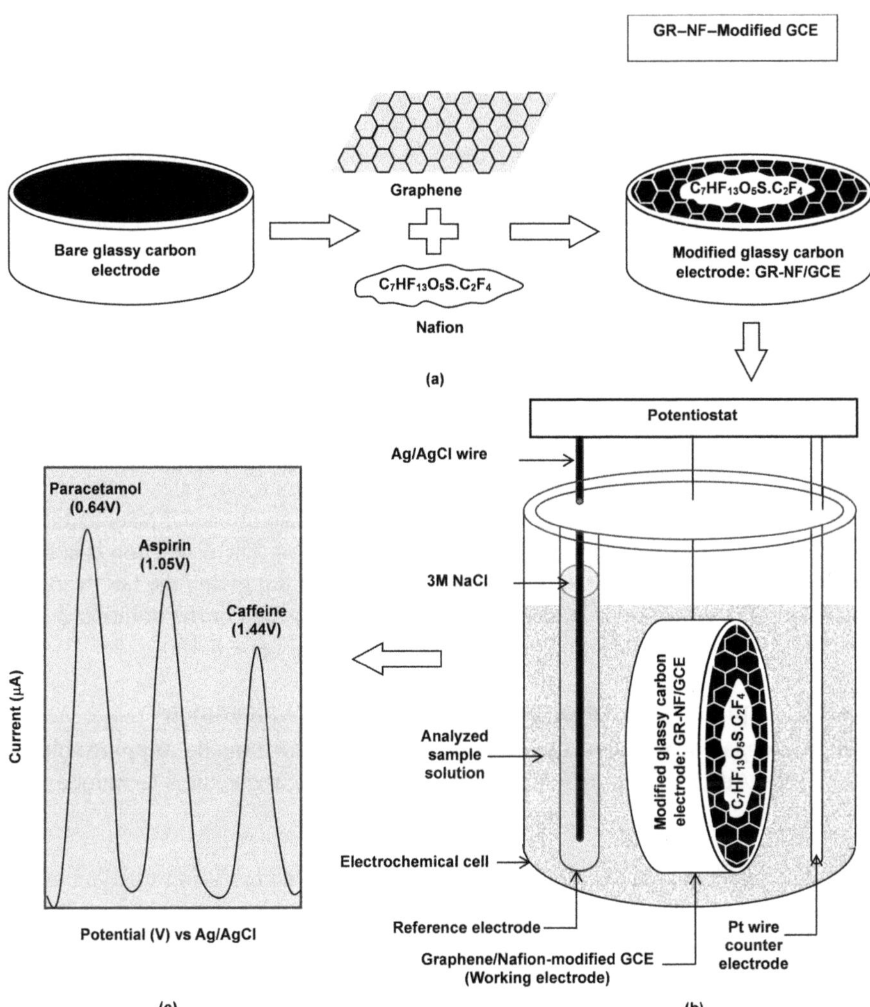

FIGURE 8.4 Graphene-Nafion (GR-NF)-modified glassy carbon electrode (GR-NF/GCE) for detecting paracetamol, aspirin, and caffeine: (a) chemical modification of GCE surface, (b) the electrochemical cell, and (c) the adsorptive anodic stripping voltammograms of an analyte sample in which paracetamol, aspirin, and caffeine are simultaneously present. Part (a) shows the treatment of a bare GCE with a {graphene + Nafion $(C_7HF_{13}O_5S.C_2F_4)$} dispersion to get a modified GCE. Part (b) shows the electrochemical cell used for adsorptive anodic stripping voltammetry. The cell is filled with the sample solution to be analyzed. It contains three electrodes termed as the reference, working, and counter electrodes. The Ag/AgCl reference electrode contains an Ag wire coated with AgCl. This wire is dipped in 3M NaCl solution. The working electrode is a graphene/Nafion-modified GCE. The counter electrode is a platinum wire. All the three electrodes are connected to a potentiostat at the upper end while their lower ends are dipped in the sample solution. Part (c) shows the graphs of current in microamperes against potential in volts with respect to Ag/AgCl reference electrode for paracetamol, aspirin, and caffeine sample solution. For all the three analytes, the current rises with voltage, reaching a peak, and then declining. The current peak for paracetamol appears at 0.64 V, for aspirin at 1.05 V, and for caffeine at 1.44 V.

8.3.4.1 Graphene Synthesis

Graphene is synthesized in two steps. First, graphene oxide (GO) is prepared by Hummers' method. Second, GO is chemically reduced to graphene (GR) or reduced graphene oxide (rGO). For reduction, the GO is mixed with water. A yellowish-brown dispersion is produced. The dispersion is sonicated until the particles become sub-visible. The clear dispersion is poured into a mixture of dimethyl formamide [$HCON(CH_3)_2$] and hydrazine hydrate [$NH_2NH_2 \cdot xH_2O$]. The blended constituents are heated at 100°C for 24 h in a reflux setup. In this setup, the solvent vapors are trapped and returned as a condensate so that no reactants are lost. Graphene is slowly precipitated as a black powder. The precipitated graphene is separated by filtration. It is washed with water and ethanol, and vacuum-dried. Black solid graphene is obtained.

8.3.4.2 Chemical Modification of GCE Surface

GCE is manually polished, sonicated in (nitric acid + ethanol + water) solution, and rinsed in water. Graphene is added to Nafion-alcohol solution. Ultrasonication is done to form a homogeneous graphene-Nafion dispersion. The dispersion is applied on GCE, and the GCE is dried at room temperature for 8 h giving the GR-NF/GCE electrode. The electrode is subjected to cyclic scanning in a buffer solution before use. The chemical modification process is illustrated in Figure 8.4(a).

8.3.4.3 Square Wave Adsorptive Anodic Stripping Voltammetry

The electroanalytical method used is square wave adsorptive anodic stripping voltammetry (SW-AdASV), Figure 8.4(b). It is a two-step electrochemical technique. The two steps are:

(i) Preconcentration step: In this step, the analyte of interest is adsorbed on the surface of the working electrode spontaneously without electrolysis, i.e., the usual application of a reducing potential is not required, rather the electrode is maintained at a potential at which adsorption increases.

(ii) Stripping step: In this step, the adsorbed analyte is stripped or removed from the working electrode by the application of an oxidation potential that reverses the redox process. During the stripping step, the current in the electrochemical cell is measured as a function of potential applied between the working and reference electrodes. The applied potential is pulsed in the forward and reverse directions at a constant frequency to minimize the capacitive background current of the working electrode and, hence, to improve sensitivity. This waveform is viewed as a square wave superimposed on a staircase potential. The current produced is proportional to the quantity of analyte on the electrode. Besides the advantage of better detection limits, the use of faster scan rates speeds up the data acquisition in square wave voltammetry than CV enabling rapid quantification and evaluation of samples.

8.3.4.4   Simultaneity of Measurements, and Effects of Interferents

In a simultaneous determination experiment (Figure 8.4(c)), the square wave stripping voltammograms show that the electrochemical oxidation peak current for PAR appears at +0.64 V, the peak current for ASA occurs at +1.05 V, and that for CAF is observed at +1.44 V, from which it is evident that:

$$\text{Difference between peak current potentials for PAR and}$$
$$\text{ASA} = 1.05\,\text{V} - 0.64\,\text{V} = 0.41\,\text{V}, \tag{8.4}$$

$$\text{Difference between peak current potentials for ASA and}$$
$$\text{CAF} = 1.44\,\text{V} - 1.05\,\text{V} = 0.39\,\text{V}, \tag{8.5}$$

and

$$\text{Difference between peak current potentials for PAR and}$$
$$\text{CAF} = 1.44\,\text{V} - 0.64\,\text{V} = 0.8\,\text{V} \tag{8.6}$$

These differences in peak current potentials indicate that the potentials are sufficiently far apart so that the three analytes can be easily distinguished from their peak current potentials when they are present in a mixture. As the concentration of PAR is increased, keeping the concentrations of ASA and CAF constant, the peak current of PAR increases regularly with concentration, while the peak currents of ASA and CAF remain unchanged. A similar pattern of variation of peak current of ASA vs. its concentration is noticed, keeping PAR and CAF concentrations fixed. The same is true for peak current changes for CAF with respect to concentration, keeping PAR and ASA concentrations static. In all the three cases, the peak current for a particular analyte varies linearly with the concentration of that analyte.

In a further simultaneous determination experiment, the concentrations of all the three analytes are varied at the same time, and their detection is confirmed when the concentrations of all analytes are changed; no concentration is fixed. The linear working range for PAR is 8.3×10^{-9} M to 5.1×10^{-7} M and its LoD is 1.2×10^{-9}M. For ASA, the linear detection range is 8.7×10^{-7} M to 1.7×10^{-5}M and LoD is 6.5×10^{-8}M. For CAF, the linear detection range is 2.6×10^{-7} M to 1.0×10^{-5}M and the LoD is 3.8×10^{-8}M.

Although there is no interference from ions such as K^+, Na^+, SO_4^{2-}, NO_3^-, and macromolecules like glucose and starch, uric acid is an interferent in PAR detection. Overall, the sensor is useful for analyzing commercial pharmaceutical formulations of these analytes (Yiğit et al. 2016).

8.3.5   Carboxylic Graphene Acetylcholinesterase Biosensor for Pesticides

An acetylcholinesterase (AChE) biosensor is designed using the cooperative action of SnO_2 nanoparticles, carboxylic graphene, and Nafion-modified glassy carbon electrode for detecting pesticides (Zhou et al. 2013).

8.3.5.1 Functions of Different Layers in the Biosensor

The biosensor has the structure: NF (Nafion)/AChE-CS (acetylcholinesterase-chitosan)/SnO_2NPs (stannic oxide nanoparticles)–CGR–NF(carboxylic graphene-Nafion)/GCE (glassy carbon electrode). It is used for the detection of methyl parathion, an organophosphate insecticide and carbofuran, a carbamate pesticide, both of which are widely used toxic agents to control insects on agricultural crops.

The top NF layer is a protective membrane for stability of the device. NF has a polytetrafluoroethylene (PTFE) backbone. Its side chains contain ether groups and a sulfonic acid unit.

AChE is an enzyme which catalyzes the breakdown of acetylcholine into choline and acetic acid. CS is used for immobilization of AChE on SnO_2NPs–CGR–NF/GCE. It also improves electronic conduction between AChE and SnO_2NPs–CGR–NF/GCE.

The CGR is carboxyl (–COOH) group-conjugated graphene. Functionalized graphene sheets are easily dispersed in polymer solvents. Their uniform dispersion produces a superior electrically conductive nanocomposite than pure graphene, which shows lower conductivity owing to aggregation effects.

A graphene–Nafion composite film is formed on the glassy carbon electrode (GR-NF/GCE). Nafion is an ionic polymer used for making high-conductivity and high-catalytic activity-modified electrodes with antifouling capacity and chemical inertness (Yiğit et al. 2016).

8.3.5.2 Fabrication of the Biosensor

8.3.5.2.1 Synthesis of Graphene Oxide and Its Functionalization with Carboxyl Groups

Hummers' method-grown graphite oxide is dispersed in water and exfoliated by ultrasonic agitation. The resulting GO solution is centrifuged, diluted, and sonicated. The -OH groups are converted into -COOH groups by the addition of sodium hydroxide (NaOH) and chloroacetic acid ($ClCH_2COOH$), and 3h-sonication. Then the suspension is separated by centrifugation, washed with DI water, and dried in an oven to get CGR.

8.3.5.2.2 Synthesis of SnO_2NPs-CGR Nanocomposites

The CGR is suspended in stannous chloride dihydrate ($SnCl_2.2H_2O$) solution by sonication. Sodium citrate [$HOC(COONa)(CH_2COONa)_2$], ethanol (CH_3CH_2OH), and DI water are added to this suspension. To the resulting mixture, freshly made, ice-cold sodium tetrahydridoborate ($NaBH_4$) solution is added until the color of the solution does not alter. The suspension is separated by centrifugation. It is washed with DI water and dried. Then the Sn nanoparticles are oxidized to SnO_2NPs by heating in air at 300°C. Thus, the SnO_2NPs-CGR nanocomposites are formed.

8.3.5.2.3 Making the AChE Biosensor Electrode

The SnO_2NPs-CGR nanocomposites are added to NF solution in ethanol and DI water and sonicated to get SnO_2NPs-CGR-NF. A small quantity of the SnO_2NPs-CGR-NF is casted on a GCE, after it has been mirror-like polished in alumina slurry, sonicated in ethanol and water, and voltage scanned to obtain a stable current–voltage curve. The casted solution is dried at room temperature to form SnO_2NPs-CGR-NF/ GCE. The AChE enzyme solution in CS is coated over the SnO_2NPs and dried overnight at 4°C to get the AChE-CS/SnO_2NPs-CGR-NF/GCE electrode. The electrode is washed with phosphate-buffered saline (PBS) and covered with an NF protective membrane to get NF/AChE-CS/SnO_2NPs-CGR-NF/GCE electrode, which is stored at 4°C.

8.3.5.3 Operation of the Biosensor

It works by enzyme inhibition, a decrease in enzyme activity in a pesticide solution. The electrode is immersed in the pesticide solution for a fixed time and transferred to the three-electrode electrochemical cell as its working electrode. The cell has a Pt foil counter electrode and a saturated calomel electrode as the reference electrode. The electrolyte is 0.1 M, pH 7.4 phosphate-buffered saline (PBS) solution. CV is performed.

The enzyme inhibition is defined as a percentage in terms of current signal recorded in the solution without the sample chemical substance ($I_{Control}$) and that in the solution containing the sample (I_{Sample})

$$\%\text{Inhibition}$$

$$= \left\{ \left(I_{Control} - I_{Sample}\right) / I_{Control} \right\} \times 100\%$$

$$= \left\{ 1 - \frac{\begin{array}{c}\text{Amperometric response of acetyl thiocholine chloride}\\ \text{(ATCl) on the electrode (NF / AChE} - \text{CS / SnO}_2\text{NPs} - \\ \text{CGR} - \text{NF / GCE) with pesticide effect}\end{array}}{\begin{array}{c}\text{Peak current of acetyl thiocholine chloride}\\ \text{(ATCl) on the electrode (NF / AChE} - \\ \text{CS / SnO}_2\text{NPs} - \text{CGR} - \text{NF / GCE)}\end{array}} \right\} \times 100\% \tag{8.7}$$

The inhibition percentage is a measure of the pesticide concentration.

8.3.5.4 Pesticide Detection Ranges and Limits

The AChE biosensor has a linear range of detection of methyl parathion from 10^{-13} M to 10^{-10} M and another linear detection range from 10^{-10} M to 10^{-8} M. The LoD is 5×10^{-14} M. For carbofuran, the linear range of detection is 10^{-12} M to 10^{-10} M and 10^{-10} M to 10^{-8} M. The LoD is 5×10^{-13} M (Zhou et al. 2013).

8.4 THE 2D MATERIAL FET BIOLOGICAL SENSORS

8.4.1 SOLUTION-GATED GRAPHENE TRANSISTOR (SGGT) FOR GLUCOSE SENSING

As a viable alternative to the commonly used electrochemical sensors for glucose detection, a whole-graphene solution-gated transistor is realized (Zhang et al. 2015).

Figure 8.5 shows the main fabrication steps of the sensor. Au/Cr electrode patterns are formed on a glass substrate by magnetron sputtering of these metals through a shadow mask. Graphene is separately synthesized by chemical vapor deposition on a copper foil and transferred over the Au/Cr electrodes. The channel and gate of the device are defined lithographically (Figure 8.5(a)). PDMS walls are fixed over the substrate ((Figure 8.5(b)).

8.4.1.1 Surface Modification of the Graphene Gate Electrode

The graphene gate electrode is modified as below (Figure 8.5(c)):

(i)   Platinum nanoparticles are deposited by electrochemical process on graphene gate electrode in hexachloroplatinic acid/hydrochloric acid (H_2PtCl_6/ HCl) aqueous solution to get PtNPs/graphene gate electrode.

(ii) 0.5wt% Nafion is dropped on graphene gate electrode and dried obtaining: Nafion/PtNPs/graphene gate electrode.

(iii) (G_{ox} Solution + 0.5wt % chitosan solution) is placed on graphene gate electrode yielding G_{ox}-chitosan/Nafion/PtNPs/graphene gate electrode.

Glucose is oxidized in the presence of G_{ox} catalyst to H_2O_2 near the graphene gate electrode (Figure 8.5(d)). Oxidation of H_2O_2 modulates the gate potential (Figure 8.6).

8.4.1.2 Glucose Response of the SGGT

Channel current changes with the concentration of glucose. The corresponding effective changes in gate voltages are related to glucose concentrations. The sensitivity of the device is 173 mV decade^{-1} in the glucose concentration range from 10^{-5} M to 5×10^{-4} M. Transconductance of whole graphene transistor is much larger (several mS) than that of silicon transistor accounting for its higher sensitivity (Zhang et al. 2015).

8.4.2 MoS$_2$ Nanosheet FET Biosensor for PSA Detection

This sensor (Lee et al. 2014) is a bottom-gated MOS$_2$ FET without any insulating oxide over MOS$_2$ surface, as shown in Figure 8.7. This becomes possible because the MOS$_2$ surface is hydrophobic. The contact angle is 75.77°. A hydrophobic surface has a high affinity for the adsorption of proteins than a hydrophilic surface. So, the physisorption of biomolecules is easy and they can directly get attached to MoS$_2$ surface. The advantage derived is that chemical treatment of the surface is not necessary and hence the process is simplified. Moreover, the biomolecules are more efficiently coupled with MOS$_2$ surface.

8.4.2.1 Fabrication of the Sensor

SiO$_2$/heavily-doped Si substrates are taken. MoS$_2$ flakes are mechanically exfoliated from bulk MoS$_2$ with the help of Scotch tape and placed over the substrates. Residues are removed by cleaning in acetone. Source and drain contact regions are defined in the photoresist by lithography. Ti/Au film is deposited by electron-beam evaporation. The photoresist is lifted off leaving Ti/Au contacts. The MoS$_2$ FET device is shown in Figure 8.7(a).

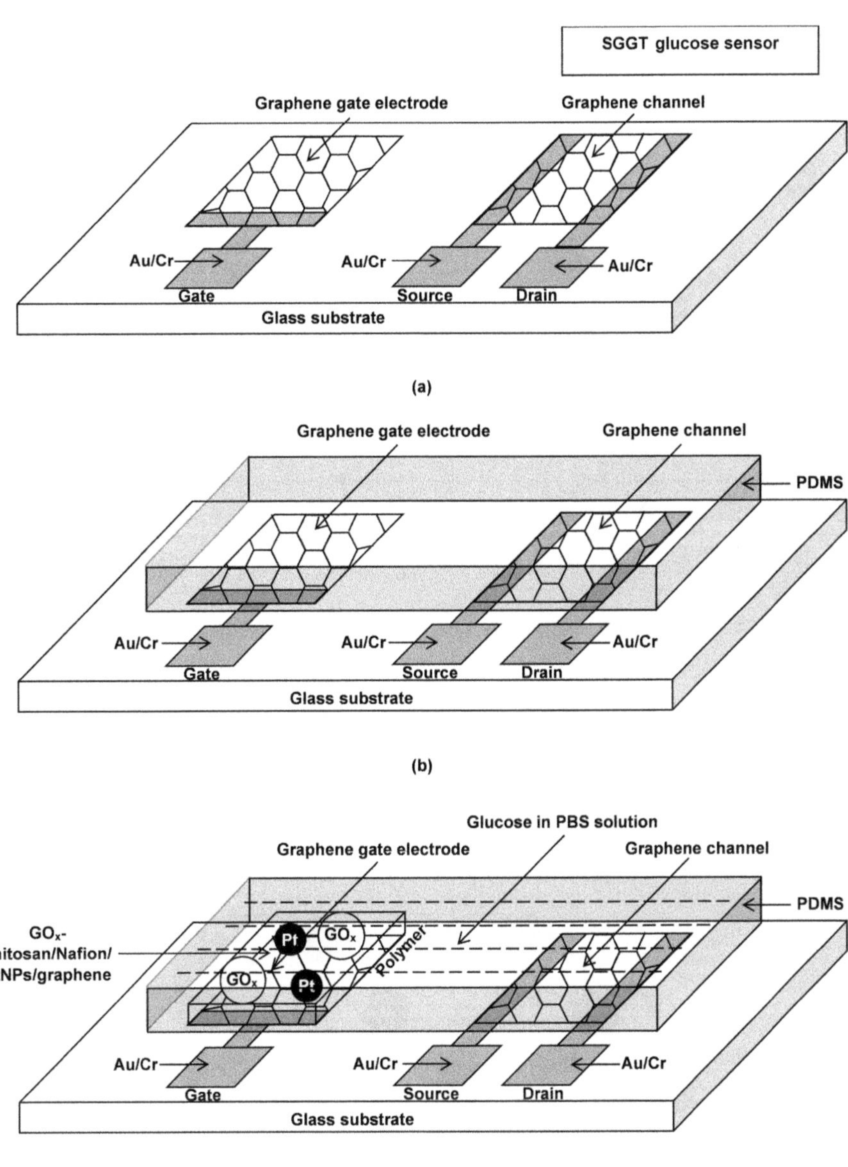

FIGURE 8.5 Fabrication steps of the solution-gated graphene transistor (SGGT) for glucose sensing: (a) defining the contact pads, channel and gate electrode, (b) fixation of PDMS walls, (c) modification of graphene gate electrode surface with PtNPs and enzyme, and (d) making the electrical connections. Part (a) shows a glass substrate over which contact pads for source, drain, and gate are defined; and the graphene film is laid down for the graphene gate electrode and channel formation. Part (b) shows the attachment of PDMS walls surrounding the graphene gate electrode and channel regions. Part (c) shows the PtNPs- and glucose oxidase enzyme-functionalization over the graphene gate electrode and glucose solution filled in the PDMS tub. A GO_x-chitosan/Nafion/PtNPs/graphene layer is formed.

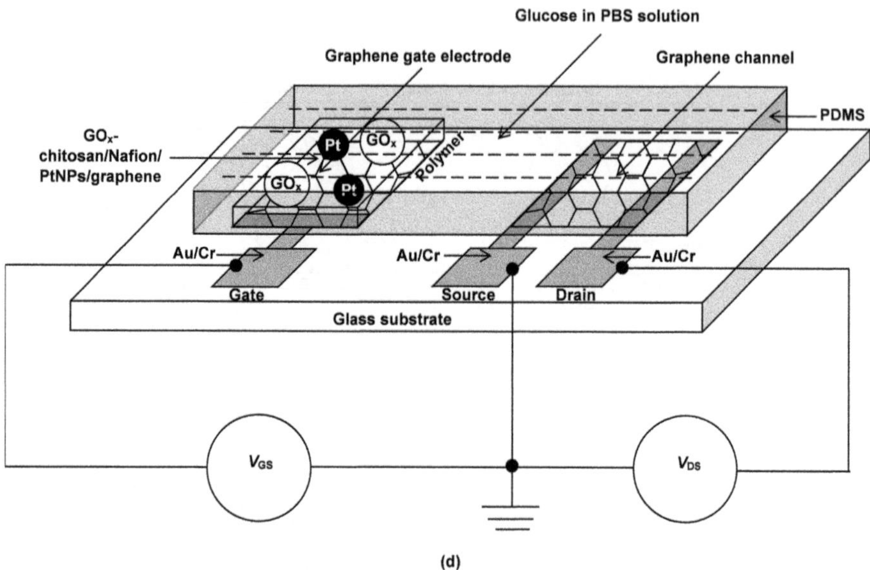

FIGURE 8.5 (Continued) Part (d) shows the electrical connections: V_{GS} is the gate–source voltage and V_{DS} is the drain–source voltage. The source terminal is grounded.

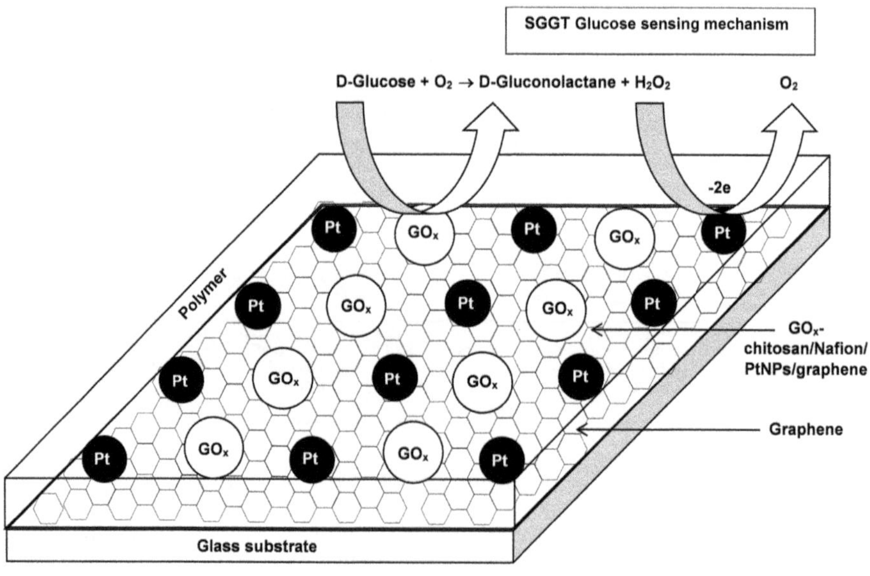

FIGURE 8.6 Glucose-sensing mechanism of SGGT. An enlarged view of the graphene gate electrode region is shown with electrodeposited PtNPs and immobilized glucose oxidase enzyme in the GO$_x$-chitosan/Nafion/PtNPs/graphene layer. Glucose reaction with oxygen is catalyzed by GO$_x$ to form gluconolactone and hydrogen peroxide. H$_2$O$_2$ loses two electrons and is oxidized to oxygen.

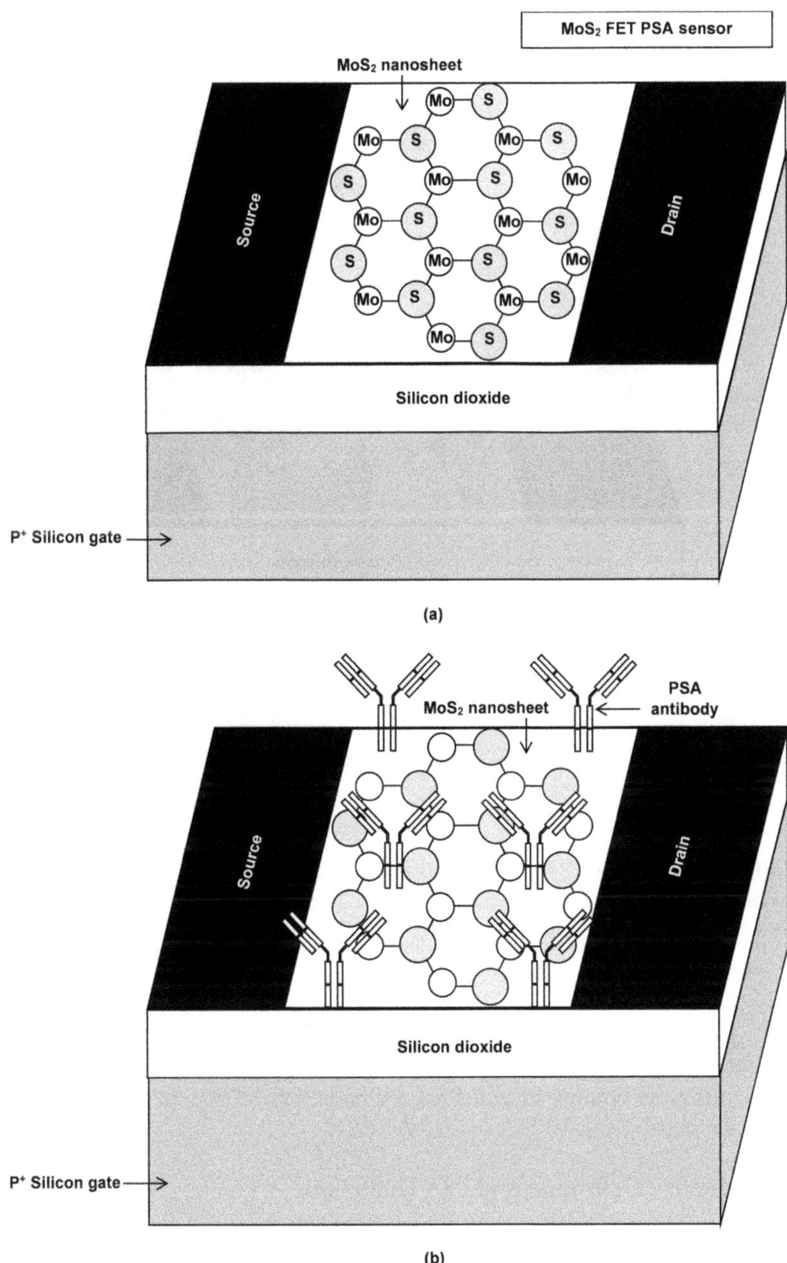

FIGURE 8.7 MoS$_2$ field-effect transistor PSA sensor: (a) basic transistor configuration, (b) with PSA antibodies, and (c) with PSA antigens bound to PSA antibodies. Part (a) shows the MOSFET consisting of a P$^+$ Si back gate covered with silicon dioxide film. The exfoliated MoS$_2$ nanosheet is laid down over the SiO$_2$ film. Source and drain contacts are formed at the two ends of the MoS$_2$ nanosheet. Part (b) shows the MOS$_2$ FET with the MOSFET surface functionalized with PSA antibodies.

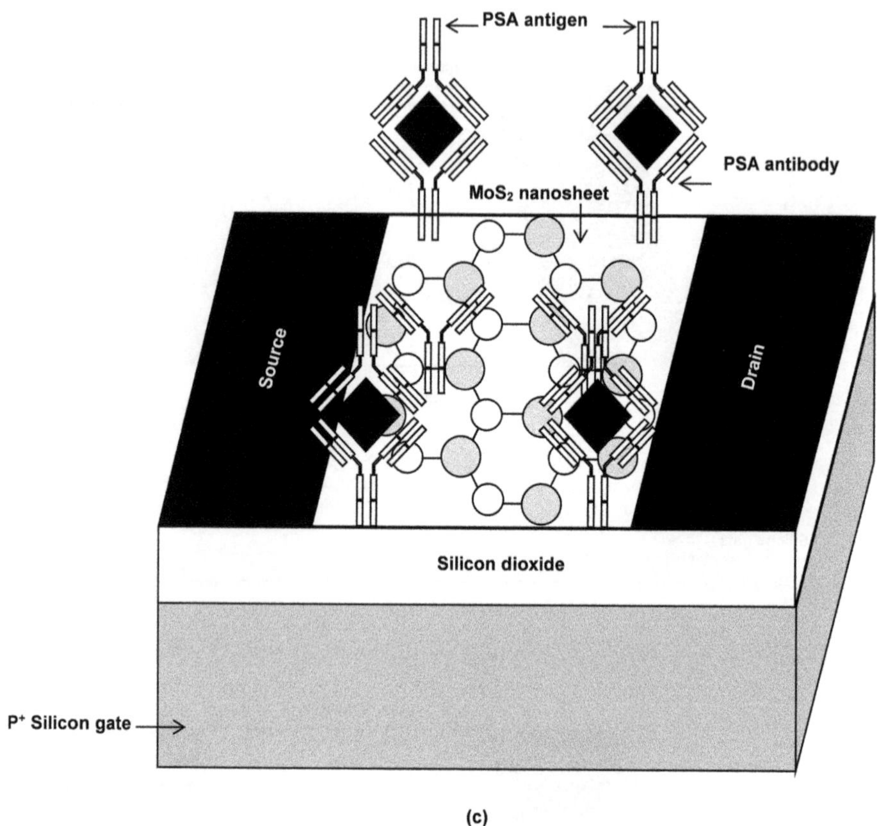

FIGURE 8.7   (Continued)   Part (c) shows the PSA-antibody-functionalized biosensor with PSA antigen sample dropped over the PSA-antibody-covered MoS_2 surface when the PSA antigens bond with the PSA antibodies.

After thermal annealing, the MoS_2 surface is functionalized with anti-prostate-specific antigen (PSA) antibodies. Figure 8.7(b) shows the antibody-functionalized device. During the binding of anti-PSA antibodies to MOS_2, the off-current of the transistor increases from 10^{-12} A to 10^{-9} A.

8.4.2.2   Device Operation and PSA Detection

The device is operated at $V_{GS} = -40$ V and $V_{DS} = 1$ V.

In this sensor, the off-state current changes are measured to determine the PSA concentration instead of threshold voltage shift. When the device is exposed to PSA solutions of different concentrations, the off-state current decreases from 2.1 nA at 1 $pgmL^{-1}$ PSA to 0.2 nA at 10 $ngmL^{-1}$ PSA as more and more PSA antigens in the solutions are bound with PSA antibodies on MOS_2 surface (Figure 8.7(c)). The lowest detectable PSA concentration is 1 pgL^{-1}. The clinical cut-off value of 4 ngL^{-1} is three orders of magnitude higher than the minimum detection limit (Lee et al. 2014).

8.4.3 MONOLAYER WSe₂ FET COVID-19 VIRUS SENSOR

Rapid and sensitive in vitro detection of SARS-CoV-2 spike proteins is accomplished using a WSe_2 2D-FET biosensor (Fathi-Hafshejani et al. 2021).

8.4.3.1 Principle of Detection

The sensor is shown in Figure 8.8 in ungated form (a), and with ionic gating (b). Monoclonal antibodies against the SAR-CoV-2 fixed on the WS_2 surface attract the antigens of coronavirus disease in a test sample. The antibody–antigen binding causes a change in the electrical characteristics of the FET dependent on the antigen concentration, enabling quantitative detection of the virus.

8.4.3.2 Fabrication and Testing of the Sensor

Fabrication and testing of the sensor are carried out as follows:

(i) Laser-assisted synthesis of monolayer WSe_2 crystals on SiO_2/Si substrate: The substrate used is silicon covered with thin thermally grown silicon dioxide. WSe_2 powder is loaded in a graphite boat. The SiO_2/Si substrate is placed on top of the boat. The WSe_2 powder is evaporated by shining light from a continuous wave CO_2 laser having a wavelength of 10.6 μm. Upon evaporation, the WSe_2 powder deposits in the form of a random dispersion of triangular-shaped monolayer crystals on the SiO_2 film over Si. The size of the crystals is 20–40 μm.

(ii) Creating a pattern of titanium interdigitated electrodes on the surface of WSe_2 film: The pattern is defined by photolithography. After developing the photoresist, titanium sputtering is done over the pattern. The titanium electrodes are formed when the photoresist is lifted off. The interdigitated electrode design increases the sensing area.

(iii) Immobilization of the monoclonal antibodies against the spike protein of SARS-COV-2: The devices are cleaned in ethanol, rinsed in DI water, and dried in nitrogen (Figure 8.9).

  (a) Surface modification: This is done using 11-mercaptoundecanoic acid (MUA), $HS(CH_2)_{10}CO_2H$. The MUA is used as a probe linker. The MUA molecule has an SH-terminated end which strongly interacts with WSe_2 surface and is adhered to it (Figure 8.9(a)).

  (b) MUA activation: The MUA is activated by dipping in N-hydroxysuccinimide (NHS), $(CH_2CO)_2NOH$ and 1-ethyl-3-(3-dimethylaminopropyl) carbodiimide hydrochloride (EDC), $C_8H_{17}N_3$·HCl solution for 3–6 h. The N-hydroxysuccinimide ester formed by exposure to NHS-EDC solution plays a vital role in the next step of antibody fixation because it reacts with amino groups of the antibodies to form an amide bond. The MUA activation is followed by water rinsing and drying in nitrogen (Figure 8.9(b)).

(iv) Addition of SARS-CoV-2 spike protein molecules: Using a micropipette, 3.3 ng μL^{-1} spike protein molecules are dropped on the modified and activated surface. Incubation is done overnight (Figure 8.9(c) and (d)).

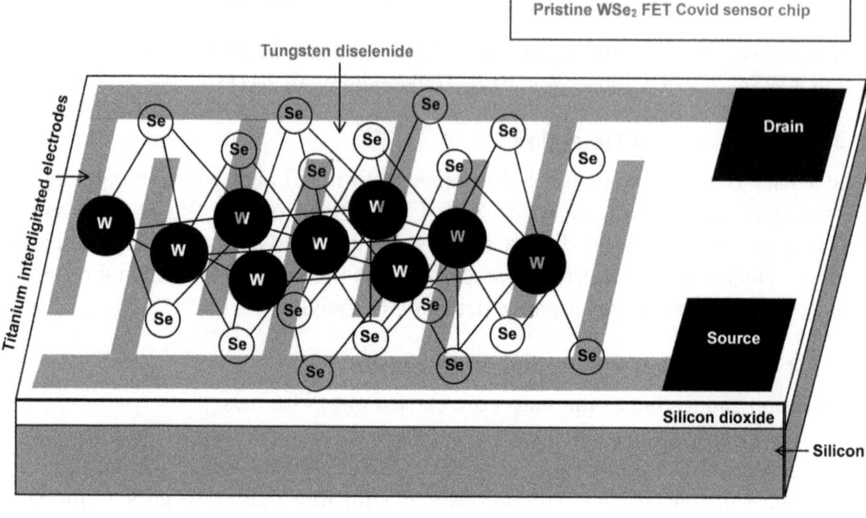

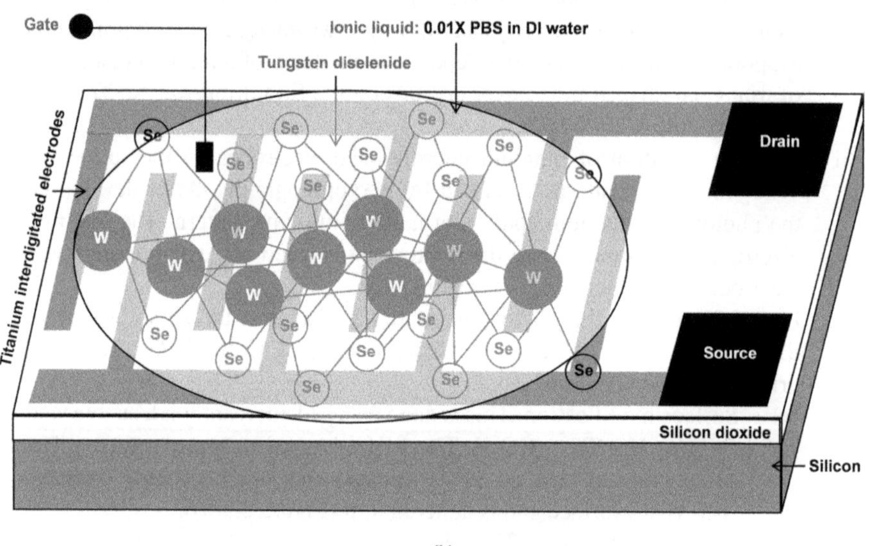

FIGURE 8.8 Pristine chip of the FET covid sensor: (a) without gate connection and (b) with ionic liquid gating. Part (a) shows a silicon dioxide-over-silicon substrate on which WSe$_2$ film is synthesized. An interdigitated pattern of titanium electrodes is formed over the WSe$_2$ film. Source and drain contact pads are seen but there is no gate electrode. Part (b) shows the same structural arrangement except that a drop of ionic liquid (0.01X phosphate buffered saline in deionized water) is poured over the interdigitated electrode pattern and a wire is immersed in this liquid to apply the gate bias. This is the gate electrode.

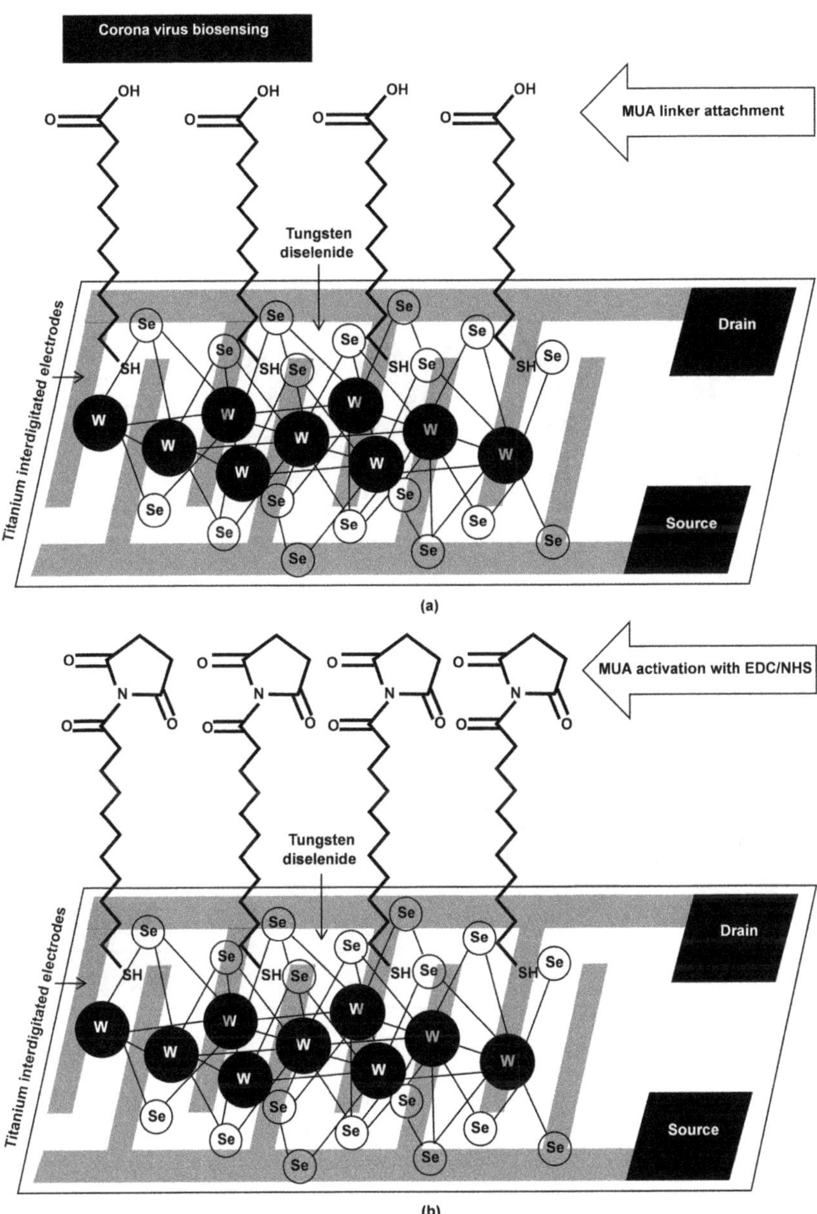

FIGURE 8.9 Steps of functionalization of the raw chip surface and coronavirus biosensing: (a) attachment of the 11-mercaptoundecanoic acid (MUA) linker molecules, (b) activation of MUA by exposing to EDC/NHS, (c) immersion in SARS-CoV-2 antibody solution to immobilize the antibodies of WS_2 film via activated MUA, and (d) immersion in a specimen solution containing SARS-CoV-2 spike proteins. Part (a) shows the SH-terminated end of MUA molecule anchored on the surface of WS_2 film with carboxyl groups at the upper extremity. Part (b) shows the formation of N-hydroxysuccinimide ester (NHS ester). EDC activates the carboxyl groups and couples NHS to carboxyls thereby producing the ester.

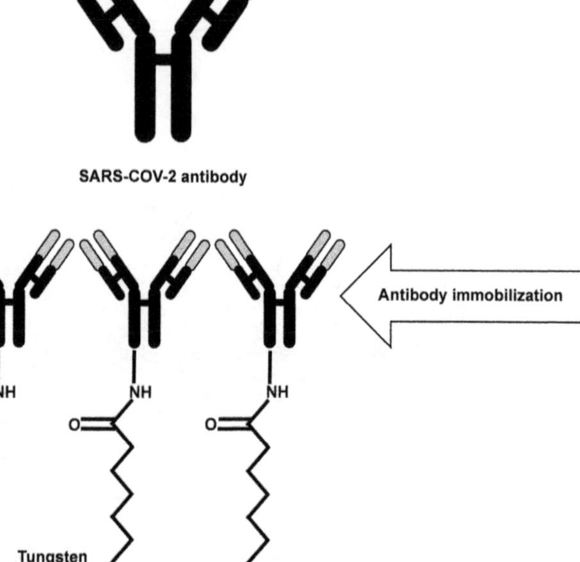

FIGURE 8.9 (Continued) Part (c) upper side shows the SARS-COV-2 antibody. Part (c) lower side shows the attachment of antibodies to the sensor surface by reaction of ester with amino groups of antibodies forming linkages via amide bonds.

(v) For measurements, the sensors are gated with ionic liquid. The liquid used is 0.01X PBS in DI water. The SARS-CoV-2 antigens are detected in real time with $V_{GS} = -0.5$ V and $V_{DS} = 1$ V and measuring $\delta I/I_0 = (I - I_0)/I_0$ with respect to time where I is the current at a given antigen concentration and I_0 is the original current. The signal is observed as rising peaks for increasing concentrations of antigen from 25 fgµL^{-1} to 10 ngµL^{-1}. The LoD is 25 fgµL^{-1} (Fathi-Hafshejani et al. 2021).

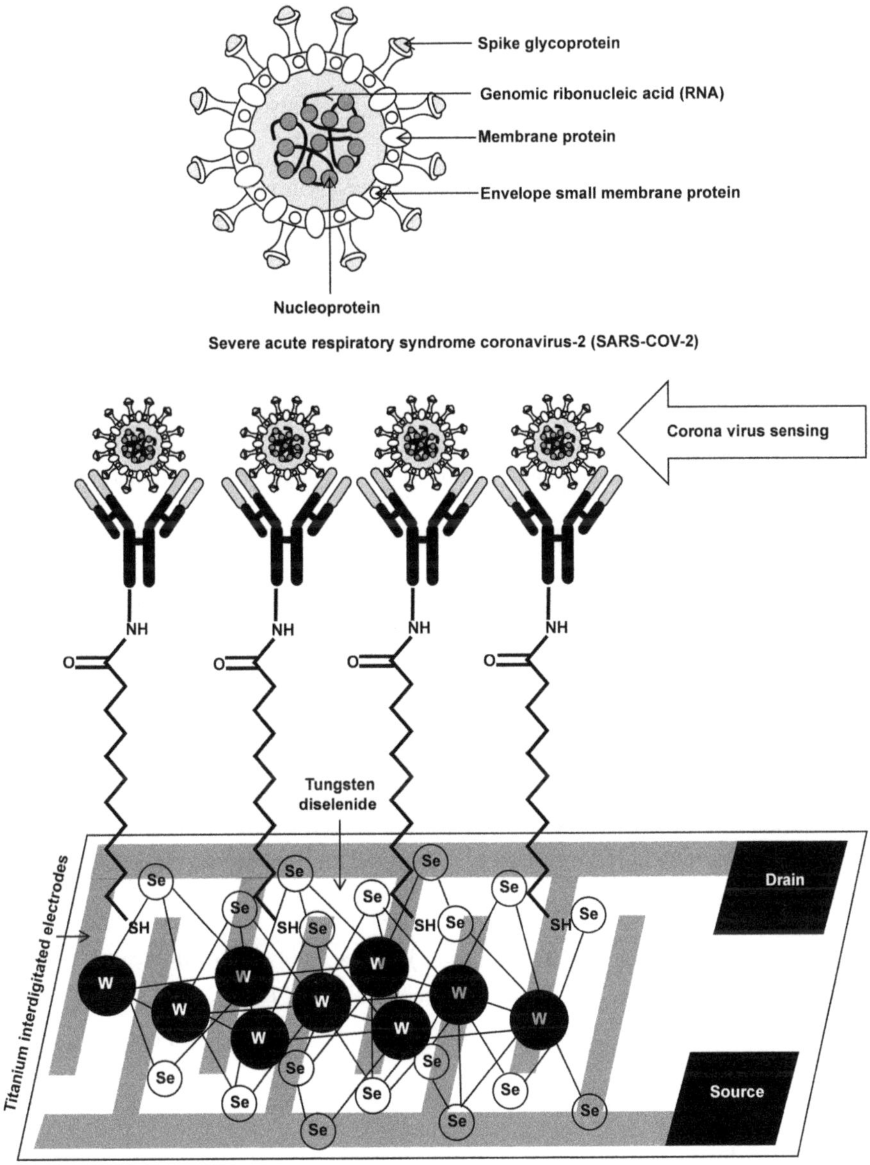

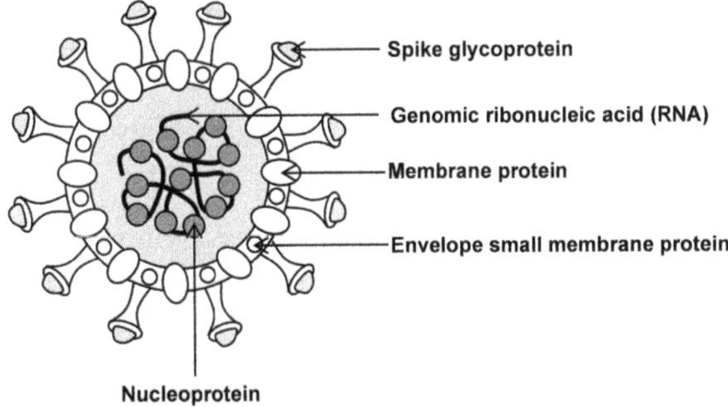

FIGURE 8.9 (Continued) Part (d) upper side shows the constituent parts of coronavirus such as spike glycoprotein, genomic ribonucleic acid (RNA), membrane protein, envelope small membrane protein, and nucleoprotein. Part (d) lower side shows the gripping of SARS-CoV-2 antigen spike protein by the SARS-CoV-2 antibodies when the sample solution is dropped over the interdigitated electrodes.

8.5 2D MATERIAL OPTICAL SENSORS FOR BIOLOGICAL ANALYSIS

8.5.1 BSA-Capped Au Nanoclusters/g-C₃N₄ Nanosheets Fluorescent Sensor for Dopamine

A dopamine-sensing fluorescent probe is developed (Guo et al. 2016).

8.5.1.1 Sensing Principle

This sensor works on the fluorescence resonance energy transfer (FRET) process (Figure 8.10), a non-radiative transfer of energy from a molecular fluorophore in an electronically excited state called the donor fluorophore to a nearby fluorophore in the ground state known as the acceptor fluorophore via long-range dipole coupling (Förster 1946). The efficiency of this radiation-less energy transfer depends on the donor-to-acceptor separation (effective only at <10 nm distance), the molecular orientations of the donor and acceptor, and the overlap integral between donor emission and acceptor absorption spectra.

In this sensor, the FRET takes place between:

(i) strong red fluorescent $g\text{-}C_3N_4$ NSs-AuNCs (graphitic carbon nitride nanosheets-gold nanoclusters) nanocomposite emitting at 420 nm and acting as the donor molecules, and

(ii) the oxidized DA (dopamine) quinine molecules formed by DA oxidation in air and serving as the acceptor molecules.

As a result, the red fluorescence shows a sharp decline depending on the dopamine concentration. This physicochemical process of decrease in intensity of light emitted by a fluorescent molecule is referred to as fluorescence quenching. The fluorescence quenching ratio is defined as

$$QR = \frac{\text{Initial fluorescence intensity}\left(F_{\text{Initial}}\right) - \text{Final fluorescence intensity}\left(F_{\text{Final}}\right)}{\text{Initial fluorescence intensity}\left(F_{\text{Initial}}\right)}$$

(8.8)

is a function of the dopamine concentration in the sample.

8.5.1.2 Roles of Nanocomposite Constituents

(i) $g\text{-}C_3N_4$: The N-containing structure $g\text{-}C_3N_4$ NSs bestows upon it a strong ability to absorb protons (H^+) enhancing the sensitivity of dopamine detection. The fluorescence emitted by pristine bulk $g\text{-}C_3N_4$ with bandgap 2.7 eV is in the blue wavelength region of the spectrum with photoluminescence peak at 470 nm. But its quantum yield, expressing the ratio of the number of photons emitted/number of photons absorbed, is poor. Besides its solubility in water is low. These disadvantages together

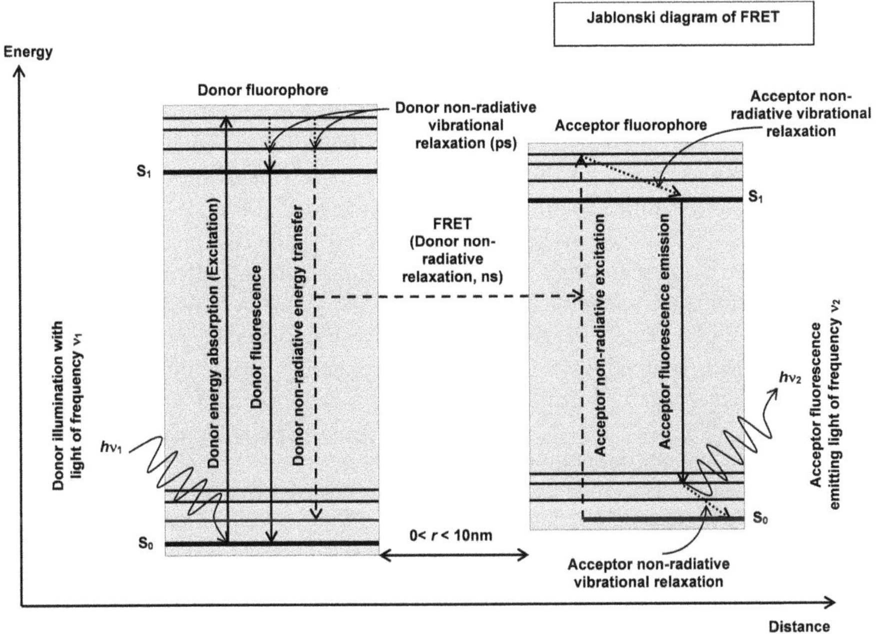

FIGURE 8.10 Jablonski diagram of the fluorescence resonance energy transfer (FRET) process. An energy-band diagram is drawn with electron energy plotted along the Y-axis and distance along the X-axis. The left-hand side of the energy band diagram represents a donor fluorophore and the right-side an acceptor fluorophore. Energy states of the donor and acceptor are indicated by horizontal lines. For both the donor and the acceptor, the sets of ground-state energy levels S_0 and the sets of excited-state energy levels S_1 are shown. A photon of energy $h\nu_1$ strikes an electron of the donor and uplifts it from the ground-state energy S_0 to an excited energy level (donor energy absorption or excitation) from which it relaxes in a few picoseconds to the lowermost excited-energy state S_1 by a radiation-less transition (donor non-radiative vibrational relaxation). From the state S_1, the electron decays radiatively to the ground state (donor fluorescence). Alternatively, it falls non-radiatively to an energy level in the ground state (donor non-radiative energy transfer). This donor non-radiative energy transfer is accompanied in a time interval of a few nano seconds by a non-radiative relaxation from the donor to the nearby-located acceptor within a distance <10 nm resulting in acceptor excitation (acceptor non-radiative excitation). The transfer of excitation energy from an excited donor fluorophore to an acceptor fluorophore through long-range non-radiative dipole–dipole intermolecular coupling constitutes the FRET. The acceptor electron next moves from its energy level in the excited state to the lowermost excited-state energy level S_1 without emitting light (acceptor non-radiative vibrational relaxation). Subsequently, it descends from the lowermost excited state to an energy level in the ground state. This is associated with the emission of a photon of energy $h\nu_2$ (acceptor fluorescence emission). The remaining portion of electron's energy is released through a non-radiative transition to the lowermost ground state S_0 as an acceptor non-radiative vibrational relaxation.

with its macroscopic size have thwarted its practical use in biosensors. Contrastingly, atomic-scale thick g-C_3N_4 nanostructures such as nanosheets engineered from bulk g-C_3N_4 show higher quantum yield of photoluminescence and also better solubility in water. Combined with the merits of their low cost, easy preparation, good biocompatibility, non-toxicity, and excellent photostability, g-C_3N_4 nanosheets act as efficient fluorescent nanoprobes for chemical and biosensing experiments (Cai et al. 2021).

(ii) AuNCs: Gold nanoclusters (AuNCs) have molecule-like properties. They emit red fluorescent light with high quantum yield and offer non-toxicity and stability advantages too. So, they are widely used in biosensors for various analytes. Hence, the composite made with Au nanoclusters and g-C_3N_4 nanosheets is an effective combination of materials for building a sensitive and selective fluorescence sensor for quantitative analysis. The fluorescence emission intensity of g-C_3N_4 nanosheets is considerably increased by grafting gold nanoclusters on their surfaces.

8.5.1.3 Fabrication of the Sensor

(i) Synthesis of g-C_3N_4 NSs:

The g-C_3N_4 NSs are obtained by exfoliation from bulk g-C_3N_4. The bulk g-C_3N_4 is prepared by pyrolysis of urea for which 50 g urea (H_2NCONH_2) is placed in an alumina crucible with cover. The urea is heated in a muffle furnace by raising its temperature from room temperature to 600°C at 15°Cmin^{-1}. It is kept at 600°C for 2 h and cooled to get bulk yellow g-C_3N_4. Figure 8.11 illustrates the fabrication and testing of the sensor. For exfoliation (Figure 8.11(a) and (b)), the bulk g-C_3N_4 is protonated by stirring with 10M HCl, followed by filtration and washing. The protonated solids are dispersed in DI water. After ultrasonic treatment and centrifugation, a dispersion of proton-functionalized g-C_3N_4 NSs in water is formed.

(ii) Amination of g-C_3N_4 NSs by carbodiimide crosslinker chemistry:

1-Ethyl-3-(-3-dimethylaminopropyl) carbodiimide hydrochloride, EDC and N-hydroxysuccinimide, NHS are dissolved in g-C_3N_4 NSs solution, ultrasonicated, and stirred (Figure 8.11(c)). Then ethanediamine, $C_4H_{13}N_3$, is added. After vigorous stirring at 80°C for 2 h and cooling, aminated g-C_3N_4 NSs solution is obtained.

(iii) Synthesis of BSA-capped AuNCs:

[Chloroauric acid ($HAuCl_4$) + bovine serum albumin (BSA)] solution is stirred and reacted for 2 min. Sodium hydroxide (NaOH) solution is added and pH is adjusted to 11. After heating at 70°C for ½ h, the color of the light-yellow mixture turns light brown. When irradiated with 365 nm UV radiation, the solution emits red light.

(iv) Synthesis of BSA-Capped AuNCs/g-C_3N_4 NSs:

The g-C_3N_4 NSs solution is mixed with BSA-capped AuNCs solution (Figure 8.11(d)). The mixture is stirred in the dark for 2 h and dialyzed in double-distilled water for 24 h after which the BSA-capped AuNCs/g-C_3N_4 NSs nanocomposite solution is collected.

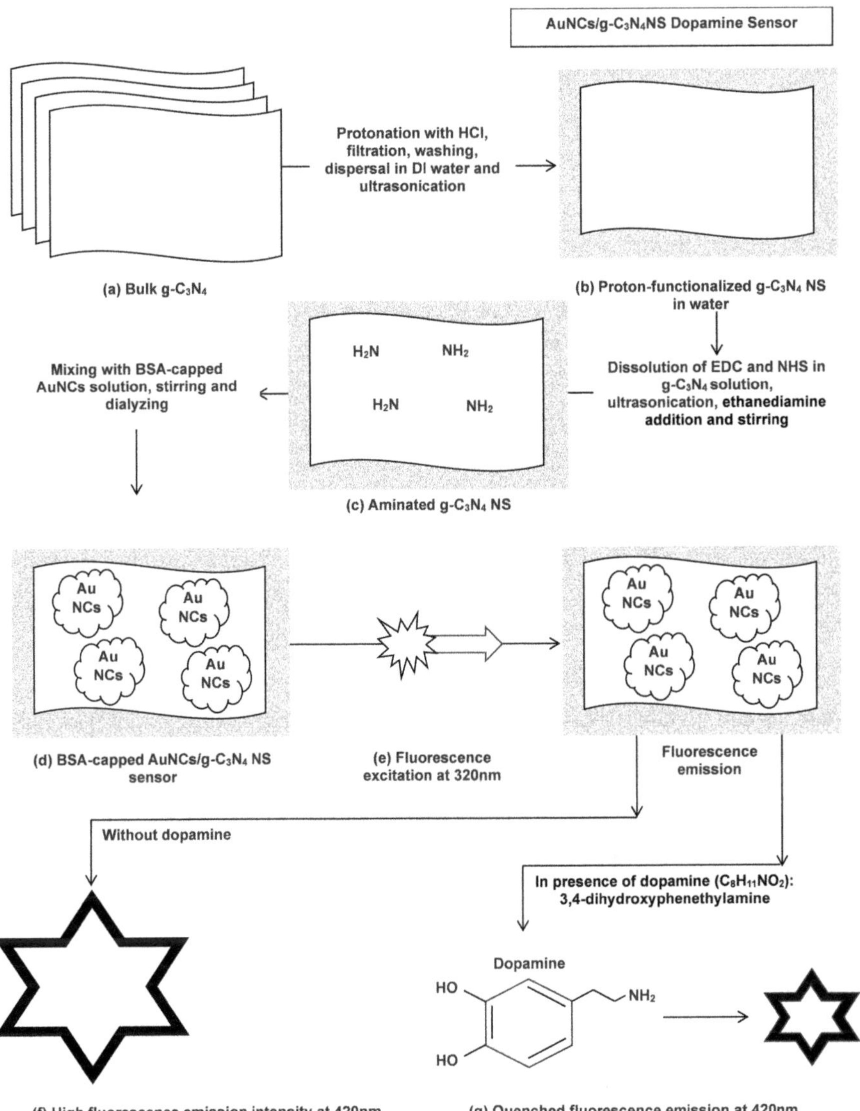

FIGURE 8.11 Fabrication and characterization of the g-C_3N_4NSs–AuNCs composite fluorescent sensor for dopamine: (a) and (b) exfoliation of bulk g-C_3N_4, (c) amination of g-C_3N_4 NSs, (d) making BSA-capped AuNCs/g-C_3N_4 NSs, (e), (f), and (g) fluorescence experiments. Part (a) shows bulk g-C_3N_4 made of four layers. Part (b) shows a single protonated g-C_3N_4 nanosheet suspended in water. Part (c) shows g-C_3N_4 nanosheet with attached NH_2 groups in water. Part (d) shows g-C_3N_4 nanosheet with appended gold nanoclusters in water. Part (e) shows the illumination of g-C_3N_4 nanosheet with gold nanoclusters using 320 nm radiation for exciting fluorescence. Part (f) shows a big star indicating that high-intensity fluorescence is observed at a wavelength of 420 nm for a sample that does not contain dopamine. Part (g) shows the chemical structure of dopamine with an arrow pointing toward a small star, which means that the intensity of fluorescence emission at 420 nm is smaller for a sample in which dopamine is present.

8.5.1.4   Characterization of the Sensor

From the fluorescence spectra recorded in the range 350–550 nm at 320 nm excitation wavelength, for different DA concentrations (Figures 8.11(e)–(g)), the fluorescence intensity of g-C_3N_4NSs-AuNCs composite at 420 nm is found to decrease with increase in DA concentration due to quenching of the red fluorescence emission of the composite by DA. The quenching ratio $(F_{Initial}$-$F_{Final})/F_{Initial}$ varies linearly with DA concentration in the DA concentration range 0.05 µmol L^{-1}–8.0 µmol L^{-1}, and the LoD is 0.018 µmol L^{-1}. The response parameters for g-C_3N_4 NSs alone at 436 nm (without AuNCs) are: linear range = 2.5–30 µmol L^{-1} and the LoD is 2.31 µmol L^{-1}. Interference effects from the presence of metal ions and organic compounds, as possible in real samples, are very small confirming the reliability of the detection of DA in human serum and urine specimens (Guo et al. 2016).

8.5.2   GRAPHENE OXIDE-UPCONVERSION NANOPARTICLES FLUORESCENCE SENSOR FOR DNA

A DNA sensor working on the FRET principle is developed (Alonso-Cristobal et al. 2015). The FRET takes place between GO and sodium yttrium tetrafluoride ($NaYF_4$)-based $NaYF_4$: Yb, Er@SiO_2 upconversion nanoparticles (UCNPs). It utilizes the fluorescence quenching property of GO.

8.5.2.1   Graphene Oxide as a Fluorescence Quencher

Graphene and graphene-like nanomaterials can be tailored to form either fluorescent emitters or efficient fluorescence quenchers, making them powerful platforms for fabricating a series of optical biosensors to sensitively detect various targets including ions, small biomolecules, deoxyribonucleic acid (DNA)/ribonucleic acid (RNA), and proteins. GO causes fluorescence quenching of various fluorophores because of its delocalized π-conjugated domains. In the quenching process, GO acts as an energy acceptor and UCNPs as the energy donor. GO has several advantages as a fluorescence quencher (Zheng and Wu 2017):

(i)   Provision of an abundance of bonding sites by the large surface area.
(ii) Ease of bioconjugation due to the oxygen-containing functional groups.
(iii) Biocompatibility, non-toxicity, intercellular penetrability, and water-solubility.

Two sets of circumstances are of interest:

(i)   DNA-functionalized $NaYF_4$: Yb, Er@SiO_2 UCNPs near GO surface: A π–π interaction occurs between sp^2 carbon atoms of GO and the nucleobases of DNA leading to a FRET fluorescence quenching due to the overlapping of absorption spectrum of GO with fluorescence emission of $NaYF_4$:YF_4: Yb, Er UCNPs. This can be expressed by the equation:

$$\text{DNA} - \text{functionalized NaYF}_4 : \text{Yb, Er @ SiO}_2 \text{UCNPs} + \text{GO} \rightarrow$$
$$\text{DNA} - \text{functionalized NaYF}_4 : \text{Yb, Er @ SiO}_2 \text{UCNPs adsorbed on GO} \rightarrow$$
$$\text{Fluorescence Quenching}$$

$$(8.9)$$

(ii) cDNA-functionalized $NaYF_4$: Yb, Er @SiO_2 UCNPs near GO surface: Hybridization occurs between DNA and cDNA strands forming double-stranded DNA, ds-DNA. The ds-DNA does not interact with the GO surface. Hence, the fluorescence of $NaYF_4$:Yb, Er@SiO_2 UCNPs remains unquenched resulting in fluorescence emission, as described by the equation:

$$cDNA + DNA - functionalized\ NaYF_4 : Yb, Er @ SiO_2\ UCNPs + GO \rightarrow$$
$$ds - DNA - functionalized\ NaYF_4 : Yb, Er @ SiO_2\ UCNPs \rightarrow$$
$$Fluorescence\ Emission \qquad (8.10)$$

The fluorescence intensity increases with increase in the number of cDNA strands. Thus, by measuring relative fluorescence emission compared to a reference emission, the concentration of cDNA is determined.

8.5.2.2 Need of Upconversion Nanoparticles

The sensor uses upconversion nanoparticles (UCNPs) in place of traditional fluorophores. The upconversion nanoparticles are new-generation fluorophores used in biosensing and bioimaging (Wang et al. 2011). They are luminescent nanomaterials made of rare-earth-based lanthanide- or actinide-doped transition metals. They absorb two or more lower energy photons (near-infrared radiation) and emit one higher energy photon (UV or visible radiation).

8.5.2.3 Limitations of Conventional Downconversion Fluorophores

The primary limitation of conventional fluorophores stems from the requirement of ultraviolet or short-wavelength excitation resulting in low signal-to-noise ratio because of autofluorescence (primary or native fluorescence), the naturally occurring fluorescence from cells in the UV or short-wavelength regions arising from endogenous fluorophores. The signal-to-noise ratio is further degraded by the intense scattering of light from biomolecules when irradiated with short-wavelength radiation. Particularly, UV radiation also damages the DNA and may cause cell death (Chen et al. 2014).

8.5.2.4 Advantages of Upconversion Nanoparticles

These nanoparticles overcome the main disadvantages of conventional fluorophores making them ideal for biosensors.

(i) The main advantages of UCNPs for biosensors are the large wavelength used for excitation of fluorescence, typically ~980 nm lying in the near-infrared region (800–2500 nm). At this wavelength, the autofluorescence from biomolecules is too small. The virtually zero autofluorescence background greatly improves the signal-to-noise (S–N) ratio. Additionally, the scattering of light is much lower than from UV–visible radiation, which further upgrades the S–N ratio. Thus, a high sensitivity of detection is achievable.

(ii) The UCNPs are highly resistive to photobleaching, meaning that they are not easily photochemically altered. They are also resistant to blinking, the propensity for fluorescence intermittency observed as a flickering effect due to temporally random discrete switching between bright and dark states. These properties make them suitable for long-term, reliable, repetitive use.

(iii)   They provide large anti-Stokes shift helping to easily separate the photoluminescence from the excitation wavelength. In anti-Stokes fluorescence, the emitted photon has a higher energy (shorter wavelength) than the absorbed photon, which has a lower energy (longer wavelength) and the anti-Stokes shift is the difference in energy between the emitted and absorbed photons. The extra energy is supplied by the dissipation of thermal phonons in the crystal lattice whereby the crystal gets cooled.

(iv)   They do not cause any photodamage to the DNA.

8.5.2.5   Making Probe ss-DNA-Functionalized $NaYF_4$: Yb, Er @SiO_2 UCNPs

The probe ss-DNA-functionalized $NaYF_4$: Yb, Er @SiO_2 UCNPs-ssDNA are made and the fluorescence measurements are performed through the following steps (Figure 8.12):

(i)   Synthesis of $NaYF_4$: Yb, Er upconversion nanoparticles ($NaYF_4$: Yb, Er UCNPs):

This is done by dissolving yttrium (III) chloride hexahydrate (YCl_3 $\cdot 6H_2O$) ytterbium (III) chloride hexahydrate, ($YbCl_3$ $\cdot 6H_2O$), and erbium (III) chloride hexahydrate ($ErCl_3 \cdot 6H_2O$) in oleic acid [$CH_3(CH_2)_7CH = CH(CH_2)_7COOH$], and 1-octadecene [$CH_3(CH_2)_{15}CH = CH_2$] followed by 1.5 h heating at 160°C in N_2 and dropwise adding [sodium hydroxide (NaOH) + ammonium fluoride (NH_4F)] solution in methanol, subsequently heating to 100°C for 2 h and then ½ h in vacuum, placing the mixture-loaded flask in a heating mantle with N_2 atmosphere, raising the temperature to 300°C, allowing the rection to take place for 1.5 h, cooling to room temperature, centrifuging and collecting the $NaYF_4$:Yb, Er nanoparticles with (hexane + ethanol + water) mixture, redispersing in ethanol, centrifuging, and redispersing purified $NaYF_4$:Yb, Er nanoparticles in hexane. A $NaYF_4$: Yb, Er nanoparticle is shown in Figure 8.12(a).

(ii)   Coating of $NaYF_4$: Yb, Er UCNPs with SiO_2 shells to form $NaYF_4$: Yb, Er @SiO_2 UCNPs:

This is done by base-catalyzed polymerization of tetraethyl orthosilicate (TEOS) [$Si (OC_2H_5)_4$] by a method for producing superfine particles known as the reverse microemulsion technique. Figure 8.12(b) shows an $NaYF_4$: Yb, Er @SiO_2 UCNP.

(iii)   Surface modification of $NaYF_4$: Yb, Er @SiO_2 UCNPs to produce carboxylic acid-functionalized $NaYF_4$: Yb, Er @SiO_2 UCNPs or, COOH-$NaYF_4$: Yb, Er @SiO_2 UCNPs:

This is done by dissolving the $NaYF_4$:Yb, Er@SiO_2 nanoparticles in ethanol, adding (3-aminopropyl) triethoxysilane (APTES) [$H_2N(CH_2)_3$ $Si(OC_2H_5)_3$] to the solution, stirring overnight, centrifuging the obtained $NaYF_4$:Yb, Er@SiO_2-NH_2 nanoparticles, dispersing in anhydrous dimethylformamide (DMF) [$HCON(CH_3)_2$], dropwise adding succinic anhydride

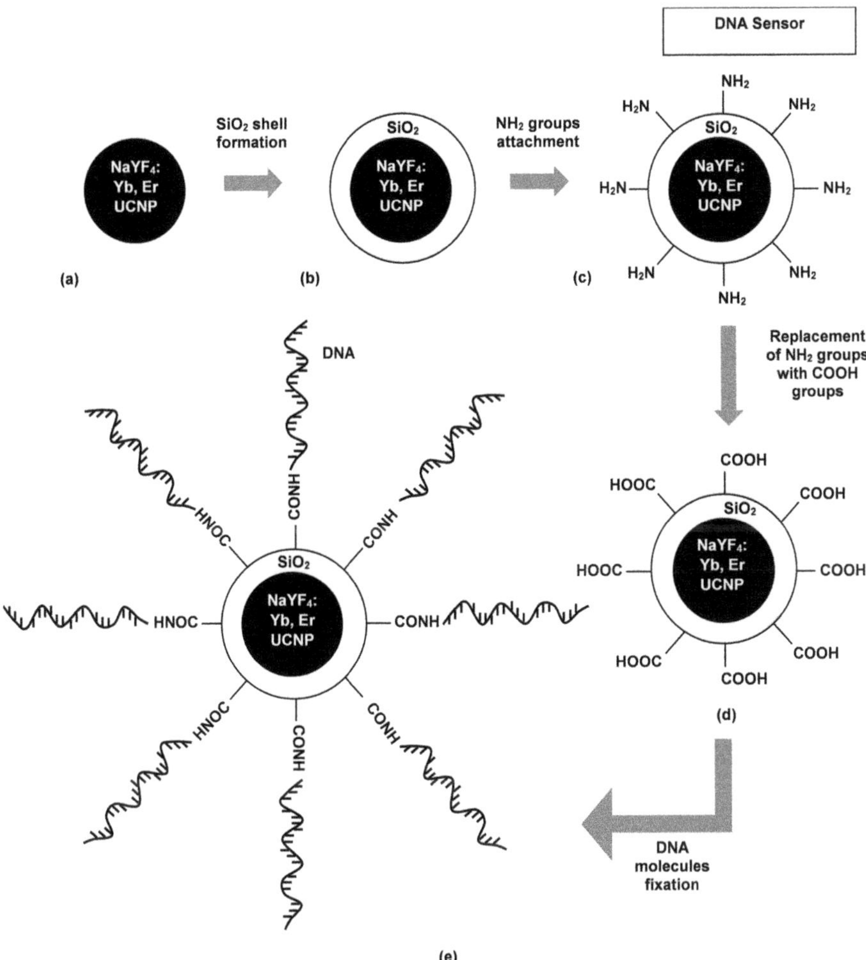

FIGURE 8.12 Preparing probe ss-DNA-functionalized NaYF$_4$: Yb, Er @SiO$_2$ UCNPs-ssDNA and investigating the effect of cDNA on fluorescence quenching: (a) NaYF:Yb, Er upconversion nanoparticle, (b) NaYF$_4$: Yb, Er @SiO$_2$ UCNP, (c) NaYF:Yb, Er@SiO$_2$-NH$_2$ nanoparticle, (d) COOH-NaYF:Yb, Er@SiO$_2$ nanoparticle, (e) ssDNA-NaYF:Yb, Er@SiO$_2$ nanoparticle, (f) fluorescence quenching by GO without cDNA, and (g) non-fluorescence quenching by GO in the presence of cDNA. Part (a) shows a black circle with NaYF:Yb, Er UCNP written inside. It represents a single upconversion nanoparticle. Part (b) shows a bigger circle surrounding the dark circle with NaYF:Yb, Er UCNP written inside. The bigger circle is the SiO$_2$ shell of the nanoparticle. Part (c) shows the black circle and bigger circle with NH$_2$ groups attached to the bigger circle for SiO$_2$ shell on all its sides. Part (d) shows the black circle and bigger circle with COOH groups attached to the bigger circle for SiO$_2$ shell on all its sides. These COOH groups replace the NH$_2$ groups of part (c). Part (e) shows the black circle and bigger circle with CONH groups attached to the bigger circle for SiO$_2$ shell on all its sides. At the opposite terminals of the CONH groups, DNA molecules are affixed.

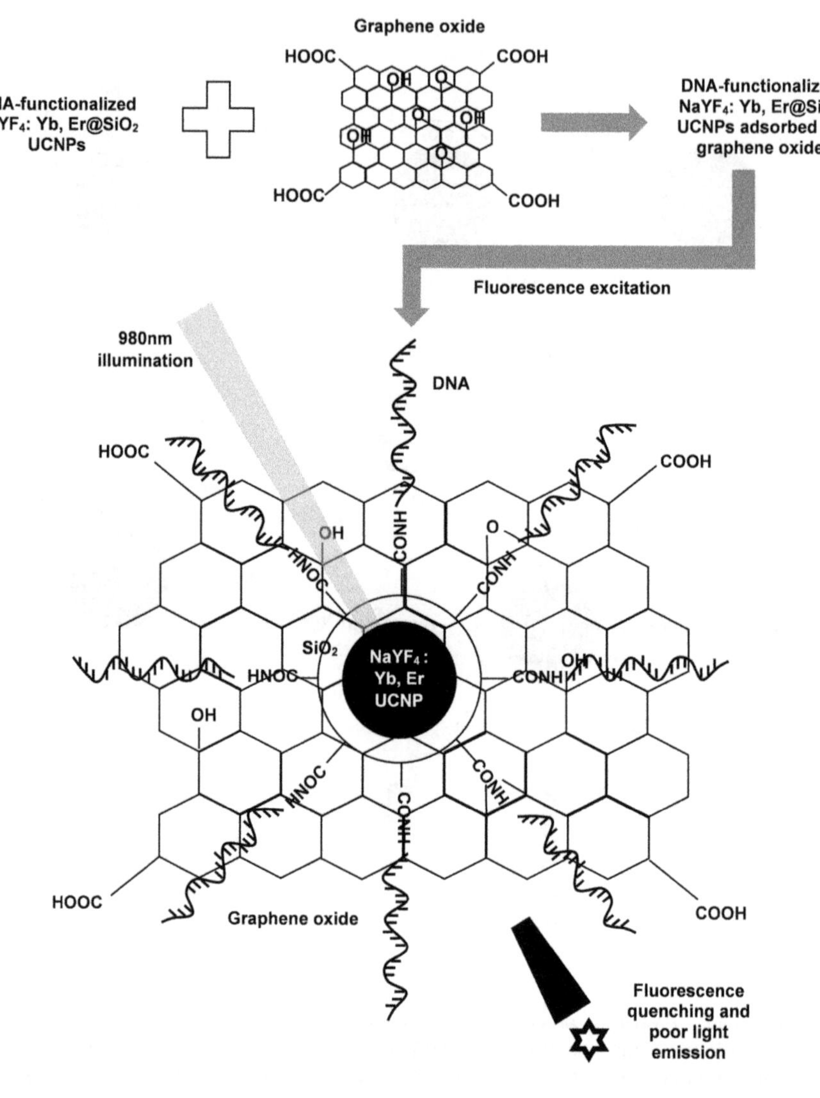

FIGURE 8.12 (Continued) Part (f) shows the addition of graphene oxide to the DNA-functionalized NaYF₄: Yb, Er@SiO₂ UCNPs of part (e) resulting in the adsorption of graphene oxide on this DNA. Fluorescence is excited on the composite structure (DNA-functionalized NaYF₄: Yb, Er@SiO₂ UCNPs-graphene oxide) thus formed. The excitation is done with 980 nm radiation. A weak light signal is emitted signifying that the fluorescence is quenched.

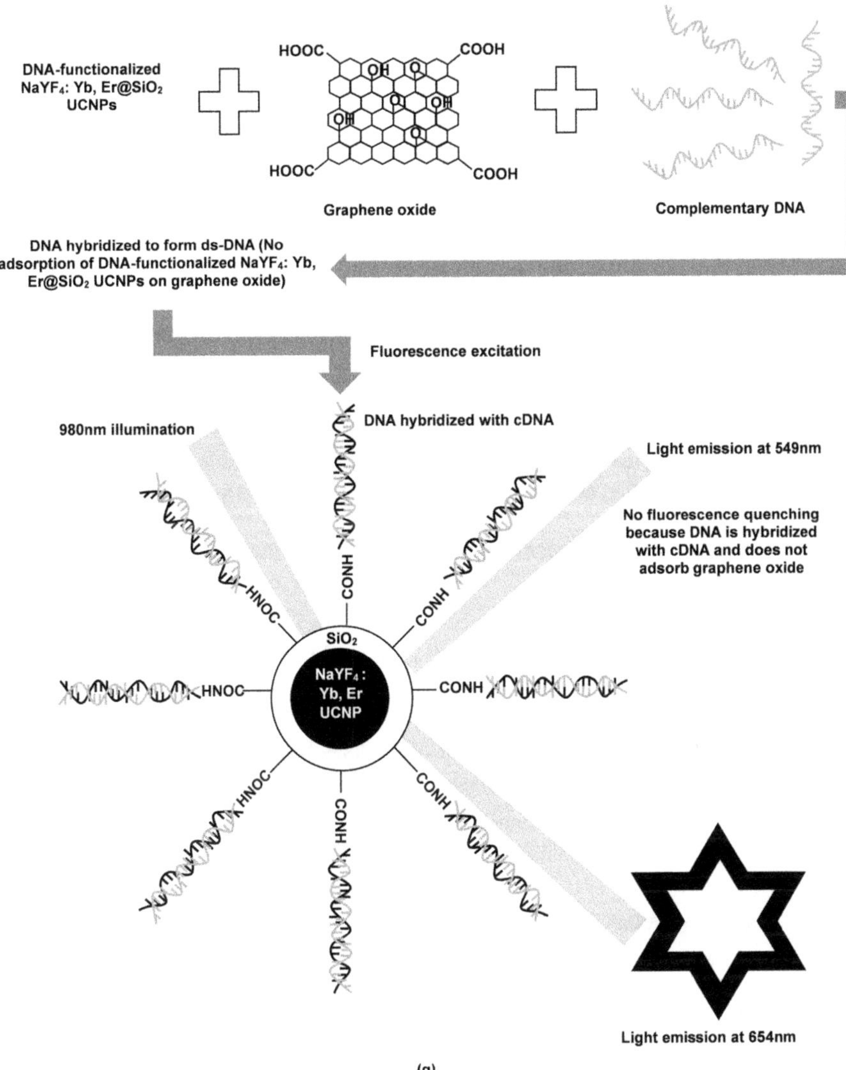

FIGURE 8.12 (Continued) Part (g) shows the addition of graphene oxide and cDNA to the DNA-functionalized NaYF$_4$: Yb, Er@SiO$_2$ UCNPs of part (e). DNA is hybridized with cDNA to form double-stranded DNA, and graphene oxide is not absorbed on ds-DNA. Fluorescence is excited at 980 nm on the composite structure (ds-DNA-functionalized NaYF$_4$: Yb, Er@ SiO$_2$ UCNPs) thus obtained. A strong light signal is emitted suggesting that the fluorescence is not quenched.

[$C_4H_4O_3$] dissolved in anhydrous DMF, stirring overnight, centrifuging and collecting NaYF$_4$:Yb, Er@SiO$_2$-COOH nanoparticles, removing the remnant DMF by centrifuging with ethanol, and dispersing COOH-NaYF$_4$:Yb, Er@SiO$_2$ nanoparticles in water. An NaYF$_4$: Yb, Er@SiO$_2$-NH$_2$ nanoparticle is shown in Figure 8.12(c) while a COOH-NaYF$_4$: Yb, Er@SiO$_2$ nanoparticle is shown in Figure 8.12(d).

(iv) Covalent attachment of probe ss-DNA to COOH-NaYF$_4$: Yb, Er @SiO$_2$ UCNPs to form ssDNA-NaYF$_4$: Yb, Er @SiO$_2$ UCNPs:

This is done by dispersing the COOH-NaYF$_4$: Yb, Er@SiO$_2$ nanoparticles in a borate-buffered solution, loading a portion of the solution to an Eppendorf tube, adding 1-ethyl-3-(3-dimethylaminopropyl) carbodiimide (EDC) in a borate buffer, then adding N-hydroxysulfosuccinimide (sulfo-NHS) [$C_4H_4NNaO_6S$] in a borate buffer, shaking, adding aqueous probe ss-DNA solution, stirring overnight, centrifuging to purify ssDNA-NaYF$_4$: Yb, Er@SiO$_2$ nanoparticles, and collecting the purified nanoparticles with PBS solution. Figure 8.12(e) shows an ssDNA-NaYF$_4$: Yb, Er@SiO$_2$ nanoparticle.

8.5.2.6 Increase in Fluorescence Intensity of UCNPs with cDNA Concentration

The fluorescence spectra (fluorescence intensity vs. wavelength) of UCNPs are recorded for cDNA concentrations from 0 to 400 nM. In the absence of cDNA, the ssDNA-UCNPs are physiosorbed on the GO surface. So, the fluorescence emission of UCNPs is quenched by GO (Figure 8.12(f)). In the presence of cDNA, dsDNA-UCNPs are formed and no physisorption of dsDNA-UCNPs takes place on the GO surface. Hence, the fluorescence emission of UCNPs becomes unquenched by GO (Figure 8.12(g)). The intensity of fluorescence emission varies with cDNA concentration. The LoD is 5 pM (Alonso-Cristobal et al. 2015).

8.5.3 Carboxyl-MoS$_2$ Nanocomposite Surface Plasmon Resonance Immunosensor for Lung Cancer Biomarker CYFRA21-1

An SPR sensing chip is fabricated using single-layer carboxyl-MoS$_2$ nanocomposite film to detect the biomarker CYFRA21-1 for lung cancer in human serum (Chiu and Yang 2020).

8.5.3.1 The Kretschmann Configuration

A BK7 (a high-purity borosilicate glass) substrate is coated with Cr/Au (2 nm/47 nm) film by thermal evaporation.

8.5.3.2 Carboxyl-MoS$_2$ Film Synthesis

MOS$_2$ powder is added to n-butyllithium (n-BuLi) [$CH_3(CH_2)_3Li$] in hexane [$CH_3(CH_2)_4CH_3$]. The mixture is heated to 100°C for 2 h in a hydrothermal synthesis reactor for 2 h. The LixMoS$_2$ (lithium intercalated MoS$_2$) thus obtained is separated by centrifugation, rinsed with anhydrous hexane solution, and centrifuged again. The resulting LixMoS$_2$ anhydrous hexane solution is heated at 70°C for drying. The addition of DI water gives a single-layer MoS$_2$ solution. To this solution, chloroacetic

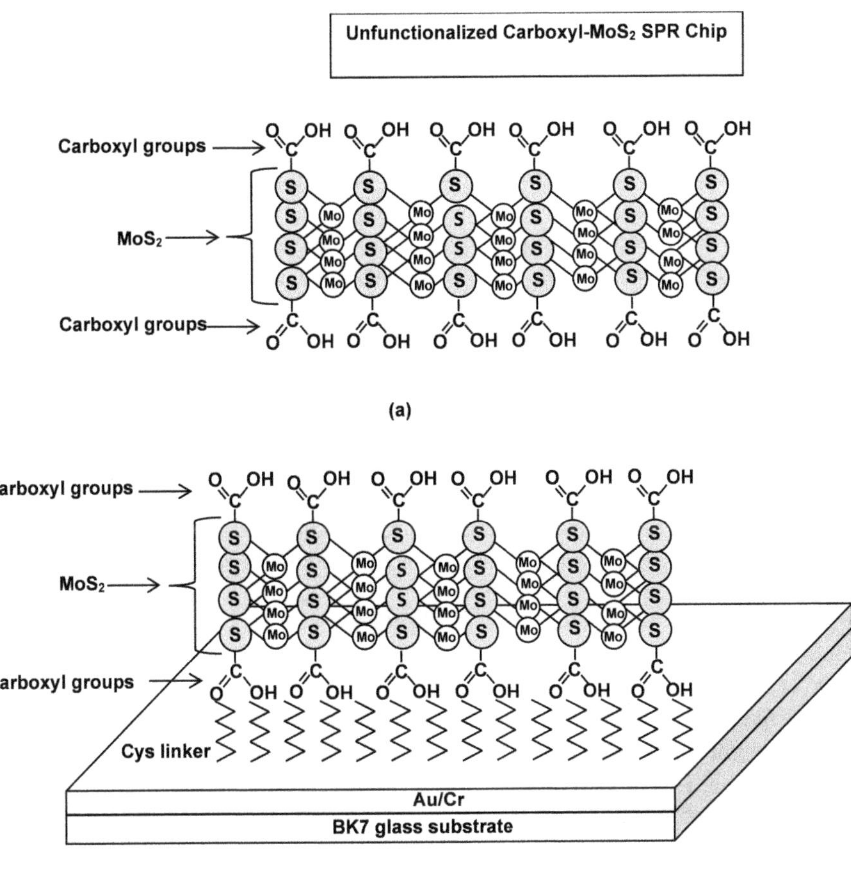

FIGURE 8.13 Single-layer carboxyl-MOS₂ sheet (a) and the sheet bound to BK7 glass/Au-Cr substrate by Cys linker to form the single-layer carboxyl-MoS₂ SPR chip (b). Part (a) shows the MoS₂ layer packed between layers of carboxyl groups on both sides. Layers of carboxyl groups and MoS₂ are indicated. Part (b) shows that the MoS₂ layer crammed between layers of carboxyl groups on opposite sides is fixed on the Au/Cr layer coated on a BK7 glass substrate through Cys linker. Layers of carboxyl groups, MoS₂, and Cys linker are marked.

acid [ClCH₂COOH] and sodium hydroxide (NaOH) are added, and vigorous ultrasonic agitation is done. After centrifuging many times and replacing the supernatant with DI water, ultrasonic shock waves are applied to MoS₂ sheet forming a single-layer carboxyl-MoS₂ sheet modified with chloroacetic acid. Figure 8.13 presents the construction of the SPR chip with the carboxyl-MOS₂ sheet. Figure 8.13(a) shows the single-layer carboxyl-MoS₂ nanocomposite sheet.

8.5.3.3 Fabrication of the Carboxyl-MoS₂ SPR Chip

A self-assembled monolayer of cystamine (NH₂CH₂CH₂SSCH₂CH₂NH₂·2HCl) amino-terminated alkanethiol (a compound containing a sulfanyl group, –SH attached to an alkyl group) is formed on the surface of the SPR chip. The self-assembled monolayer (SAM) is a one-molecule thick crystalline-like layer spontaneously formed on the surface of a solid by chemisorption of organic molecules. A robust SAM is created on gold immersed in alkanethiol. Although the monolayer is formed within a few seconds/minutes, it is not well ordered in the incipient stage but becomes more ordered and organized with the passage of time. The molecular assembly time is 24 h.

In the carboxyl-MoS_2 sheets solution, the carboxyl-MoS_2 sheet ties up with the -NH_2 terminus of cystamine and so gets adhered to the SAM through covalent bonds. Figure 8.13(b) shows the carboxyl-MoS_2 SPR chip.

8.5.3.4 Immobilization of Lung Cancer Antibodies

Various surface modification treatment steps of the sensor are displayed in Figure 8.14. The carboxyl group surface layer (Figure 8.14(a)) of carboxyl-MoS_2 SPR chip is activated with the help of EDC {N-ethyl3-(3-dimethylaminopropyl) carbodiimide}/NHS (N-hydroxysuccinimide) solution (Figure 8.14(b) and (c)). Then recombinant anti-CYFRA21-1 rat immunoglobulin (Ig) monoclonal antibodies are coupled with -COOH terminal groups through covalent bonds (Figure 8.14(d)).

8.5.3.5 Prevention of Non-specific Binding

Unbound anti-CYFRA21-1 proteins are removed with a regeneration buffer. The –COOH terminus of the still leftover unbound anti-CYFRA21-1 proteins is blocked with a blocking buffer employing a high concentration of bovine serum albumin (BSA) in 1×PBS (Figure 8.14(e)). The unbound BSA proteins are removed with NaCl solution. Ethanolamine (EA) ($NH_2CH_2CH_2OH$) solution is used to get rid of the remnant activated –COOH groups.

8.5.3.6 Sensorgrams

The sensorgrams are recorded in spiked human serum specimens where the CYFRA21-1 antigens are trapped by the anti-CYFRA21-1 antibodies (Figure 8.14(f)). The experimental arrangement consists of the Kretschmnn configuration including the prism, and the BK7 glass substrate coated with Au/Cr film along with a laser as a source of light and a photodetector for monitoring the reflected light, as displayed in Figure 8.15. The modified chip surface and the attached antibodies and antigens can be seen.

The sensorgrams reveal that at 0 $pgmL^{-1}$ CYFRA21-1 concentration, the shift of SPR angle is 0.02 m° (millidegree= 10^{-3} degree). At 0.05 $pgmL^{-1}$ CYFRA21-1 concentration, it increases to 0.64 m°. As the CYFRA21-1 concentration rises to higher values, the SPR angle shift also increases. The SPR angle shifts are 1.54 m° at 0.5 $pgmL^{-1}$, 3.48 m° at 5.0 $pgmL^{-1}$, 6.37 m° at 50 $pgmL^{-1}$, 8.96 m° at 0.5 $ngmL^{-1}$, 14.23 m° at 5.0 $ngmL^{-1}$, 18.19 m° at 50 $ngmL^{-1}$, and 21.77 m° at 100 $ngmL^{-1}$ CYFRA21-1 concentration. The LoD of the carboxyl-MoS_2 SPR chip is 0.05 $pgmL^{-1}$ CYFRA21-1 concentration. Since the LoD of enzyme-linked immunosorbent assay (ELISA) is 600 $pgmL^{-1}$, a 600/0.05 = 12000-fold enhancement in LoD is achieved by the carboxyl-MoS_2 SPR chip with respect to ELISA (Chiu and Yang 2020).

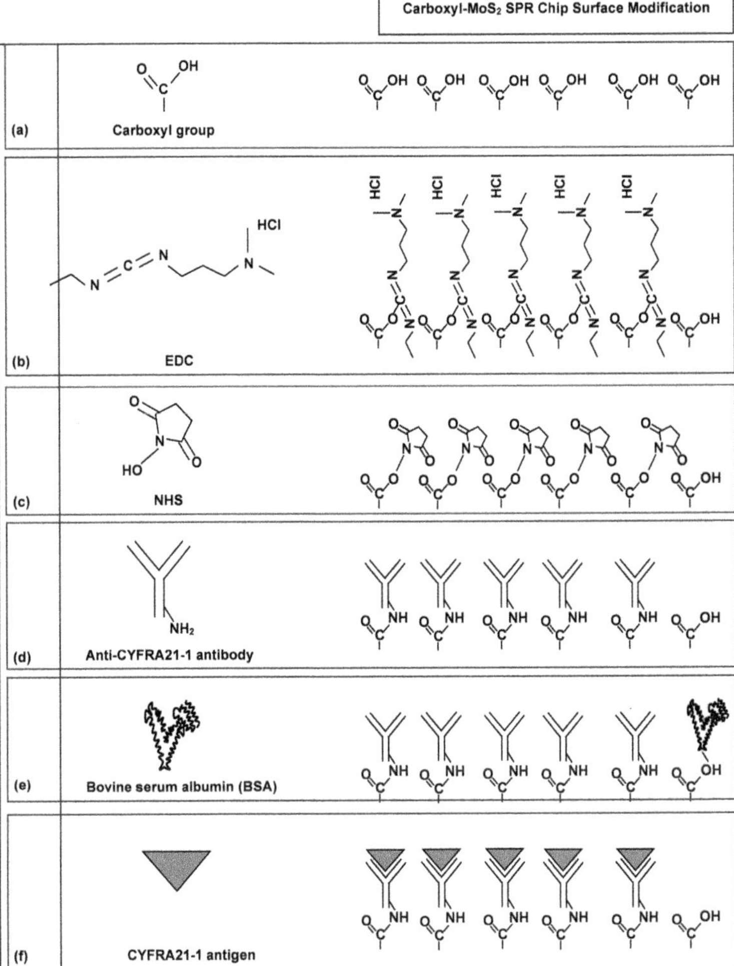

FIGURE 8.14 Different stages of surface modification of single-layer carboxyl-MoS$_2$ SPR chip. Since the surface modification affects the top carboxyl layer, only this layer is shown. The underlying MoS$_2$ layer, carboxyl layer, Cys liker, Au/Cr layer, and the BK7 glass substrate are not shown to avoid repetition but they are present:(a) the top carboxyl layer, (b) and (c) surface activation using EDC/NHS, (d) anti-CYFRA21-1 antibody immobilization, (e) blocking of remaining carboxylic groups having unbound antibodies with bovine serum albumin, and (f) detection of the antigen CYFRA21-1. Part (a) shows the structure and bonds of the carboxyl group on the left side and the carboxyl group layer on the right side. Part (b) shows the molecular structure of EDC on the left side and its binding to carboxyl groups on the right side. Part (c) shows the molecular constitution of NHS on the left side and its effect on carboxyl groups on the right side. Part (d) shows the anti-CYFRA21-1 antibody on the left side and its attachment to the carboxyl groups on the right side. Part (e) shows the bovine serum albumin on the left side and its fixation on leftover COOH group on which antibody is not fastened on the right side. Part (f) shows the CYFRA21-1 antigen on the left side. On the right side are shown the CYFRA21-1 antigens captured and trapped by the anti-CYFRA21-1 antibodies upon exposure of the sensor to an antigen-containing solution.

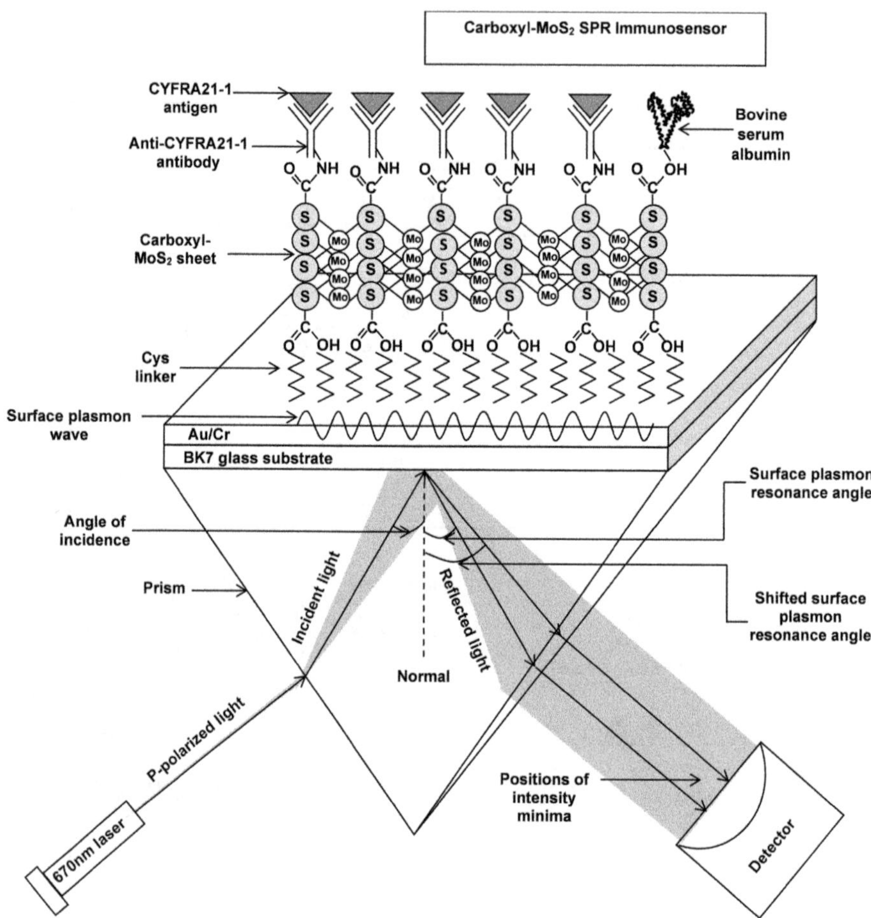

FIGURE 8.15 Depiction of surface-plasmon resonance setup using the Kretschmann configuration for lung cancer biomarker measurement. The diagram shows a BK7 glass substrate coated with Au/Cr film. To the Au/Cr film a three-layer structure comprising a carboxyl group layer, a MoS$_2$ layer and another carboxyl group layer, is fixed by Cys linker. The surface of the top carboxyl group layer is chemically treated to bind anti-CYFRA21-1 antibodies. Bovine serum albumin blocks the carboxyl group site at which antibody is not tied. The CYFRA21-1 antigens are seen grabbed by the anti-CYFRA21-1 antibodies when the chip is immersed in antigen solution. P-polarized light from a 670-nm laser is pointed through a prism at a certain angle on the metal-coated surface exciting surface plasmon waves. The reflected light is received by the detector. At the resonance angle, the light is absorbed by the electrons in the Au/Cr film. Then the intensity of reflected light is minimum. This is the SPR angle. On binding the CYFRA21-1 antigens of a given concentration to anti-CYFRA21-1 antibodies on the surface of the chip, the refractive index of the medium at the metal–dielectric interface is altered thereby shifting the angle of absorption of light. This is the shifted SPR angle at that antigen concentration.

8.5.4   Black Phosphorous-Tilted Fiber Grating (BP-TFG) Biosensor for Neuron-Specific Enolase (NSE) Cancer Biomarkers

A black phosphorus–fiber optic biosensor for diagnoses of human neuron-specific enolase cancer biomarkers is proposed (Zhou et al. 2019).

8.5.4.1   Principle, Optical Interrogation and Role of BP Nanosheets

In this sensor (Figure 8.16), a tilted fiber grating acts as an optical transducer. It couples light from the core of the fiber to its cladding. An evanescent field is produced at the cladding and the boundary region surrounding it. Any affinity binding in the surrounding medium causes a localized alteration in the concentration of the analyte inducing a change in local refractive index.

The sensor is optically interrogated using a broadband source of light and the transmission spectrum is recorded with an optical spectrum analyzer. The wavelength of the transmission/attenuation bands of the spectrum is found to shift with the concentration of the analyte. Hence, the analyte concentration can be measured by tracking the wavelength shift. The transverse magnetic resonance shows higher sensitivity to refractive index variation than transverse electric resonance. So, the transverse magnetic mode is used in biosensing experiments.

Neuron-specific enolase, the neuronal form of enzyme enolase, is a reliable biomarker of small-cell lung cancer or neuroblastoma (Isgrò et al. 2015). Elevated levels of NSE have been noticed with malignant proliferation in patients suffering from renal cell carcinoma, seminoma, melanoma, etc. For sensitive detection of NSE biomarkers, anti-human neuron-specific enolase (anti-NSE) antibodies are immobilized on black phosphorous nanosheets deposited on an optical fiber with inscribed tilted fiber grating. The shift in wavelength of the optical spectrum of the fiber caused by interactions between anti-NSE bioreceptors and target NSE biomarkers is measured with respect to the concentration of NSE biomarkers.

The BP nanosheets enhance the interactions of light with matter. Owing to their large surface area-to-volume ratio and favorable molecular adsorption energy, BP nanosheets enlarge the sensing area and raise the number of binding sites, resulting in enormous amplification of the optical signal.

8.5.4.2   Fabrication and Response Evaluation

(i)   Inscription of TFG: A hydrogenated BGe single-mode fiber is used. The 82° tilted fiber grating is inscribed using a frequency-doubled argon laser (wavelength 244 nm).

(ii)   Synthesis of BP nanosheets: Liquid ultrasonication-based exfoliation method is used. Small pieces of black phosphorous obtained by grinding bulk BP are dissolved in isopropanol, $(CH_3)_2CHOH$. The dispersion is sonicated in an ultrasonic bath for breaking apart the BP layers stacked together by van der Waals forces and then centrifuged for removing the remnant unexfoliated black phosphorous. The supernatant liquid contains few-layer BP nanosheets. It is collected.

(iii)   Deposition of black phosphorous nanosheets on the optical fiber: The portion of the fiber with TFG is cleaned with acetone, rinsed with water, and dried

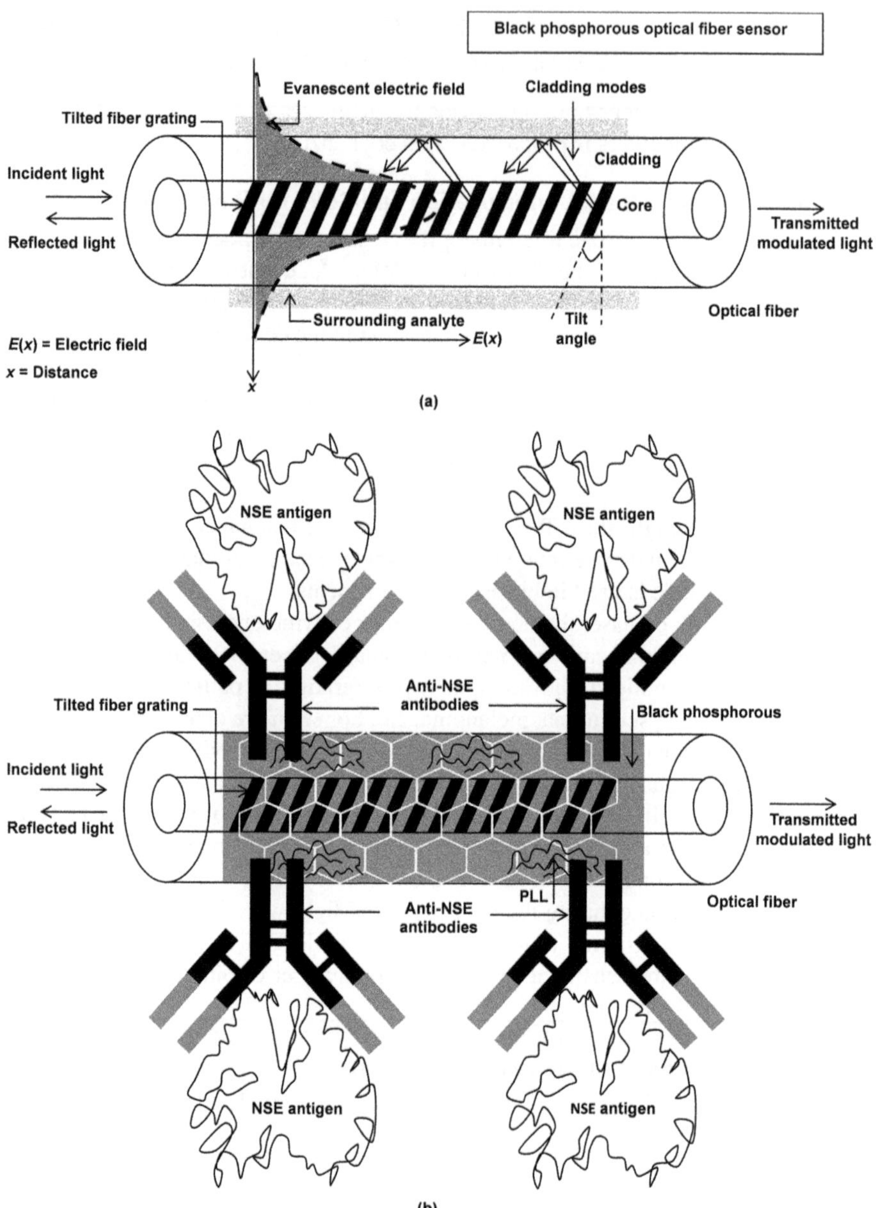

FIGURE 8.16 Black phosphorous-based optical fiber biosensor in: (a) un-biofunctionalized state and (b) biofunctionalized state and exposed to neuron-specific enolase biomarker. Part (a) shows an optical fiber consisting of a concentric core surrounded by a cladding of lower refractive index material than the core. Inside the core of the fiber a tilted fiber grating is seen. The tilt angle of the grating is marked. Directions of incident, reflected, and transmitted modulated light beams are indicated. The fiber is dipped in the analyte solution which is in contact with the outer surface of the fiber. The cladding modes are shown; these are modes confined to the cladding due to coupling of light from the core to the cladding. The evanescent electric field is seen.

FIGURE 8.16 (Continued)   The evanescent field arises because light striking the interface between higher and lower refractive index media suffers total internal reflection. However, it also penetrates through the interface and moves farther into the lower refractive index medium. It is this part of the electromagnetic field which is known as evanescent field. Its primary feature is that it decays exponentially with distance. The graph between electric field $E(x)$ and distance x in the direction perpendicular to the direction of propagation of light is sketched showing the variation of electric field along this direction. Part (b) shows the use of the surface-modified and functionalized optical fiber as a biosensor. On the outer surface of the fiber, black phosphorous nanosheets are deposited. Also seen are PLL (poly-L-lysine) molecules having amino groups for attaching bioactive molecules. Anti-NSE antibodies are immobilized on the black phosphorous surface with PLL molecules sticking to it. NSE antigens are seen bound with anti-NSE antibodies through antibody–antigen interaction when the sensor is exposed to a specimen solution containing NSE antigens.

(Figure 8.17). For enrichment of OH groups on the surface of the fiber, it is immersed in 1.0 M sodium hydroxide (NaOH) solution. The alkali-treated fiber is incubated in APTES-ethanol solution. APTES [(3-Aminopropyl) triethoxysilane, $H_2N(CH_2)_3Si\ (OC_2H_5)_3$], reacts with the OH groups on the fiber surface forming Si–O–Si bonds and making the fiber surface positively charged. After washing in water, the fiber is baked in an oven. After subjecting the fiber to above treatments, it is kept in a microchannel container. To this container, a solution of BP in isopropanol is added. Evaporation of the solvent is accompanied by binding of the negatively charged BP nanosheets with the positively charged surface of the fiber. After several cycles of pipetting the BP nanosheets solution in the microchannel container and evaporation of the solvent together with binding of BP nanosheets to the fiber, the fiber is placed in a vacuum oven to improve the binding strength.

(iv) Biofunctionalization of the BP nanosheets by anti-NSE immobilization: The black phosphorous-coated fiber is incubated in poly-L-lysine, $(C_6H_{12}N_2O)_n$ solution. The poly-L-lysine has a large number of amino groups. It acts as a cross-linker via amide groups. The cross-linking is established by the attachment of the positively charged poly-L-lysine to black phosphorous. Anti-NSE antibody is mixed with EDC and NHS in PBS for the activation of carboxyl groups. Covalent immobilization of anti-NSE over poly-L-lysine-functionalized black phosphorous coating on the optical fiber is performed by incubating it in the (anti-NSE+EDC+NHS) solution. After washing with 1×PBS buffer solution, the fiber is dipped in 1% bovine serum albumin (BSA) solution for blocking the remaining active carboxylic groups and hence avoiding non-specific absorption.

(v) NSE biomarker sensing: Different concentrations of NSE biomarkers in 1×PBS buffer are applied on the biosensor and the transverse magnetic wavelength shifts are recorded. Before each measurement, the sensor is washed with PBS buffer, and then it is dipped in the NSE biomarker solution of a particular concentration. The wavelength shift is 210 pm at 0.01 $ngmL^{-1}$ NSE concentration. It is 240 pm at 0.1 ng mL^{-1} NSE concentration. The wavelength shifts at 1.0, 10, and 100 $ngmL^{-1}$ NSE concentrations are 290, 320, and 370 pm respectively. The LoD of NSE is 1.0 $pgmL^{-1}$ (Zhou et al. 2019).

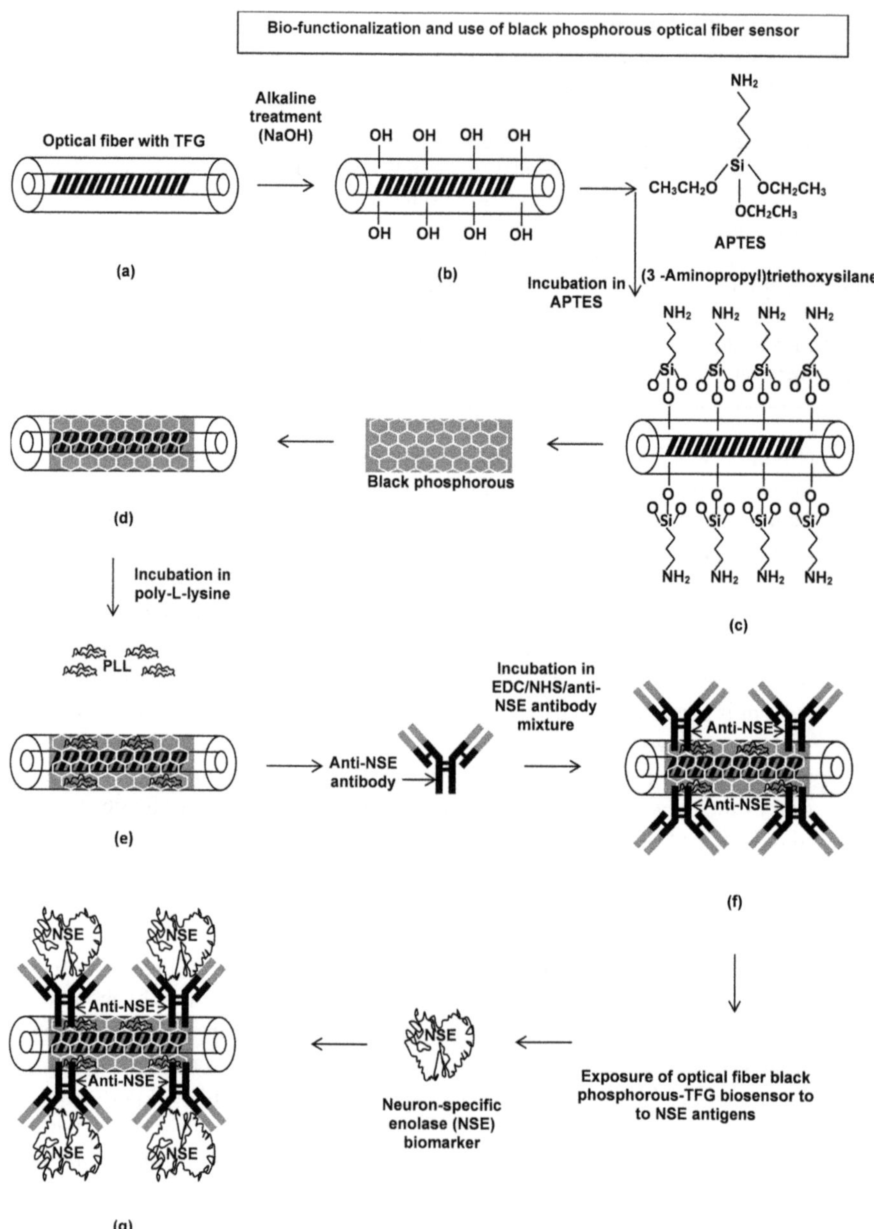

FIGURE 8.17 Biofunctionalization of black phosphorous-based optical fiber biosensor, and its exposure to neuron-specific enolase: (a) as fabricated optical fiber with tilted fiber grating, (b) after treatment with NaOH, (c) after incubation in APTES, (d) after black phosphorous nanosheets deposition, (e) after incubation in poly-L-lysine, (f) after immobilization of anti-NSE antibodies via bioconjugation with EDC/NHS, and (g) after bio-affinity binding of NSE biomarkers with anti-NSE antibodies. Part (a) shows a single-mode optical fiber consisting of a core and cladding with tilted fiber grating inscribed in the core of the fiber.

FIGURE 8.17 (Continued) Part (b) shows that hydroxyl groups are attached to the outer surface of the optical fiber after treating with NaOH. Part (c) shows the structural formula of APTES and the silanized surface of the optical fiber with affixed alkoxysilane molecules. Part (d) shows the fiber with black phosphorous deposited on its outer surface. Black phosphorous has a layered graphene-like structure with an interlinked six-membered ring on which phosphorous atoms are placed at the joining points of the lines in the ring. Part (e) shows the black phosphorous after it is surface functionalized with poly-L-lysine molecules serving as cross-linkers. Part (f) shows the black phosphorous surface with immobilized anti-NSE antibodies. Part (g) shows the NSE antigens bound to anti-NSE antibodies on the fiber surface when the sensor is immersed in a sample solution.

8.6 DISCUSSION AND CONCLUSIONS

This chapter presented glimpses of the expanding landscape of 2D material biological sensors (Table 8.1). Like any other biosensor, the surface of a 2D material biosensor has to interact repeatedly with biological fluids during its use. This interaction is likely to alter the calibration characteristics of the biosensor. Further, the biomolecules used in these sensors have a limited life causing a drift in sensor characteristics with time. Preservation of long-term stability and prevention of drifts over extended periods of storage are issues of paramount importance confronting biosensors. The drift mechanisms must be thoroughly understood and suitable strategies must be devised for using biosensors considering these limitations. Biological sensors which do not use bio-recognition elements are impervious to such deterioration mechanisms.

TABLE 8.1
Variety of 2D Material Biological Sensors

S. No.	Type of 2D Material Sensor	Sensing Device	Analyte (es) Detected	Reference
1.	2D material-modified electrode electrochemical biosensors	Au/MXene nanocomposite-GCE enzymatic sensor	Glucose	Rakhi et al. (2016)
		Au NPs-MoS$_2$ nanosheet-GCE sensor	Ascorbic acid, dopamine, and uric acid	Sun et al. (2014)
		2D conductive MOF nanosheets sensor	Cancer biomarker H$_2$O$_2$ in live cells	Huang et al. (2022)
		Graphene/Nafion-modified electrode	Paracetamol, aspirin, and caffeine	Yiğit et al. (2016)
		Carboxylic graphene acetylcholinesterase sensor	Pesticides	Zhou et al. (2013)

(Continued)

TABLE 8.1 (Continued)

S. No.	Type of 2D Material Sensor	Sensing Device	Analyte (es) Detected	Reference
2.	2D material FET biological sensors	Solution-gated graphene transistor	Glucose	Zhang et al. (2015)
		MoS_2 nanosheet sensor	PSA	Lee et al. (2014)
		Monolayer WSe_2 FET sensor	COVID-19 virus	Fathi-Hafshejani et al. (2021)
3.	2D material optical sensors for biological analysis	BSA-capped Au nanoclusters/g-C_3N_4 nanosheets fluorescent sensor	Dopamine	Guo et al. (2016)
		Graphene oxide-upconversion nanoparticles fluorescence sensor	DNA	Alonso-Cristobal et al. (2015)
		Carboxyl-MoS_2 nanocomposite surface plasmon resonance immunosensor	Lung cancer biomarker CYFRA21-1	Chiu and Yang (2020)
		Black phosphorous-tilted fiber grating sensor	Neuron-specific enolase cancer biomarkers	Zhou et al. (2019)

REFERENCES

Alonso-Cristobal P., P. Vilela, A. El-Sagheer, E. Lopez-Cabarcos, T. Brown, O. L. Muskens, J. Rubio-Retama, and A. G. Kanaras 2015 Highly sensitive DNA sensor based on upconversion nanoparticles and graphene oxide, *ACS Applied Materials & Interfaces*, 7(23), pp. 12422–12429.

Bhalla N., P. Jolly, N. Formisano and P. Estrela 2016 Introduction to biosensors, *Essays in Biochemistry*, 60(1), pp. 1–8.

Cai Z., J. Chen, S. Xing, D. Zheng, and L. Guo 2021 Highly fluorescent g-C_3N_4 nanobelts derived from bulk g-C_3N_4 for NO_2 gas sensing, *Journal of Hazardous Materials*, 416, p. 126195.

Chen G., H. Qiu, P. N. Prasad, and X. Chen 2014 Upconversion nanoparticles: Design, nanochemistry, and applications in theranostics, *Chemical Reviews*, 114(10), pp. 5161–5214.

Chiu N.-F. and H.-T. Yang 2020 High-sensitivity detection of the lung cancer biomarker CYFRA21-1 in serum samples using a carboxyl-MoS_2 functional film for SPR-based immunosensors, *Frontiers in Bioengineering and Biotechnology*, 8, 234, pp. 1–14.

Fathi-Hafshejani P., N. Azam, L. Wang, M. A. Kuroda, M. C. Hamilton, S. Hasim, and M. Mahjouri-Samani 2021 Two-dimensional-material-based field-effect transistor biosensor for detecting covid-19 virus (SARS-CoV-2), *ACS Nano*, 15(7), pp. 11461–11469.

Filice M., J. A. Marchal, and F. Gamiz 2021 Chapter 7 – Biosensors Based on Two-Dimensional Materials, In: Y.M. Jhon, and J. H. Lee (Eds.), *2D Materials for Nanophotonics*, Elsevier, Amsterdam, The Netherlands, pp. 245–312.

Förster T. 1946 Energiewanderung und Fluoreszenz, *Naturwissenschaften*, 33(6), pp. 166–175; 2012 Energy migration and fluorescence, *Journal of Biomedical Optics*, 17(1), pp. 011002-1 to 011002-10.

Guo X., F. Wu, Y. Ni, and S. Kokot 2016 Synthesizing a nano-composite of BSA-capped Au nanoclusters/graphitic carbon nitride nanosheets as a new fluorescent probe for dopamine detection, *Analytica Chimica Acta*, 942, pp. 112–120.

Huang W., Y. Xu, Z. Wang, K. Liao, Y. Zhang, and Y. Sun 2022 Dual nanozyme based on ultrathin 2D conductive MOF nanosheets intergraded with gold nanoparticles for electrochemical biosensing of H_2O_2 in cancer cells, *Talanta*, 249, p. 123612.

Isgrò M.A., P. Bottoni, and R. Scatena 2015 Neuron-specific enolase as a biomarker: Biochemical and clinical aspects. In: Scatena, R. (Eds) *Advances in Cancer Biomarkers*. Advances in Experimental Medicine and Biology, Vol. 867. Springer Nature Switzerland AG, pp. 125–143.

Lee J., P. Dak, Y. Lee, H. Park, W. Choi, M. A. Alam and S. Kim 2014 Two-dimensional layered MoS_2 biosensors enable highly sensitive detection of biomolecules, *Scientific Reports*, 4, 7352, pp. 1–7.

Rakhi R. B., P. Nayak, C. Xia and H. N. Alshareef 2016 Novel amperometric glucose biosensor based on MXene nanocomposite, *Scientific Reports*, 6, 36422, pp. 1–9.

Sun H., J. Chao, X. Zuo, S. Su, X. Liu, L. Yuwen, C. Fan and L. Wang 2014 Gold nanoparticle-decorated MoS_2 nanosheets for simultaneous detection of ascorbic acid, dopamine and uric acid, *RSC Advances*, 4, pp. 27625–27629.

Wang M., G. Abbineni, A. Clevenger, C. Mao, and S. Xu 2011 Upconversion nanoparticles: Synthesis, surface modification and biological applications, *Nanomedicine*, 7(6), pp. 710–729.

Yiğit A., Y. Yardım, M. Çelebi, A. Levent, and Z. Şentürk 2016 Graphene/Nafion composite film modified glassy carbon electrode for simultaneous determination of paracetamol, aspirin and caffeine in pharmaceutical formulations, *Talanta*, 158, pp. 21–29.

Zhang M., C. Liao, C. H. Mak, P. You, C. L. Mak and F. Yan 2015 Highly sensitive glucose sensors based on enzyme-modified whole-graphene solution-gated transistors, *Scientific Reports*, 5, 8311, pp. 1–6.

Zheng P. and N. Wu 2017 Fluorescence and sensing applications of graphene oxide and graphene quantum dots: A review, *Chemistry: An Asian Journal*, 12(18), pp. 2343–2353.

Zhou L., C. Liu, Z. Sun, H. Mao, L. Zhang, X. Yu, J. Zhao, and X. Chen 2019 Black phosphorus based fiber optic biosensor for ultrasensitive cancer diagnosis, *Biosensors and Bioelectronics*, 137, pp. 140–147.

Zhou Q., L. Yang, G. Wang, and Y. Yang 2013 Acetylcholinesterase biosensor based on SnO_2 nanoparticles–carboxylic graphene–nafion modified electrode for detection of pesticides, *Biosensors and Bioelectronics*, 49, pp. 25–31.

9 Challenges and Perspectives of 2D Materials-Based Sensors

This chapter is devoted to an introspection of advancements in 2D material-based sensors vis-à-vis the problems that need to be immediately addressed to accelerate the pace of progress. It goes without saying that the issues beleaguering sensors are essentially linked to the process technologies of 2D materials. These are the roadblocks of paramount importance which must be surmounted to move ahead from the present stage of development. So, it is time to look back and assess the scenario. The scenario is unfolded beginning with graphene.

9.1 THE GRAPHENE REVOLUTION

Graphene, the wonder material of the 21st century, was discovered by Andre Konstantin Geim, a Russian-born Dutch–British physicist, and Konstantin Sergeevich Novoselov, a Russian–British physicist, in 2004 (Novoselov et al. 2004). Graphene quickly revealed a cornucopia of new physics and prospective applications (Geim and Novoselov 2007). Geim and Novoselov were honored with the Nobel Prize in Physics in 2010 for their groundbreaking work on graphene.

9.2 FASCINATING 2D MATERIALS BEYOND GRAPHENE

The discovery of mechanically exfoliated graphene was a defining moment for the emergence of a new field in materials science, the birth of two-dimensional materials, extending to non-graphene materials, including transition-metal dichalcogenides (TMDCs), black phosphorous, monoelement 2D materials (silicene, phosphorene), transition metal carbide- and carbon nitride-based MXenes, and many more (Bhimanapati et al. 2015).

9.3 THE ROSY PICTURE AND ITS FLIP SIDE

The tremendous research interest in 2D materials sparked by graphene has motivated relentless progress in this field since 2004. All this work unequivocally confirms the immense potential of these materials for developing new and enhanced-capability sensors. On the other hand, it is equally true that not many 2D materials-based sensors have been commercialized, which indicates that the technology-readiness level

DOI: 10.1201/9781003330585-9

has not reached the stage at which wide-scale proliferation of these sensors can be expected (Sulleiro et al. 2022). The laboratory-scale experimental process demonstrations need to be upscaled. There is no doubt about the exotic qualities of 2D materials and the achieved super-sensitivity of sensors. But there are intervening steps between the laboratory-level technology development and its upgradation for mass manufacturing. There must be reasons for the impediments to the industrialization of 2D materials devices. Let us enquire into and identify the probable causes in this chapter and look for possible solutions.

9.4 LARGE-AREA SYNTHESIS OF 2D MATERIALS

This is an area of extensive ongoing research to hasten the industrial uptake of developed technologies (Zavabeti et al. 2020). Process engineers are still confronted with common challenges in 2D material fabrication and device performance regarding the garnering of high crystal quality and uniformity, scaling up the growth, controlling thickness and morphology, optimizing device structures, and improving repeatability (Yao et al. 2019). Batch production of 2D material sensors has to grow and burgeon rapidly from the small-scale prototyping.

9.4.1 MECHANICAL EXFOLIATION

Among top-down synthesis methods, mechanical exfoliation is a cost-effective approach to produce high-quality 2D material sheets measuring greater than half a millimeter laterally. But it suffers from scalability and yield constraints.

9.4.2 CHEMICAL VAPOR DEPOSITION (CVD) AND METAL-ORGANIC CHEMICAL VAPOR DEPOSITION (MOCVD)

Considering, bottom-up synthesis methods, a meter-sized single-crystal graphene film has been epitaxially grown in a CVD furnace at 1030°C on copper (111) substrate (Xu et al. 2017b). The graphene film has a size of (5×50) cm^2. It has >99% ultra-highly oriented grains. The graphene film has a mobility of up to ~23,000 cm^2 V^{-1} s^{-1} at 4 K. It shows a sheet resistance of ~230 Ω/square at room temperature.

High-mobility TMDCs films are required for FET channels. CVD of crystalline MoS$_2$ film of >½ mm size has been done on molten glass substrates. Single-crystal monolayer MoS$_2$ with a domain size of up to 563 μm is obtained by an optimized atmospheric pressure CVD process. It shows room temperature mobility ~24 cm^2 V^{-1} s^{-1} which increases to 84 cm^2 V^{-1} s^{-1} at 20 K (Zhang et al. 2018).

Wafer-scale uniform monolayer 2H-TaSe$_2$ films are directly synthesized on gold foils by CVD (Shi et al. 2018).

CVD is applied for synthesizing highly crystalline MoTe$_2$ films extending over ~ 1 cm scale lengths. These films are uniform across the whole area showing electronic properties similar to exfoliated MoTe$_2$ flakes (Zhou et al. 2015).

MOCVD technique, a variant of CVD, is employed for the growth of monolayer molybdenum disulfide (MoS$_2$) and tungsten disulfide (WS$_2$) films of 4-inch wafer-scale dimensions on SiO_2 substrates. For MoS_2, the electron mobility is 30 cm^2 V^{-1} s^{-1} at room temperature and 114 cm^2 V^{-1} s^{-1} at 90 K. MoS_2 FETs are demonstrated with 99% yield, advancing toward atomically thin integrated circuits (Kang et al. 2015).

CVD is a big-budget and time-consuming method than the mechanical exfoliation technique. Whole-hearted efforts, engineering knowledge, and proficiency are essential.

9.4.3 Molecular Beam Epitaxy (MBE)

MBE is even more complicated and resource-intensive than CVD/MOCVD but the quality of MBE-deposited 2D material films is unquestionably far superior to CVD films, e.g., two-fold higher quantum yield (QY) of photoluminescence (PL) with lower photobleaching is achieved in MBE MoS_2 film than CVD MoS_2 film (Singh and Gupta 2022).

9.5 DOPING OF 2D MATERIALS

9.5.1 Inapplicability of Standard Doping Techniques to 2D Materials

The standard substitutional doping methods of silicon technology, namely, thermal diffusion and ion implantation cannot be used for 2D materials because of their damaging effects on the crystal lattice and copious defect generation caused in the thin 2D materials by their use.

9.5.2 Graphene Doping

Considering the dopants and the doping processes, let us take the example of graphene. Nearly equal atomic sizes of nitrogen and boron atoms to carbon atoms make them suitable dopants for graphene. Nitrogen atom acts as a donor impurity introducing free electrons and therefore producing N-type graphene. The boron atom acts as an acceptor impurity creating holes and hence giving P-type graphene. A few methods for doping graphene are given in subsections below (Guo et al. 2011).

9.5.2.1 Heteroatom Doping

It is a process, involving the replacement of some carbon atoms in the graphitic structure with heteroatoms. The heteroatom is any atom except carbon and hydrogen.

An arc discharge is struck at a high current between graphite electrodes. The presence of H_2+NH_3 during arc discharge gives N-graphene while the presence of $H_2+B_2H_6$ leads to P-graphene (Panchakaria et al. 2009).

Nitrogen-doped N-graphene is synthesized by CVD, using a Cu film (as a catalyst) on a Si substrate in an (H_2 +20% Ar) atmosphere, and $CH_4 + NH_3$, with CH_4 as a carbon source and NH_3 as a nitrogen source (Wei et al. 2009).

Boron-doped P-graphene is made using a bubbler-assisted CVD method with precursors toluene ($C_6H_5CH_3$) as the carbon source and (9,10-dimesityl-9,10-diboraanthracene (DBA(Mes)$_2$) as the boron source (Wu et al. 2017).

9.5.2.2 Chemical Modification

The chemically modified graphene is covalently doped graphene. Organic molecules are used (Farmer et al. 2009):

(i) For N-doping, molecules with electron-donor groups, e.g., 9,10-dimethyl-anthracene (An-CH$_3$), 1,5 naphthalenediamine (Na-NH$_2$), are used.
(ii) For P-doping, molecules with electron-acceptor groups, e.g., 9,10-Dibromo-anthracene (An-Br), 2,3,5,6-tetrafluoro-7,7,8,8-tetracyanoquinodimethane (F4-TCNQ), etc., are used (Lee et al. 2016).

9.5.2.3 Electrostatic Field Tuning

This is done on top gate and back gate FETs (Castro Neto et al. 2009). The Fermi level of pristine graphene is tuned between conduction and valence bands by changing the gate voltage. In P-type graphene, it lies in the valence band, while in N-type graphene, it is in the conduction band.

9.6 OHMIC CONTACTS FOR 2D MATERIALS

9.6.1 Mechanisms of Deviations of Contacts from Ohmic Character and Resulting Effects

Departure in behavior of 2D material/metal contacts from Ohm's law occurs through different routes (Zheng et al. 2021):

(i) Damages inflicted by processing: The fragile, atomically thin 2D materials are prone to damages by semiconductor process techniques such as electron-beam lithography and physical deposition. These damages result in the creation of interfacial states or Schottky barriers, causing the degradation of 2D material/metal contacts. The height of the Schottky barrier is determined by the difference between the work function of the metal and electron affinity of the semiconductor.
(ii) States introduced by adsorption of molecules: Interfacial states are also induced due to the ultra-sensitivity of 2D materials to adsorbed molecules from the environment. Adsorbate-induced states too are detrimental for contacts.

The interfacial states and the Schottky barriers increase the contact resistance adversely affecting the performance of 2D material devices, e.g., low ON/OFF ratio of currents and poor mobility in an FET device. So, processes ensuring ideal ohmic contacts with linear current–voltage characteristics and low contact resistances are necessary for making high-performance sensors.

9.6.2 The 2D Material/2D Material Ohmic Contacts

Novel strategies have been devised to overcome the problems of contacts (Chuang et al. 2016). An example is the use of 2D material/2D material ohmic contacts to solve the contact problems. The performance of metal-contacted field-effect transistors using TMDCs such as MoS_2, $MoSe_2$, and WSe_2 is usually impaired by the generation of a Schottky barrier at the metal–semiconductor junction. In silicon microelectronics, Si is degenerately doped at the contact regions to reduce the barrier height for making ohmic contacts but this practice is not recommended for 2D materials due to its damaging effects on them. Graphene contact electrodes on MoS_2 have yielded nearly zero barrier contacts but the same is not true for WSe_2.

The concept of 2D material/2D material contacts has been successfully applied to MoS_2, $MoSe_2$, and WSe_2 FETs. Low contact resistance values ~0.3 kΩ μm could be attained in few-layered WSe_2 FETs with this method. Here, the drain/source contact regions are substitutionally doped while the channel regions are not doped, followed by the van der Waals assembly of the two regions. Niobium (rhenium) dopants are used for P-doping (N-doping). The process steps are:

(i)   Exfoliated TMDC channel 2D material is placed over the previously transferred hBN layer on an SiO_2/Si substrate.

(ii) The channel region is covered by transferring an exfoliated hBN layer over it.

(iii) Degenerately doped TMDC drain/source electrodes are stacked over the exposed contact regions of the TMDC channel.

(iv) Metal electrodes are deposited over the TMDC drain/source electrodes.

Note that in step (iii), the doped contact-2D material is laid over the undoped channel 2D material establishing a 2D material/2D material contact to resolve a major bottleneck for TMDCs FETs (Chuang et al. 2016).

9.7   INSULATORS FOR 2D NANOELECTRONICS

9.7.1   PROBLEM OF INTEGRATION OF 2D SEMICONDUCTORS WITH 3D DIELECTRICS

Ill-defined interfaces and defects of amorphous oxides used in silicon technology render them impractical for 2D material electronics. The associated defect states and imperfections produced worsen the device's characteristics. Therefore, the integration of 2D semiconductors with 3D dielectrics seriously impacts the interface quality.

9.7.2   INSULATORS EXAMINED FOR 2D FETS

Insulators investigated for 2D FETs include thermally grown SiO_2 as a substrate or back-gate dielectric; high-dielectric constant (high-κ) materials: aluminum oxide (Al_2O_3) and hafnium dioxide (HfO_2), generally used in top-gate structures; crystalline calcium fluoride (CaF_2) and the 2D crystalline insulator hexagonal boron nitride (hBN). CaF_2 is found to outperform hBN for tunnel-thin gate insulators (Illarionov et al. 2019, 2020). Although hBN is the most promising 2D dielectric for FETs, exceedingly high leakage currents tarnish its reputation in ultra-scaled electronics for high-density integration (Knobloch et al. 2021).

9.8   DEFECTS ENGINEERING IN 2D MATERIALS

9.8.1   Types of Defects Created

Common crystalline defects observed in 2D materials are (Jiang et al. 2019):

(i)   Vacancies, lattice sites in a crystal structure where host atoms are absent.
(ii)   Antisites, defects caused by switching of positions of different types of atoms in the crystal structure.
(iii) Adatoms, which are atoms from lattice sites lying on the surface by gaining sufficient energy to leave their lattice positions but having insufficient energy to escape from the surface.
(iv)   Substitution defects arising from replacement of original lattice atoms by different kinds of atoms.
(v)   Grain boundaries, which are planar defects separating regions of different crystalline orientations.

9.8.2   Influence of Defects on 2D Material Properties

These defects act either as donors of charge carriers or as trapping, scattering, or recombination centers for carriers, influencing the electronic, optical, and optoelectronic properties of 2D materials. The desired properties are realized by eradication of unwanted defects and by incorporation of favorable defects. The methods used for this purpose constitute defect engineering, which must be properly understood and mastered to fabricate sensors with the required response characteristics.

9.8.3   Tools and Processes for Defect Engineering

Defects are generated and manipulated through various processes, notably e-beam irradiation, plasma, chemical, and ozone treatments; laser illumination, alloying, etc. They are also tailored by substitutional doping during crystal growth and ion implantation.

9.8.3.1   E-Beam Irradiation

The bombardment of 2D materials with electrons causes knock-on damage or displacement of an atom from a lattice site, creating point defects in the materials. It also assists in the migration and agglomeration of point defects. Hence, structural manipulations at nanoscale are possible (Xu et al. 2017a).

Healing of vacancies with foreign atoms is applied for substitutional doping by providing the source of foreign atoms. Suppose B and N vacancies are produced in the honeycomb-like lattice of a boron nitride (BN) nanosheet by electron irradiation, and paraffin wax is present as a source of carbon. In this condition, carbon atoms from paraffin wax fill these vacancies, converting the electrically insulating BN nanostructure to conducting BCN (Wei et al. 2011).

Further, the transformation between phases is accomplished by e-beam irradiation, e.g., the semiconducting 2H phase of MoS_2 is converted into its metallic 1T phase, a 2H/1T phase transition (Lin et al. 2014). Thus, innovative device structures are fabricated with atomic precision.

Shortcomings of e-beam irradiation are its cost and efficiency limitations.

9.8.3.2 Plasma Treatment

The main advantages of plasma irradiation are:

(i) It is easily controlled by varying the power, pressure, and time.
(ii) It is not susceptible to atmospheric contamination because plasma irradiation is a vacuum-based process.

Plasma is used for:

(i) Tuning of electronic properties of 2D materials: Plasma treatment can increase carrier mobility (Nan et al. 2017), adjust the N/P ambipolar behavior of materials (Giannazzo et al. 2017), and reduce contact resistances (Schulman et al. 2018).
(ii) Modulating the optical properties of 2D materials: a key photophysical parameter of a semiconductor is its photoluminescence (PL) quantum yield (QY), which determines the maximum efficiency of the device. This parameter is gravely affected by defects. The higher the defect density, the lower this parameter (Lien et al. 2019).

The role of defects is exemplified with PL of MoS_2. Irradiation of monolayer MoS_2 with mild oxygen plasma produces a strong enhancement of its PL intensity (Nan et al. 2014). The underlying reason for this PL improvement can be understood from the fact that the low PL intensity of monolayer MoS_2 arises from the N-type doping effect of sulfur vacancy defects leading to trion formation, which furnish a non-radiative recombination pathway. A trion is a quasiparticle in condensed matter physics analogous to an exciton. It is not a particle in the customary sense. It is looked upon as a localized excitation comprising three charged particles, e.g., two electrons and one hole (negatively charged trion), or two holes and one electron (positively charged trion). Oxygen atoms delivered by plasma treatment are seated on the sulfur vacancy defects causing transference of electrons from MoS_2 to O_2. As a result, P-doping of MoS_2 takes place and the trions are converted into excitons which have radiative properties unlike the nonradiative trions. Additionally, the excitons have a much larger binding energy to the defect sites. Hence, thermally assisted nonradiative recombination is inhibited and a high PL QY is attained. The PL intensity increases by around two orders of magnitude.

9.8.3.3 Chemical Treatment

Efficient doping capability of chemical treatments with less damage than plasma treatments makes it suitable for the modulation of trap states in optoelectronic devices, especially for photodetectors whose performance is crucially dependent on defects.

The photocurrent of a photoconductor is proportional to its photoconductivity gain, which in turn varies with carrier mobility and lifetime. Defects produce trap states, lengthening the minority-carrier lifetime. The minority carriers remain trapped for extended periods so that the lifetime of minority carriers becomes larger and a higher photocurrent is observed. The response time increases because the trapped carriers take more time for liberation from traps by thermal excitation.

To illustrate with an example, the as-prepared intrinsic ReS_2 photoconductor transistor shows a high responsivity of $88,600\,AW^{-1}$ with a long response time, around tens of seconds indicating the presence of deep trap states due to sulfur vacancies (Liu et al. 2016). But when ReS_2 devices are decorated with protoporphyrin IX, H_2PP, 8,13-divinyl-3,7,12,17-tetramethyl-21H-23H-porphine-2,18-dipropionic acid molecules, most of the sulfur vacancies are filled by chemical bonding of H_2PP molecules. Reduction of the number of sulfur vacancies and the associated trap states leads to severalfold diminution of decay time. At the same time, charge transfer between ReS_2 and H_2PP molecules significantly improves the specific detectivity of the device (Jiang et al. 2018).

PL of a 2D material can be modulated by chemical doping, e.g., treatment of MoS_2 (Amani et al. 2015) and WS_2 monolayers (Amani et al. 2016) with bis(trifluoromethane)sulfonimide (TFSI), $C_2F_6NO_4S_2^-$, yielded near-unity PL QY.

9.9 DISCUSSION AND CONCLUSIONS: LOOKING FORWARD

A major chunk of the experimental work on 2D materials has been performed by exfoliation of the monoatomic layer/a few layers and the transfer of its flake to the relevant substrate. The process is irreconcilable with large-scale bulk manufacturing of nanosensors. Consistently synthesizing good quality 2D materials over large areas is a hot topic.

The road to commercialization of 2D materials-based sensors will become smoother if research efforts are concentrated on the development of easy solution-based methods for scaling up production of large-area 2D materials while retaining their laboratory-scale properties and on formulating reliable surface functionalization chemistries to make drift-free chemical and biosensors using 2D materials. In both fields, chemists play a central role (Editorial 2023).

Undeterred by the various obstacles, earnest endeavors by scientists and technologists worldwide are paving the way to hasten the process upscaling for the transfer of lab-scale fabrication methods for manufacturing marketable sensors, as evidenced by the unabated progress and the ongoing deluge of research papers on this subject. The 2D materials-based sensors continue to march forward from laboratory to industry with ever-increasing confidence and enthusiasm. Judging from the state-of-the-art picture, the realization of the dual goals of industrialization and the commercialization of 2D material-based sensors are well on the horizon. The dream to reach the stage where the social impact of 2D material nanosensors will be felt is approaching fast.

Scientific research advances societies and economies and builds civilization
By translating ideas from laboratory investigations and innovations to industrialization
Through technological cooperation and product commercialization
Knowledge exchanges, mainstreaming, integration and globalization!

REFERENCES

Amani M., D.-H. Lien, D. Kiriya, J. Xiao, A. Azcatl, J. Noh, S.R. Madhvapathy, R. Addou, K. C. Santosh, et al. 2015 Near-unity photoluminescence quantum yield in MoS_2, *Science*, 350(6264), pp. 1065–1068.

Amani M., P. Taheri, R. Addou, G. H. Ahn, D. Kiriya, D.-H. Lien and J.W. Ager, et al. 2016 Recombination kinetics and effects of superacid treatment in sulfur- and selenium-based transition metal dichalcogenides, *Nano Letters*, 16(4), pp. 2786–2791.

Bhimanapati G. R., Z. Lin, V. Meunier, Y. Jung, J. Cha, S. Das, D. Xiao, and Y. Son, et al. 2015 Recent advances in two-dimensional materials beyond graphene, *ACS Nano*, 9(12), pp. 11509–11539.

Castro Neto A. H., F. Guinea, N. M. R. Peres, K. S. Novoselov, and A. K. Geim 2009 The electronic properties of graphene, *Reviews of Modern Physics*, 81(1), pp. 1–55.

Chuang H.-J., B. Chamlagain, M. Koehler, M. M. Perera, J. Yan, D. Mandrus, D. Tománek and Z. Zhou 2016 Low-resistance 2D/2D ohmic contacts: A universal approach to high-performance WSe_2, MoS_2, and $MoSe_2$ transistors, *Nano Letters*, 16(3), 1896–1902.

Editorial 2023 2D materials, a matter for chemists, *Nature Nanotechnology*, 18, p. 535.

Farmer D. B., Y.-M. Lin, A. Afzali-Ardakani, and P. Avouris 2009 Behavior of a chemically doped graphene junction, *Applied Physics Letters*, 94(21), 213106-1 to 213106-3.

Geim A. and K. Novoselov 2007 The rise of graphene, *Nature Materials*, 6, 183–191.

Giannazzo F., G. Fisichella, G. Greco, S. D. Franco, I. Deretzis, A. L. Magna, C. Bongiorno, G. Nicotra, et al. 2017 Ambipolar MoS_2 transistors by nanoscale tailoring of Schottky barrier using oxygen plasma functionalization, *ACS Applied Materials & Interfaces*, 9(27), pp. 23164–23174.

Guo B., L. Fang, B. Zhang, J. R. Gong 2011 Graphene doping: A review, *Insciences Journal*, 1(2), pp. 80–89.

Illarionov Y. Y., A. G. Banshchikov, D. K. Polyushkin, S. Wachter, T. Knobloch, M. Thesberg, M. I. Vexler, M. Waltl, et al. 2019 Reliability of scalable MoS_2 FETs with 2 nm crystalline CaF_2 insulators, *2D Materials*, 6(4), pp. 1–11.

Illarionov Y. Y., T. Knobloch, M. Jech, M. Lanza, D. Akinwande, M. I. Vexler, T. Mueller, et al. 2020 Insulators for 2D nanoelectronics: The gap to bridge, *Nature Communications*, 11, pp. 1–15.

Jiang J., C. Ling, T. Xu, W. Wang, X. Niu, A. Zafar, Z. Yan, X. Wang, Y. You, L. Sun, J. Lu, J. Wang and Z. Ni 2018 Defect engineering for modulating the trap states in 2D photoconductors, *Advanced Materials*, 30(40), 1804332.

Jiang J., T. Xu, J. Lu, L. Sun, and Z. Ni 2019 Defect engineering in 2D materials: Precise manipulation and improved functionalities, *Research (Washington, DC)*, 2019, pp. 1–14.

Kang K., S. Xie, L. Huang, Y. Han, P. Y. Huang, K. F. Mak, C.-J. Kim, D. Muller and J. Park 2015 High-mobility three-atom-thick semiconducting films with wafer-scale homogeneity, *Nature*, 520, pp. 656–660.

Knobloch T., Y. Y. Illarionov, F. Ducry, C. Schleich, S. Wachter, K. Watanabe, T. Taniguchi, et al. 2021 The performance limits of hexagonal boron nitride as an insulator for scaled CMOS devices based on two-dimensional materials, *Nature Electronics*, 4, pp. 98–108.

Lee J. H., S. Yoon, M. S. Ko, N. Lee, I. Hwang and M. J. Lee 2016 Improved performance of organic photovoltaic devices by doping F4TCNQ onto solution-processed graphene as a hole transport layer, *Organic Electronics*, 30, pp. 302–311.

Lien D. H., S. Z. Uddin, M. Yeh, M. Amani, H. Kim, J. W. Ager III, E. Yablonovitch and A. Javey 2019 Electrical suppression of all nonradiative recombination pathways in monolayer semiconductors, *Science*, 364(6439), pp. 468–471.

Lin Y.-C., D. O. Dumcenco, Y.-S. Huang and K. Suenaga 2014 Atomic mechanism of the semiconducting-to-metallic phase transition in single-layered MoS_2, *Nature Nanotechnology*, 9(5), pp. 391–396.

Liu E., M. Long, J. Zeng, W. Luo, Y. Wang, Y. Pan, W. Zhou, B. Wang et al. 2016 High responsivity phototransistors based on few-layer ReS_2 for weak signal detection, *Advanced Functional Materials*, 26(12), pp. 1938–1944.

Nan H., Z. Wang, W. Wang, Z. Liang, Y. Lu, Q. Chen, D. He, P. Tan, et al. 2014 Strong photoluminescence enhancement of MoS_2 through defect engineering and oxygen bonding, *ACS Nano*, 8(6), pp. 5738–5745.

Nan H., Z. Wu, J. Jiang, A. Zafar, Y. You, and Z. Ni 2017 Improving the electrical performance of MoS_2 by mild oxygen plasma treatment, *Journal of Physics D: Applied Physics*, 50(15), 154001.

Novoselov K. S., A. K. Geim, S. V. Morozov, D. Jiang, Y. Zhang, S.V. Dubonos, I. V.Grigorieva, and A. A. Firsov 2004 Electric field effect in atomically thin carbon films, *Science*, 306, pp. 666–669.

Panchakaria L. S., K. S. Subrahmanyam, S. K. Saha, A. Govindaraj, H. R. Krishnamurthy, U. V. Waghmare and C. N. R. Rao. 2009 Synthesis, structure and properties of boron and nitrogen doped graphene, *Advanced Materials*, 21(46), pp. 4726–4730.

Schulman D. S., A. J. Arnold and S. Das 2018 Contact engineering for 2D materials and devices, *Chemical Society Reviews*, 47(9), pp. 3037–3058.

Shi J., X. Chen, L. Zhao, Y. Gong, M. Hong, Y. Huan, Z. Zhang, P. Yang, et al. 2018 Chemical vapor deposition grown wafer-scale 2D tantalum diselenide with robust charge-density-wave order, *Advanced Materials*, 30(44), p. 1804616.

Singh D. K. and G. Gupta 2022 van der Waals epitaxy of transition metal dichalcogenides *via* molecular beam epitaxy: looking back and moving forward, *Materials Advances*, 3, pp. 6142–6156.

Sulleiro M. V., A. Dominguez-Alfaro, N. Alegret, A. Silvestri, and I. J. Gómez 2022 2D Materials towards sensing technology: From fundamentals to applications, *Sensing and Bio-Sensing Research*, 38, (100540), pp. 1–21.

Wei D., Y. Liu, Y. Wang, H. Zhang, L. Huang and G. Yu 2009 Synthesis of N-doped graphene by chemical vapor deposition and its electrical properties, *Nano Letters*, 9(5), pp. 1752–1758.

Wei X., M.-S. Wang, Y. Bando and D. Golberg 2011 Electron-beam-induced substitutional carbon doping of boron nitride nanosheets, nanoribbons, and nanotubes. *ACS Nano*, 5(4), pp. 2916–2922.

Wu T.-L., C.-H. Yeh, W.-T. Hsiao, P.-Y. Huang, M.-J. Huang, Y.-H. Chiang, C.-H. Cheng, R.-S. Liu and P.-W. Chiu 2017 High-performance organic light-emitting diode with substitutionally boron-doped graphene anode, *ACS Applied Materials & Interfaces*, 9, 17, pp. 14998–15004.

Xu T., K. Yin and L. Sun 2017a *In-situ* study of electron irradiation on two-dimensional layered materials, *Chinese Science Bulletin*, 62(25), pp. 2919–2930.

Xu X., Z. Zhang, J. Dong, D. Yi, J. Niu, M. Wu, L. Lin, R. Yin, M. Li, et al. 2017b Ultrafast epitaxial growth of meter-sized single-crystal graphene on industrial Cu foil, *Science Bulletin*, 62(15), pp. 1074–1080.

Yao J.D., Z.Q. Zheng and G.W. Yang 2019 Production of large-area 2D materials for high-performance photodetectors by pulsed-laser deposition, *Progress in Materials Science*, 106(100573), 1–64.

Zavabeti A., A. Jannat, L. Zhong, A. A. Haidri, Z. Yao, and J. Z. Ou 2020 Two-dimensional materials in large-areas: Synthesis, properties and applications, *Nano-Micro Letters*,12(66), pp. 1–34.

Zhang Z., X. Xu, J. Song, Q. Gao, S. Li, Q. Hu, X. Li and Y. Wu 2018 High-performance transistors based on monolayer CVD MoS_2 grown on molten glass, *Applied Physics Letters*, 113(20), p. 202103.

Zheng Y., J. Gao, C. Han and W. Chen 2021 Ohmic contact engineering for two-dimensional materials, *Cell Reports Physical Science*, 2(1), pp. 1–27.

Zhou L., K. Xu, A. Zubair, A. D. Liao, W. Fang, F. Ouyang, Y.-H. Lee, K. Ueno, et al. 2015 Large-area synthesis of high-quality uniform few-layer $MoTe_2$, *Journal of the American Chemical Society*, 137, 11892–11895.

Index

Pages in *italics* refer to figures, pages in **bold** refer to tables.

For Product Safety Concerns and Information please contact our EU
representative GPSR@taylorandfrancis.com
Taylor & Francis Verlag GmbH, Kaufingerstraße 24, 80331 München, Germany